FOURIER SERIES IN SEVERAL VARIABLES WITH APPLICATIONS TO PARTIAL DIFFERENTIAL EQUATIONS

Published Titles

CHAPMAN & HALL/CRC APPLIED MATHEMATICS
AND NONLINEAR SCIENCE SERIES

FOURIER SERIES IN SEVERAL VARIABLES WITH APPLICATIONS TO PARTIAL DIFFERENTIAL EQUATIONS

Victor L. Shapiro

Department of Mathematics
University of California, Riverside
USA

CRC Press
Taylor & Francis Group
Boca Raton London New York

CRC Press is an imprint of the
Taylor & Francis Group, an **informa** business

A CHAPMAN & HALL BOOK

CRC Press
Taylor & Francis Group
6000 Broken Sound Parkway NW, Suite 300
Boca Raton, FL 33487-2742

First issued in paperback 2019

© 2011 by Taylor & Francis Group, LLC
CRC Press is an imprint of Taylor & Francis Group, an Informa business

No claim to original U.S. Government works

ISBN-13: 978-1-4398-5427-3 (hbk)
ISBN-13: 978-0-367-38292-6 (pbk)

Visit the Taylor & Francis Web site at
http://www.taylorandfrancis.com

and the CRC Press Web site at
http://www.crcpress.com

To Flo, my wife and dancing partner

Preface

The primary purpose of this book is to show the great value that Fourier series methods provide in solving difficult problems in nonlinear partial differential equations. We illustrate these methods in three different cases.

Probably the most important of these three cases are the results that we present for the stationary Navier-Stokes equations. In particular, we show how to obtain the best possible results for periodic solutions of the stationary Navier-Stokes equations when the driving force is nonlinear. We also present the basic theorem for the distribution solutions of said equations. The ideas for this material come from a paper published by the author in the Journal of Differential Equations.

Also, we show how to obtain classical solutions to the stationary Navier-Stokes equations by applying the Calderon-Zygmund C^a-theory developed for multiple Fourier series earlier in the book. This technique using the Calderon-Zygmund C^a-theory does not appear to be in any other text dealing with this subject and is based on a paper that appeared in the Transactions of the AMS.

The second case we consider handles nonlinear reaction-diffusion systems and uses a technique involving multiple Fourier series to strongly improve on a theorem previously introduced by Brezis and Nirenberg. The idea for doing this comes from a recent (2009) paper published by the author in the Indiana University Math Journal. Reaction-diffusion systems are important in many areas of applied mathematics including mathematical biology. The main reason we were able to improve on the results of Brezis and Nirenberg is because the use of multiple Fourier series enables one to make sharper estimates and thus obtain a better compactness lemma. The second theorem we present in this area involves a conventional result involving weak solutions to the reaction-diffusion system.

The third case we consider is in the area of quasilinear elliptic partial differential equations and resonance theory. We deal with an elliptic operator of the form

$$Qu = -\sum_{i,j=1}^{N} D_i[a^{ij}(x,u)D_j u] + \sum_{j=1}^{N} b^j(x,u,Du)D_j u$$

and establish a resonance result based on the work of Defigueredo and Gossez in a Journal of Differential Equations paper and on the work of the author in a Transactions of the AMS manuscript. The resonance result obtained is the best possible and is proved via a Galerkin type argument that illustrates once again the power of Fourier analysis in handling tough problems in nonlinear PDE. The second and third theorems that we present give necessary and sufficient conditions for the solution of certain other equations at a resonance involving the above operator Qu.

Another aim of this book is to establish the connection between multiple Fourier series and number theory. We present an N-dimensional, $N \geq 2$, number theoretic result, which gives a necessary and sufficient condition that

$$C(\xi_1) \times \cdots \times C(\xi_N)$$

be a set of uniqueness for a class of distributions on the N-torus, T_N. The ideas behind this result come from a paper published in the Journal of Functional Analysis.

Here, $C(\xi_j)$ is the familiar symmetric Cantor set on $[-\pi, \pi]$ depending on the real number ξ_j where $0 < \xi_j < 1/2$. The condition is that each ξ_j^{-1} be an algebraic integer called a Pisot number. What is important about this result is that the considered class of distributions, labeled $\mathcal{A}(T_N)$, does not necessarily have Fourier coefficients that go to zero as the spherical norm $|m| = (m_1^2 + \cdots + m_N^2)^{1/2} \to \infty$ but as $\min(|m_1|, ..., |m_N|) \to \infty$. This gives rise to a wider class of distributions; when it appeared, it was the first result of this nature in the mathematical literature.

As a corollary to the result just mentioned, we have the following:

Let p and q be positive relatively-prime integers with $p < 2q$. Then a necessary and sufficient condition that

$$C(\frac{p_1}{q_1}) \times \cdots \times C(\frac{p_N}{q_N})$$

be a set of uniqueness for the class $\mathcal{A}(T_N)$ is that $p_j = 1$ for $j = 1, ..., N$.

An additional aim of this book is to present the periodic C^α-theory of Calderon and Zygmund. We deal with a Calderon-Zygmund kernel of spherical-harmonic type, called $K^*(x)$, and show that it has a principal-valued Fourier coefficient $\widehat{K^*}(m)$. We set $\widetilde{f} = f * K^*$ and show that the following very important theorem prevails:

$$f \in C^\alpha(T_N), 0 < \alpha < 1, \Rightarrow \widetilde{f} \in C^\alpha(T_N).$$

We also give an application of this theorem to a periodic boundary value problem involving the Laplace operator and later use it to obtain the regularity result mentioned above for the stationary Navier-Stokes equations.

Another aim of this book is to present the recent (2006) article in the Proceedings of the AMS, which extends Fatou's famous work on anti-derivatives and nontangential limits to higher dimensions. The big question answered is "How does an individual handle a concept that depends on the one-dimensional notion of the anti-derivative in dimension $N \geq 2$?" Our answer to the question is

> "Generalize the notion of the Lebesgue point set and show
>
> that the concepts are the same in one-dimension."

Chapter 1 of the book deals with four different summability methods used in the study of multiple Fourier series, namely the methods of (i) iterated Fejer, (ii) Bochner-Riesz, (iii) Abel, and (iv) Gauss-Weierstrass. The iterated Fejer method in §2 gives a global uniform approximation for continuous periodic functions as well as a global L^p approximation theorem. In §3, the classical Bochner theorem for pointwise Bochner-Riesz summability of multiple Fourier series is established. To understand the proof of this theorem, a knowledge of various Bessel identities and estimates is essential. This Bessel background material is presented in §1 and §2 of Appendix A.

Several Abel summability theorems, which are important in the study of harmonic functions including the nontangential result discussed above, are also presented in Chapter 1, §4. In §5 of Chapter 1, the Gauss-Weierstrass summability method, which is fundamental in the study of the heat equation, is developed; it includes a theorem necessary for a subsequent number theoretic result appearing later in the book.

Chapter 2 is devoted to the study of conjugate multiple Fourier series where the conjugacy is defined by means of periodic Calderon-Zygmund kernels that are of spherical harmonic type. In particular, the periodic Calderon-Zygmund kernel, $K^*(x)$, is defined, and it is proved that its principal-valued Fourier coefficient $\widehat{K^*}(m)$ exists. The conjugate function of f is designated by $\widetilde{f}$, and it is shown that if things are good, $\widehat{\widetilde{f}}(m) = \widehat{K^*}(m)\widehat{f}(m)$, which is similar to the one-dimensional situation. The main result established is the following: If $f \in C^\alpha(T_N)$, then $\widetilde{f} \in C^\alpha(T_N)$. This C^α- theorem is presented in complete detail in §4 of Chapter 2 and is based on a paper published by Calderon and Zygmund in the Studia Mathematica.

In §5 of Chapter 2, an application of this C^α- result to a periodic boundary value problem involves the Laplace operator. Also, a Tauberian convergence theorem for conjugate multiple Fourier series motivated by an interesting one-dimensional result of Hardy and Littlewood is given in §3 of Chapter 2. The Tauberian background material is developed in Appendix B.

Chapter 3 contains the details of the solution to *a one hundred year old problem*, namely

> Establish the two-dimensional analogue of Cantor's famous
>
> uniqueness theorem dealing with the convergence
>
> of one-dimensional trigonometric series.

The solution depends upon an elegant paper published by Roger Cooke in the Proceedings of the AMS establishing the two-dimensional Cantor-Lebesgue lemma joined with a manuscript of the author that appeared in the Annals of Mathematics.

Chapter 3 also contains the N-dimensional number theoretic theorem discussed above giving a necessary and sufficient condition that

$$C(\xi_1) \times \cdots \times C(\xi_N)$$

be a set of uniqueness for the class of distributions $\mathcal{A}(T_N)$ on the N-torus. In addition, Chapter 3 contains the recent (2004) article about fractal sets called generalized carpets that are not Cartesian product sets but are sets of uniqueness for a smaller class of distributions on the N-torus labeled $\mathcal{B}(T_N)$. These fractal results come from a paper published in the Proceedings of the AMS.

The analogous problem to Cantor's uniqueness theorem for a series of two-dimensional surface spherical harmonics on S_2 is still open and is presented in complete detail in Chapter 3, §2. This problem has been open now for 140 years. The background material in spherical harmonics, which plays an important role throughout this monograph, is presented in Appendix A, §3.

The material in Chapter 4 is motivated by Schoenberg's theorem involving positive definite functions on S_2 and surface spherical harmonics published in the Duke Journal of Math. It turns out that part of Schoenberg's theorem is highly useful in studying the kissing problem, $k(3)$, in discrete geometry, as Musin's 2006 result shows. Here, $k(3)$ is the largest number of white billiard balls that can simultaneously kiss (touch) a black billiard ball and represents a problem going back to Isaac Newton's time in 1694.

Chapter 4 presents Schoenberg's theorem on S_{N-1}, then on T_N, and finally on $S_{N_1-1} \times T_N$. The proof on $S_{N_1-1} \times T_N$ makes use of a number of different concepts that occur in this monograph.

Chapter 5 presents five theorems dealing with periodic solutions of nonlinear partial differential equations. As mentioned earlier, the methods employed illustrate the huge power of Fourier analysis in solving seemingly impenetrable problems in a nonlinear analysis. Chapter 5, §1 presents, in particular, periodic solutions in the space variables to a system of nonlinear reaction-diffusion equations of the form

$$\begin{cases} \frac{\partial u_j}{\partial t} - \Delta u_j = f_j(x, t, u_1, ..., u_J) & \text{in } T_N \times (0, T) \\ \\ u_j(x, 0) = 0 \end{cases}$$

$j = 1, ..., N.$

Two theorems are established with respect to this nonlinear parabolic system. The first theorem deals with one-sided conditions placed on the f_j, and the second deals with two-sided conditions on the f_j. As discussed above, the first theorem strongly improves (for periodic solutions) on a one-sided classical theorem previously established by Brezis and Nirenberg.

In §2 of Chapter 5, we deal with the equation

$$Qu = f(x, u)$$

where Qu is the partial differential operator discussed above. We set

$$\mathcal{F}_{\pm}(x) = \limsup_{s \to \pm\infty} f(x,s)/s,$$

and show that if

$$\int_{T_N} \mathcal{F}_+(x)\,dx < 0 \text{ and } \int_{T_N} \mathcal{F}_-(x)\,dx < 0$$

and certain other conditions are met, then a distribution solution $u \in W^{1,2}(T_N)$ of $Qu = f(x,u)$ exists. We also show that this is the best possible result.

In §2 of Chapter 5, we also handle the equation

$$Qu = g(u) - h(x)$$

and define

$$\lim_{s \to \infty} g(s) = g(\infty) \text{ and } \lim_{s \to -\infty} g(s) = g(-\infty).$$

We show that if certain other assumptions are met, then the condition

$$(2\pi)^N g(\infty) < \int_{T_N} h(x)\,dx < (2\pi)^N g(-\infty)$$

is both necessary and sufficient that a distribution solution $u \in W^{1,2}(T_N)$ of $Qu = g(u) - h(x)$ exists.

In §1 of Chapter 6, we handle the stationary Navier-Stokes equations with a nonlinear driving force:

$$-\nu\Delta\mathbf{v}(x) + (\mathbf{v}(x) \cdot \nabla)\mathbf{v}(x) + \nabla p(x) = \mathbf{f}(x, \mathbf{v}(x))$$

$$(\nabla \cdot \mathbf{v})(x) = 0$$

where ν is a positive constant, and $\mathbf{v}$ and $\mathbf{f}$ are vector-valued functions. In particular, $\mathbf{f} = (f_1, ..., f_N) : T_N \times \mathbf{R}^N \to \mathbf{R}^N$. We set

$$E_j(\mathbf{f}) = \{x \in T_N : \limsup_{|s_j| \to \infty} f_j(x, \mathbf{s})/s_j < 0$$

uniformly for $s_k \in \mathbf{R}, k \neq j,\ k = 1, ..., N\}$

and show that if certain other assumptions are met, then

$$|E_j(\mathbf{f})| > 0 \quad for \quad j = 1, ..., N,$$

is a sufficient condition for the pair $(\mathbf{v}, p)$ to be a distribution solution of the stationary Navier-Stokes equations with $v_j \in W^{1,2}(T_N)$ and $p \in L^1(T_N)$. Here, $|E_j(\mathbf{f})|$ represents the Lebesgue measure of $E_j(\mathbf{f})$. We also demonstrate that this is the best possible result.

Another theorem that we establish in §1 of Chapter 6 handles the situation when

$$f_j(x, \mathbf{s}) = g_j(s_j) - h_j(x).$$

In particular, we prove that if certain other conditions are met, then

$$(2\pi)^N g_j(\infty) < \int_{T_N} h_j(x)\,dx < (2\pi)^N g_j(-\infty)$$

for $j = 1,...N$, is both a necessary and sufficient condition that the pair $(\mathbf{v}, p)$ be a distribution solution of the stationary Navier-Stokes equations with $v_j \in W^{1,2}(T_N)$ and $p \in L^1(T_N)$.

In §2 of Chapter 6, we deal with the classical solutions of the stationary Navier-Stokes equations. The main tool for proving the theorem involved is the C^α-theory of Calderon and Zygmund established earlier in Chapter 2.

Given $\mathbf{f} \in [C(T_N)]^N$, we will say the pair $(\mathbf{v}, p)$ is a periodic classical solution of the stationary Navier-Stokes system provided:

$$\mathbf{v} \in \left[C^2(T_N)\right]^N \text{ and } p \in C^1(T_N)$$

and

$$-\nu\Delta\mathbf{v}(x) + (\mathbf{v}(x)\cdot\nabla)\mathbf{v}(x) + \nabla p(x) = \mathbf{f}(x) \qquad \forall x \in T_N$$

$$(\nabla\cdot\mathbf{v})(x) = 0 \qquad\qquad \forall x \in T_N.$$

To obtain the classical solutions of the Navier-Stokes system, we require slightly more for the driving force $\mathbf{f}$ than periodic continuity. In particular, we say $f_1 \in C^\alpha(T_N)$, $0 < \alpha < 1$, provided the following holds:

(i) $f_1 \in C(T_N)$;

(ii) $\exists\, c_1 > 0$ s. t. $|f_1(x) - f_1(y)| \le c_1 |x - y|^\alpha \qquad \forall x, y \in \mathbf{R}^N$.

Working in dimension $N = 2$ or 3, we show in §2 of Chapter 6 that if

$$f_j \in C^\alpha(T_N), 0 < \alpha < 1 \text{ for } j = 1,...N,$$

then there is a pair $(\mathbf{v}, p)$ which is a periodic classical solution of the stationary Navier-Stokes system with $v_j \in C^{2+\alpha}(T_N)$ and $p \in C^{1+\alpha}(T_N)$.

I have lectured on the mathematics developed in this book at various mathematical seminars at the University of California, Riverside, where I have been a professor for the last 45 years. Also, I would like to thank my colleague James Stafney for the many discussions that we have had about spherical harmonics and related matters.

I had the good fortune to write my doctoral thesis with Antoni Zygmund at the University of Chicago. Also, I did post-doctoral work with Arne Beurling at the Institute for Advanced Study and with Salomon Bochner from Princeton University. My subsequent mathematical work was backed by Marston Morse from the Institute for Advanced Study. I am indebted to these four outstanding mathematicians.

Victor L. Shapiro
Riverside, California
January, 2010

Contents

CHAPTER 1

Summability of Multiple Fourier Series

1. Introduction

We shall operate in real N-dimensional Euclidean space, $\mathbf{R}^N$, $N \geq 1$, and use the following notation:

$$
\begin{aligned}
x &= (x_1, ..., x_N) \qquad y = (y_1, ..., y_N) \\
\alpha x + \beta y &= (\alpha x_1 + \beta y_1, ..., \ \alpha x_N + \beta y_N) \\
x \cdot y &= x_1 y_1 + ... + x_N y_N \ , \qquad |x| = (x \cdot x)^{\frac{1}{2}}.
\end{aligned}
$$

With T_N, the N-dimensional torus,

$$
T_N = \{x : -\pi \leq x_j < \pi, \ j = 1, ..., N\} \,,
$$

we shall say $f \in L^p(T_N), 1 \leq p < \infty$, provided f is a real-valued (unless explicitly stated otherwise) Lebesgue measurable function defined on $\mathbf{R}^N$ of period 2π in each variable such that

$$
\int_{T_N} |f|^p \, dx < \infty.
$$

A similar definition prevails for $f \in L^\infty(T_N)$.

With m as an integral lattice point in $\mathbf{R}^N$ and Λ_N representing the set of all such points, we shall designate the series

$$
\sum_{m \in \Lambda_N} \widehat{f}(m) e^{im \cdot x}
$$

by $S[f]$ and call it the Fourier series of f where

$$
\widehat{f}(m) = (2\pi)^{-N} \int_{T_N} e^{-im \cdot x} f(x) dx.
$$

In this chapter, we study the relationship between f and its Fourier series $S[f]$.

To begin, we let $\Delta = \partial^2/\partial x_1^2 + \cdots + \partial^2/\partial x_N^2$ be the usual Laplace operator and observe that $\Delta e^{im \cdot x} = -|m|^2 \, e^{im \cdot x}$. Consequently, from an eigenvalue point of view, it is natural to ask, "In what manner does the series

$$
(1.1) \qquad \sum_{n=0}^{\infty} \left(\sum_{|m|^2 = n} \widehat{f}(m) e^{im \cdot x} \right)
$$

approximate f?" Bearing in mind the classical counter-examples of both Fejer and Lebesgue concerning the convergence of one-dimensional Fourier series, [Zy1, Chapter 8], we see that the answer to the previous question should involve some spherical summability method of the series given in (1.1).

The two most natural methods involving spherical summability are those of Bochner-Riesz and Abel. In particular, we say that $S[f]$ is Bochner-Riesz summable of order α, henceforth designated by $(B - R, \alpha)$ to $f(x)$ if

$$(1.2) \qquad \lim_{R \to \infty} \sum_{|m| \leq R} \widehat{f}(m) e^{im \cdot x} (1 - |m|^2 / R^2)^\alpha = f(x).$$

Bochner-Riesz summability plays the same role for multiple Fourier series that Cesaro summability plays for one-dimensional Fourier series. In §3 of this chapter, we shall establish a fundamental result for Bochner-Riesz summability of Fourier series.

$S[f]$ is Abel summable to $f(x)$, this means that the

$$(1.3) \qquad \lim_{t \to 0} \sum_{m \in \Lambda_N} \widehat{f}(m) e^{im \cdot x - |m|t} = f(x).$$

The reason for calling this method of summability Abel summability is motivated by the fact that the series

$$\sum_{m \in \Lambda_N} \widehat{f}(m) e^{im \cdot x - |m|t}$$

is harmonic in $\mathbf{R}_+^{N+1}$, i.e., in the variables (x,t) for $t > 0$.

We shall discuss Abel summability in detail in §4 of this chapter. Also, in Chapter 2, we shall deal with the Abel summability of conjugate multiple Fourier series. But first, it turns out that we can get some very good global results connecting f and $S[f]$ by iterating well-known one-dimensional results involving the Fejer kernel, and we will now show this iteration.

2. Iterated Fejer Summability of Fourier Series

We leave $D_n(t)$ as the well-known one-dimensional Dirichlet kernel

$$(2.1) \qquad D_n(t) = \sum_{j=-n}^{n} e^{ijt} = \frac{\sin(n + \frac{1}{2})t}{\sin(t/2)},$$

and $K_n(t)$ as the well-known one-dimensional Fejer kernel [Ru1, p. 199],

$$(2.2) \qquad K_n(t) = \frac{1}{n+1} \sum_{j=0}^{n} D_j(t) = \frac{1}{n+1} \frac{1 - \cos(n+1)t}{1 - \cos t}.$$

We also observe from [Ru1, p. 199] that $K_n(t)$ has the following three properties:

$$(a)\ K_n(t) \geq 0 \quad \forall t \in \mathbf{R},$$

(2.3)
$$(b)\ \frac{1}{2\pi} \int_{-\pi}^{\pi} K_n(t)dt = 1,$$

$$(c)\ K_n(t) \leq \frac{1}{n+1} \frac{2}{1-\cos\delta} \quad \text{if}\ \ 0 < \delta \leq |t| \leq \pi.$$

It follows from (2.1) and (2.2) that

(2.4)
$$K_n(t) = \sum_{j=-n}^{n} e^{ijt}(1 - \frac{|j|}{n+1}),$$

and we shall refer to

(2.5)
$$K_n^{\Diamond}(x) = K_n(x_1) \cdots K_n(x_N)$$

as the iterated N-dimensional Fejer kernel.

For $f \in L^1(T_N)$ with $S[f]$ as its Fourier series, we shall prove three global theorems involving $K_n^{\Diamond}(x)$ and the iterated Fejer summability of $S[f]$. In particular, we call $\sigma_n^{\Diamond}(f,x)$ the iterated Fejer partial sum of $S[f]$ where $m = (m_1, ..., m_N)$ and

(2.6) $\quad \sigma_n^{\Diamond}(f,x) = \sum_{m_1=-n}^{n} \cdots \sum_{m_N=-n}^{n} \widehat{f}(m)e^{im \cdot x}(1 - \frac{|m_1|}{n+1}) \cdots (1 - \frac{|m_N|}{n+1}).$

With $f \in C(T_N)$ signifying that f is a real-valued continuous function defined on $\mathbf{R}^N$ of period 2π in each variable and with $B(x,r)$ designating the open N-ball with a center x and radius r, the first theorem we shall prove is the following:

Theorem 2.1. Let $f \in C(T_N)$ and suppose $\sigma_n^{\Diamond}(f,x)$ is defined as in (2.6). Then
$$\lim_{n\to\infty} \sigma_n^{\Diamond}(f,x) - f(x) \quad \text{uniformly for } x \in T_N.$$

Proof of Theorem 2.1. We observe from (2.3)-(2.6) that

(2.7) $\quad \sigma_n^{\Diamond}(f,x) - f(x) = (2\pi)^{-N} \int_{T_N} [f(x-y) - f(x)]K_n^{\Diamond}(y)dy.$

Let $\varepsilon > 0$ be given. Choose $\delta > 0$ so that $|f(x-y) - f(x)| < \varepsilon$ for $y \in B(0,\delta)$ uniformly for $x \in T_N$. Now it is clear that $Cu(0, \frac{\delta}{N}) \subset B(0,\delta)$ where $Cu(0, \frac{\delta}{N})$ is the open N-cube with center 0 and a half-side δ/N. So

(2.8) $\ |f(x-y) - f(x)| < \varepsilon \quad \text{for} \quad y \in Cu(0, \frac{\delta}{N}) \quad \text{uniformly for}\ \ x \in T_N.$

Designating $P_{1,\delta}^{+}$ as the rectangular parallelopiped

$$P_{1,\delta}^{+} = \{x : \delta \leq x_1 \leq \pi,\ \text{-}\pi \leq x_j \leq \pi,\ j = 2, ..., N.\},$$

we see from (2.3) that $\lim_{n\to\infty}\int_{P^+_{1,\delta/N}}\left|K_n^\Diamond(y)\right|dy=0$. Since $T_N\backslash Cu(0,\frac{\delta}{N})$ is covered by a finite number of parallelopipeds similar to $P^+_{1,\delta/N}$, we conclude that

$$\lim_{n\to\infty}\int_{T_N\backslash Cu(0,\frac{\delta}{N})}\left|K_n^\Diamond(y)\right|dy=0.$$

Since $f(x)$ is uniformly bounded on $\mathbf{R}^N$, we also see from this last limit that n_0 can be chosen so large that

$$(2.9)\quad(2\pi)^{-N}\int_{T_N\backslash Cu(0,\frac{\delta}{N})}|f(x-y)-f(x)|\left|K_n^\Diamond(y)\right|dy\le\varepsilon\quad\text{for}\ \ n\ge n_0$$

uniformly for $x\in T_N$.

Next, returning to (2.8), we obtain from (2.3)(b) that

$$\int_{Cu(0,\frac{\delta}{N})}|f(x-y)-f(x)|\left|K_n^\Diamond(y)\right|dy\ \le\ \varepsilon\int_{T_N}\left|K_n^\Diamond(y)\right|dy$$

$$\le\ \varepsilon\,(2\pi)^N\quad\forall n$$

uniformly for $x\in T_N$.

Hence, (2.7) and this last fact joined with (2.9) shows that

$$\left|\sigma_n^\Diamond(f,x)-f(x)\right|\le2\varepsilon\quad\text{for}\ \ n\ge n_0\quad\text{uniformly for}\ x\in T_N,$$

which gives the conclusion to the theorem.   ∎

The second summability theorem that we obtain using the N-dimensional iterated Fejer kernel is the following:

Theorem 2.2.   *Let* $f\in L^p(T_N)$, $1\le p<\infty$ *and suppose* $\sigma_n^\Diamond(f,x)$ *is defined as in (2.6). Then*

$$\lim_{n\to\infty}\int_{T_N}\left|\sigma_n^\Diamond(f,x)-f(x)\right|^p dx=0.$$

Proof of Theorem 2.2.   We prove this for the case $1<p<\infty$, with a similar proof prevailing for the case $p=1$. From (2.7) with $p^{-1}+p'^{-1}=1$, we see that

$$\left|\sigma_n^\Diamond(f,x)-f(x)\right|\le(2\pi)^{-N}\int_{T_N}|f(x-y)-f(x)|\left|K_n^\Diamond(y)\right|^{p^{-1}+p'^{-1}}dy,$$

and hence from Holder's inequality and (2.3)(b) that
(2.10)
$$\int_{T_N}\left|\sigma_n^\Diamond(f,x)-f(x)\right|^p dx\le\int_{T_N}\left|K_n^\Diamond(y)\right|[\int_{T_N}|f(x-y)-f(x)|^p dx]dy.$$

Now $f \in L^p(T_N)$ and is also periodic of period 2π in each variable. Therefore, it follows that given $\varepsilon > 0$, $\exists \delta > 0$,

$$\int_{T_N} |f(x-y) - f(x)|^p \, dx \leq \varepsilon (2\pi)^{-N} \quad \text{for} \quad y \in B(0, \delta).$$

Consequently, we obtain from (2.10) that

$$\int_{T_N} \left| \sigma_n^{\Diamond}(f, x) - f(x) \right|^p dx \leq$$
$$\int_{T_N - B(0,\delta)} \left| K_n^{\Diamond}(y) \right| \left[\int_{T_N} |f(x-y) - f(x)|^p \, dx \right] dy + \varepsilon$$

But $f \in L^p(T_N)$ implies that the inner integral on the right-hand side of the above inequality is uniformly bounded. Therefore, since $Cu(0, \frac{\delta}{N}) \subset B(0, \delta)$, we infer from the limit above (2.9) and the above inequality that

$$\limsup_{n \to \infty} \int_{T_N} \left| \sigma_n^{\Diamond}(f, x) - f(x) \right|^p dx \leq \varepsilon.$$

Since $\varepsilon > 0$ is arbitrary, this gives the conclusion to the theorem. ∎

Theorem 2.2 has three important corollaries, the first of which is the following:

Corollary 2.3. $\left\{ e^{im \cdot x} \right\}_{m \in \Lambda_N}$, *the trigonometric system, is a complete orthogonal system for* $L^1(T_N)$, *i.e., if* $f, g \in L^1(T_N)$ *and* $\widehat{f}(m) = \widehat{g}(m)$ *for every integral lattice point* m, *then* $f(x) = g(x)$ *a.e. in* T_N.

Proof of Corollary 2.3. Since $f, g \in L^1(T_N)$ and $\widehat{f}(m) = \widehat{g}(m)$ for every integral lattice point m, it implies that $\sigma_n^{\Diamond}(f, x) = \sigma_n^{\Diamond}(g, x)$ $\forall x \subset T_N$ and $\forall n$. Consequently, it follows from Theorem 2.2 that

$$\int_{T_N} |f(x) - g(x)| \, dx = 0,$$

which establishes the corollary. ∎

The next corollary that we shall prove is called the Riemann-Lebesgue lemma and is the following:

Corollary 2.4 *If* $f \in L^1(T_N)$, *then* $\lim_{|m| \to \infty} \widehat{f}(m) = 0$.

Proof of Corollary 2.4. Let $\varepsilon > 0$ be given. Using Theorem 2.2, choose an n sufficiently large so that $\int_{T_N} \left| \sigma_n^{\Diamond}(f, x) - f(x) \right| dx < \varepsilon$. Then, it follows from the definition of $\widehat{f}(m)$ given above (1.1) that

$$\left|\widehat{f}(m)\right| \leq (2\pi)^{-N}\{\int_{T_N} \left|\sigma_n^\diamond(f,x) - f(x)\right| dx + \left|\int_{T_N} e^{-im\cdot x}\sigma_n^\diamond(f,x)\, dx\right|\},$$

$$\leq \varepsilon + (2\pi)^{-N}\left|\int_{T_N} e^{-im\cdot x}\sigma_n^\diamond(f,x)\, dx\right|.$$

Since $\sigma_n^\diamond(f,x)$ is a fixed trigonometric polynomial, it follows that there is a positive number s_0 such that the integral in the second inequality is zero for $|m| \geq s_0$. We conclude that $\left|\widehat{f}(m)\right| \leq \varepsilon$ for $|m| \geq s_0$, which establishes the corollary. ■

The third corollary that we can obtain from Theorem 2.2 is called Parsevaal's theorem and is the following:

Corollary 2.5. *If* $f \in L^2(T_N)$, *then*

$$\lim_{n\to\infty} (2\pi)^N \sum_{m_1=-n}^{n} \cdots \sum_{m_N=-n}^{n} \left|\widehat{f}(m)\right|^2 = \|f\|_{L^2}^2.$$

Proof of Corollary 2.5. From Theorem 2.2, we see that

$$\lim_{n\to\infty} \left\|\sigma_n^\diamond\right\|_{L^2}^2 = \|f\|_{L^2}^2.$$

Also, we have that $\left\{\left\|\sigma_n^\diamond\right\|_{L^2}^2\right\}_{n=1}^{\infty}$ is an increasing sequence, and the proof follows easily from this last observation. ■

The third summability theorem that we get using the N-dimensional iterated Fejer kernel is the following:

Theorem 2.6. *Let* $f \in L^\infty(T_N)$ *and suppose* $\sigma_n^\diamond(f,x)$ *is defined as in* (2.6). *Then* $\sigma_n^\diamond(f,x) \to f(x)$ *in the weak** L^∞-*topology, i.e.,*

$$\lim_{n\to\infty} \int_{T_N} \sigma_n^\diamond(f,x)\, h(x)dx = \int_{T_N} f(x)\, h(x)dx \quad \forall h \in L^1(T_N).$$

Proof of Theorem 2.6. Let h be a given function in $L^1(T_N)$. Then it follows from Theorem 2.2 that

(2.11) $$\lim_{n\to\infty} \int_{T_N} \left|\sigma_n^\diamond(h,x) - h(x)\right| dx = 0.$$

Next, we set

$$I_n = (2\pi)^{-N} \int_{T_N} \sigma_n^\diamond(f,x)h(x)dx,$$

and observe from (2.6) that

$$
\begin{aligned}
I_n &= \sum_{m_1=-n}^{n} \cdots \sum_{m_N=-n}^{n} \widehat{f}(m)\widehat{h}(-m)\left(1 - \frac{|m_1|}{n+1}\right)\cdots\left(1 - \frac{|m_N|}{n+1}\right) \\
&= \sum_{m_1=-n}^{n} \cdots \sum_{m_N=-n}^{n} \widehat{f}(-m)\widehat{h}(m)\left(1 - \frac{|m_1|}{n+1}\right)\cdots\left(1 - \frac{|m_N|}{n+1}\right).
\end{aligned}
$$

Consequently,

$$
\int_{T_N} \sigma_n^{\Diamond}(f,x)h(x)dx = \int_{T_N} \sigma_n^{\Diamond}(h,x)f(x)dx.
$$

But then

$$
\int_{T_N} [\sigma_n^{\Diamond}(f,x) - f(x)]h(x)dx = \int_{T_N} [\sigma_n^{\Diamond}(h,x) - h(x)]f(x)dx.
$$

Hence,

$$
\left| \int_{T_N} [\sigma_n^{\Diamond}(f,x) - f(x)]h(x)dx \right| \leq \|f\|_{L^\infty(T_N)} \int_{T_N} \left| \sigma_n^{\Diamond}(h,x) - h(x) \right| dx,
$$

and the conclusion to the theorem follows immediately from the limit in (2.11). ■

Exercises.

1. With $D_n(t) = \sum_{j=-n}^{n} e^{ijt}$, use the well-known formula for geometric progressions and prove that

$$
D_n(t) = \frac{\sin(n + \frac{1}{2})t}{\sin(t/2)}.
$$

2. With $K_n(t) = \frac{1}{n+1}\sum_{j=0}^{n} D_j(t)$, use the familiar formula $1 - \cos\phi = 2\sin^2(\phi/2)$ and prove that

$$
K_n(t) = \frac{1}{n+1}\frac{1 - \cos(n+1)t}{1 - \cos t}.
$$

3. Prove that $K_n(t)$ has the following properties:

    (a) $K_n(t) \geq 0 \quad \forall t \in \mathbf{R}$,

    (b) $\frac{1}{2\pi}\int_{-\pi}^{\pi} K_n(t)dt = 1$,

    (c) $K_n(t) \leq \frac{1}{n+1}\frac{2}{1-\cos\delta}$ if $0 < \delta \leq |t| \leq \pi$.

4. Complete the proof of Corollary 2.5.

3. Bochner-Riesz Summability of Fourier Series

As we observed in the introduction to this chapter, $\Delta e^{im \cdot x} = -|m|^2 e^{im \cdot x}$ where Δ is the usual Laplace operator. Hence, from an eigenvalue point of view, since the eigenfunctions with the same eigenvalue have their integral lattice points lying on spheres, it is a good idea to study multiple Fourier series using spherical techniques. One of the most effective spherical technique is the method of Bochner-Riesz summation, defined previously in (1.2). With $B(x, r)$ representing the open N-ball with center x and radius r, the first theorem for this method of summation that we shall prove is the following due to Bochner [Boc1]:

Theorem 3.1. *Let* $f \in L^1(T_N)$ *and set*

$$(3.1) \qquad \sigma_R^\alpha(f, x) = \sum_{|m| \leq R} \widehat{f}(m) e^{im \cdot x} (1 - |m|^2 / R^2)^\alpha.$$

Suppose that $|B(0, \rho)|^{-1} \int_{B(0, \rho)} |f(x_0 + x) - f(x_0)| \, dx \to 0$ *as* $\rho \to 0$. *Then*

$$\lim_{R \to \infty} \sigma_R^\alpha(f, x_0) = f(x_0) \quad \text{for} \quad \alpha > (N - 1)/2.$$

We refer to $\sigma_R^\alpha(f, x)$ on the left-hand side of (3.1) as the R-th Bochner-Riesz mean of order α. Also, $|B(0, \rho)|$ designates the volume of the N-ball of radius ρ, which we shall now compute.

In order to make this computation, we introduce the N-dimensional spherical coordinate notation

$$
\begin{aligned}
x_1 &= r \cos \theta_1 \\
x_2 &= r \sin \theta_1 \cos \theta_2 \\
x_3 &= r \sin \theta_1 \sin \theta_2 \cos \theta_3 \\
&\;\;\vdots \\
x_{N-1} &= r \sin \theta_1 \sin \theta_2 \cdots \sin \theta_{N-2} \cos \phi \\
x_N &= r \sin \theta_1 \sin \theta_2 \cdots \sin \theta_{N-2} \sin \phi
\end{aligned}
$$

where $0 \leq r < \rho$, $0 \leq \theta_j \leq \pi$ for $j = 1, \dots, N - 2$, and $0 \leq \phi < 2\pi$.

We label the Jacobian of this transformation, $\mathcal{J}_N(r, \theta_1, \dots, \theta_{N-2}, \phi)$. For example,

$$\mathcal{J}_3(r, \theta_1, \phi) = r^2 \begin{vmatrix} \cos \theta_1 & -\sin \theta_1 & 0 \\ \sin \theta_1 \cos \phi & \cos \theta_1 \cos \phi & -\sin \theta_1 \sin \phi \\ \sin \theta_1 \sin \phi & \cos \theta_1 \sin \phi & \sin \theta_1 \cos \phi \end{vmatrix},$$

and an easy computation shows that $\mathcal{J}_3(r, \theta_1, \phi) = r^2 \sin \theta_1$.

In a similar manner, we see that $\mathcal{J}_4(r,\theta_1,\theta_2,\phi)/r^3$ is going to be the determinant of the following array:

$$\begin{array}{llll}
\cos\theta_1 & -\sin\theta_1 & 0 & 0 \\
\sin\theta_1\cos\theta_2 & \cos\theta_1\cos\theta_2 & -\sin\theta_1\sin\theta_2 & 0 \\
\sin\theta_1\sin\theta_2\cos\phi & \cos\theta_1\sin\theta_2\cos\phi & \sin\theta_1\cos\theta_2\cos\phi & -\sin\theta_1\sin\theta_2\sin\phi \\
\sin\theta_1\sin\theta_2\sin\phi & \cos\theta_1\sin\theta_2\sin\phi & \sin\theta_1\cos\theta_2\sin\phi & \sin\theta_1\sin\theta_2\cos\phi.
\end{array}$$

Expanding this determinant using the first row, we observe that

$$\begin{aligned}
\mathcal{J}_4(r,\theta_1,\theta_2,\phi)/r^3 &= \cos^2\theta_1\sin^2\theta_1\mathcal{J}_3(r,\theta_2,\phi)/r^2 + \sin^4\theta_1\mathcal{J}_3(r,\theta_2,\phi)/r^2 \\
&= \sin^2\theta_1\mathcal{J}_3(r,\theta_2,\phi)/r^2 \\
&= \sin^2\theta_1\sin\theta_2.
\end{aligned}$$

Hence, $\mathcal{J}_4(r,\theta_1,\theta_2,\phi) = r^3\sin^2\theta_1\sin\theta_2$.

Continuing in this manner, we compute $\mathcal{J}_N(r,\theta_1,\ldots,\theta_{N-2},\phi)$ using induction and obtain

(3.2) $\mathcal{J}_N(r,\theta_1,\ldots,\theta_{N-2},\phi) = r^{N-1}(\sin\theta_1)^{N-2}\cdots(\sin\theta_{N-3})^2(\sin\theta_{N-2}).$

Now, is well known,

(3.3)
$$|B(0,\rho)| = \int_0^\rho\int_0^\pi\cdots\int_0^\pi\int_0^{2\pi}\mathcal{J}_N(r,\theta_1,\ldots,\theta_{N-2},\phi)d\phi d\theta_1\cdots d\theta_{N-2}dr.$$

Also, it is easy to see that

$$|B(0,\rho)| = \int_0^\rho r^{N-1}|S_{N-1}|\,dr = \rho^N|S_{N-1}|/N,$$

where S_{N-1} is the unit (N-1)-sphere in $\mathbf{R}^N$ and $|S_{N-1}|$ is its $(N-1)$-dimensional volume.

In particular, we see from (3.2) and (3.3) that

$$\begin{aligned}
|S_{N-1}| &= 2\pi\int_0^\pi\cdots\int_0^\pi(\sin\theta_1)^{N-2}\cdots(\sin\theta_{N-2})d\theta_1\cdots d\theta_{N-2} \\
&= 2\pi\prod_{j=1}^{N-2}\int_0^\pi(\sin\theta)^j\,d\theta.
\end{aligned}$$

From [Ti1, p. 56], we obtain

$$\int_0^\pi(\sin\theta)^j\,d\theta = \Gamma(\frac{j+1}{2})\Gamma(\frac{1}{2})/\Gamma(\frac{j+2}{2}).$$

Consequently, it follows from this last calculation that

$$|S_{N-1}| = 2\pi[\Gamma(\frac{1}{2})]^{N-2}/\Gamma(\frac{N}{2}) = 2(\pi)^{N/2}/\Gamma(\frac{N}{2}),$$

and therefore that

(3.4) $$|B(0,\rho)| = \frac{2(\pi)^{N/2}}{N\Gamma(\frac{N}{2})}\rho^N.$$

$|S_{N-1}|$ can also be computed from the following observation:

$$
\begin{aligned}
\int_{\mathbf{R}^N} e^{-|x|^2} dx &= |S_{N-1}| \int_0^\infty r^{N-1} e^{-r^2} dr \\
&= |S_{N-1}| 2^{-1} \int_0^\infty s^{\frac{N}{2}-1} e^{-s} ds \\
&= |S_{N-1}| \Gamma(\frac{N}{2})/2.
\end{aligned}
$$

Since, is well known, $\int_0^\infty e^{-t^2} dt = \frac{\pi^{\frac{1}{2}}}{2}$, the left-hand side of the above equality is $(\pi)^{N/2}$, and we obtain the same value for $|S_{N-1}|$ as we did before.

In order to prove Theorem 3.1, we shall need some lemmas. The first of such lemmas is concerned with the Bochner-Riesz summability of Fourier integrals. In particular, if $g \in L^1(\mathbf{R}^N)$ and is complex-valued, we designate the Fourier transform of g by $\widehat{g}$ and define it in a manner analogous to the one used for the Fourier coefficients of a function in $L^1(T_N)$, namely,

$$
\widehat{g}(y) = (2\pi)^{-N} \int_{\mathbf{R}^N} e^{-iy \cdot x} g(x) dx.
$$

The first lemma we prove is the following:

Lemma 3.2. *Let $g \in L^1(\mathbf{R}^N)$ and be complex-valued. Set*

$$
(3.5) \qquad \tau_R^\alpha(g, x) = \int_{B(0,R)} \widehat{g}(y) e^{ix \cdot y} (1 - |y|^2/R^2)^\alpha \, dy.
$$

Suppose that $|B(0,\rho)|^{-1} \int_{B(0,\rho)} |g(x_0 + x) - g(x_0)| \, dx \to 0$ as $\rho \to 0$. Then

$$
\lim_{R \to \infty} \tau_R^\alpha(g, x_0) = g(x_0) \quad \text{for} \quad \alpha > (N-1)/2.
$$

Proof of Lemma 3.2. We will first prove a special case of the lemma, namely, when $g(x) = e^{-|x-x_0|^2}$. We start out by observing once again that $\int_0^\infty e^{-s^2} ds = \frac{\pi^{\frac{1}{2}}}{2}$, and from (1.12) in Appendix A that

$$
\int_0^\infty e^{-s^2} \cos 2ts \, ds = \frac{\pi^{\frac{1}{2}}}{2} e^{-t^2}.
$$

Hence, $\int_{-\infty}^\infty e^{-s^2} e^{-ist} ds = \pi^{\frac{1}{2}} e^{-\frac{t^2}{4}}$, and consequently

$$
(3.6) \qquad \underbrace{\int_{-\infty}^\infty \cdots \int_{-\infty}^\infty}_{N} e^{-(x_1^2 + \cdots + x_N^2)} e^{-ix \cdot y} dx = \pi^{N/2} e^{-|y|^2/4} \quad \text{for } y \in \mathbf{R}^N.
$$

On setting $2x = u$ and $y = 0$ in this last equation, we see that

$$
(3.7) \qquad \int_{\mathbf{R}^N} e^{-|u|^2/4} du = \pi^{N/2} 2^N.
$$

We are now able to establish the lemma in the particular case when $g(x) = e^{-|x-x_0|^2}$. From (3.6), we obtain that

$$
\begin{aligned}
\widehat{g}(y) &= (2\pi)^{-N} \int_{\mathbf{R}^N} e^{-ix\cdot y} e^{-|x-x_0|^2} dx \\
&= (2\pi)^{-N} \int_{\mathbf{R}^N} e^{-iy\cdot(x+x_0)} e^{-|x|^2} dx \\
&= (2\pi)^{-N} e^{-iy\cdot x_0} \pi^{N/2} e^{-|y|^2/4}.
\end{aligned}
$$

Hence, $\widehat{g}(y) \in L^1(\mathbf{R}^N)$, and the equality in (3.7) together with this last value of $\widehat{g}(y)$ then implies that

$$
\begin{aligned}
\lim_{R\to\infty} \int_{B(0,R)} \widehat{g}(y) e^{iy\cdot x_0} (1 - |y|^2/R^2)^\alpha \, dy &= (2\pi)^{-N} \pi^{N/2} \int_{\mathbf{R}^N} e^{-|y|^2/4} dy \\
&= (2\pi)^{-N} \pi^{N/2} \pi^{N/2} 2^N \\
&= g(x_0).
\end{aligned}
$$

Therefore, the lemma is proved in the special case $g(x) = e^{-|x-x_0|^2}$.

From what we have just established, we can prove the lemma. Without loss of generality we can assume from the start that

$$(3.8) \qquad\qquad g(x_0) = 0.$$

Otherwise, we could work with the function

$$h(x) = g(x) - g(x_0) e^{-|x-x_0|^2}.$$

In order to prove the lemma, we will need two estimates concerning Bessel functions that are established in Appendix A. The first estimate we need is

$$(3.9) \qquad |J_\nu(t)| \leq K_\nu t^\nu \text{ for } 0 < t \leq 1 \text{ and } \nu > -\frac{1}{2},$$

and the second is

$$(3.10) \qquad |J_\nu(t)| \leq K_\nu t^{-\frac{1}{2}} \text{ for } 1 \leq t < \infty \text{ and } \nu > -1,$$

where K_ν is a positive constant. The estimates (3.9) and (3.10) correspond respectively to (2.1) and (2.2) in Appendix A.

Continuing with the proof of the lemma, we set

$$(3.11) \qquad (2\pi)^N H_R^\alpha(x) = \int_{B(0,R)} e^{iy\cdot x} (1 - |y|^2/R^2)^\alpha \, dy,$$

and observe from (3.5) and Fubini's theorem that

$$\tau_R^\alpha(g, x_0) = \int_{\mathbf{R}^N} g(x) H_R^\alpha(x - x_0) dx.$$

Hence,

$$(3.12) \qquad \tau_R^\alpha(g, x_0) = \int_{\mathbf{R}^N} g(x + x_0) H_R^\alpha(x) dx.$$

In (1.11) in Appendix A, it is shown that

$$(3.13) \qquad H_R^\alpha(x) = c(N, \alpha) J_{\frac{N}{2}+\alpha}(R\,|x|) R^{\frac{N}{2}-\alpha} / |x|^{\frac{N}{2}+\alpha},$$

where $c(N, \alpha) = (2\pi)^{-N} \omega_{N-2} 2^\alpha \Gamma(\alpha + 1) = 2^\alpha \Gamma(\alpha + 1)/(2\pi)^{N/2}$. We therefore conclude from (3.12) and (3.13) that

$$(3.14) \quad \tau_R^\alpha(g, x_0) = c(N, \alpha) R^N \int_{\mathbf{R}^N} g(x + x_0) J_{\frac{N}{2}+\alpha}(R\,|x|)/(R\,|x|)^{\frac{N}{2}+\alpha} dx.$$

Next, we set

$$G(r) = \int_{B(0,r)} |g(x + x_0)|\, dx,$$

and observe from (3.8) and the hypothesis of the lemma that

$\qquad (i)\ G(r) = o(r^N)$ as $r \to 0$,

$\qquad (ii)\ G(r)$ is uniformly bounded for $0 < r < \infty$,

(3.15)

$\qquad (iii)\ G(r)$ is absolutely continuous on $0 < r < \infty$,

$\qquad (iv)\ dG(r)/dr \geq 0$ a.e. on $0 < r < \infty$.

From the definition of $G(r)$ above, we see from (3.14) and (3.15)(iii) and (iv) that

$$(3.16) \quad |\tau_R^\alpha(g, x_0)| \leq c(N, \alpha) R^N \int_0^\infty \frac{dG(r)}{dr} \left| J_{\frac{N}{2}+\alpha}(Rr) \right| /(Rr)^{\frac{N}{2}+\alpha}\, dr.$$

Also, we see that the statements in (3.15) together with $\alpha > (N-1)/2$ imply that for any $\delta > 0$,

$$R^{N/2-(\alpha+\frac{1}{2})} \int_\delta^\infty r^{-(\frac{N}{2}+\alpha+\frac{1}{2})} dG(r)/dr\, dr = o(1) \text{ as } R \to \infty.$$

Hence, we obtain from (3.12) and (3.16) that

$$(3.17) \quad \limsup_{R \to \infty} |\tau_R^\alpha(g, x_0)|/c(N, \alpha) \leq R^N \int_0^\delta \frac{dG(r)}{dr} \frac{\left| J_{\frac{N}{2}+\alpha}(Rr) \right|}{(Rr)^{\frac{N}{2}+\alpha}}\, dr.$$

Next, given $\varepsilon > 0$ and using (3.15)(i), we choose δ, with $0 < \delta < 1$, so that

$$|G(r)| < \varepsilon r^N \qquad \text{for } 0 < r < \delta,$$

and observe after an integration by parts that

$$\limsup_{R \to \infty} R^{N/2-(\alpha+\frac{1}{2})} \int_{R^{-1}}^\delta r^{-(\frac{N}{2}+\alpha+\frac{1}{2})} dG(r)/dr\, dr \leq \varepsilon \frac{\alpha + \frac{1}{2} + \frac{N}{2}}{\alpha + \frac{1}{2} - \frac{N}{2}}.$$

So using (3.10) and this last computation, we obtain

$$\limsup_{R \to \infty} R^N \int_{R^{-1}}^\delta \frac{dG(r)}{dr} \frac{\left| J_{\frac{N}{2}+\alpha}(Rr) \right|}{(Rr)^{\frac{N}{2}+\alpha}}\, dr \leq \varepsilon K_{\frac{N}{2}+\alpha} \frac{\alpha + \frac{1}{2} + \frac{N}{2}}{\alpha + \frac{1}{2} - \frac{N}{2}}.$$

Also, using (3.9) and (3.15) (iii) and (iv), we see that

$$R^N \int_0^{R^{-1}} \frac{dG(r)}{dr} \left| J_{\frac{N}{2}+\alpha}(Rr) \right| / (Rr)^{\frac{N}{2}+\alpha} \, dr \leq \varepsilon K_{\frac{N}{2}+\alpha}$$

for R sufficiently large.

Hence, on writing the integral on the right-hand side of the inequality in (3.17) in the form $\int_0^\delta = \int_0^{R^{-1}} + \int_{R^{-1}}^\delta$, we see from these last two inequalities that

$$\limsup_{R\to\infty} |\tau_R^\alpha(g,x_0)| / c(N,\alpha) \leq \varepsilon K_{\frac{N}{2}+\alpha} \left(\frac{\alpha + \frac{1}{2} + \frac{N}{2}}{\alpha + \frac{1}{2} - \frac{N}{2}} + 1 \right).$$

Since ε is an arbitrary positive number, we conclude that

$$\lim_{R\to\infty} |\tau_R^\alpha(g,x_0)| = 0,$$

which finishes the proof of the Lemma 3.2 because $g(x_0) = 0$. ∎

The next lemma that we need for the proof of Theorem 3.1 is the following:

Lemma 3.3. *Let $S(x)$ be the trigonometric polynomial $\sum_{|m|\leq R_1} b_m e^{im\cdot x}$, i.e., $S(x) = \sum_{m\in\Lambda_N} b_m e^{im\cdot x}$ where $b_m = 0$ for $|m| > R_1$. For $R>0$, set*

$$\sigma_R^\alpha(S,x) = \sum_{|m|\leq R} b_m e^{im\cdot x} (1 - |m|^2/R^2)^\alpha.$$

Then for $\alpha > (N-1)/2$,

$$(3.18) \qquad \sigma_R^\alpha(S,x) = c(N,\alpha) R^{N/2-\alpha} \int_{\mathbf{R}^N} S(y) \frac{J_{\frac{N}{2}+\alpha}(R|x-y|)}{|x-y|^{\frac{N}{2}+\alpha}} dy$$

where $c(N,\alpha)$ is the constant in (3.13).

Proof of Lemma 3.3. Define $\phi(t) = (1-t^2)^\alpha$, $0 \leq t \leq 1$, and $\phi(t) = 0$ for $t \geq 1$. Then since $S(x)$ is a finite linear combination of exponentials, it is clear that the lemma will follow if we can show that for fixed x and every $u \in \mathbf{R}^N$,

$$(3.19) \qquad e^{iu\cdot x} \frac{\phi(|u|/R)}{c(N,\alpha)} = R^{N/2-\alpha} \int_{\mathbf{R}^N} e^{iu\cdot y} \frac{J_{\frac{N}{2}+\alpha}(R|x-y|)}{|x-y|^{\frac{N}{2}+\alpha}} dy.$$

Set $g(u) = e^{iu\cdot x}\phi(|u|/R)$. Then $g(u)$ is a continuous function which is also in $L^1(\mathbf{R}^N)$. If $\widehat{g}(y)$ is also in $L^1(\mathbf{R}^N)$, it follows from Lemma 3.2 and the Lebesgue dominated convergence theorem that

$$g(u) = \int_{\mathbf{R}^N} e^{iu\cdot y} \widehat{g}(y) dy.$$

For fixed x, (3.9) and (3.10) let

$$J_{\frac{N}{2}+\alpha}(R\,|x-y|)/\,|x-y|^{\frac{N}{2}+\alpha} \in L^1(\mathbf{R}^N) \text{ with respect to } y.$$

So (3.19) will be established if we show that

$$\frac{\widehat{g}(y)}{c(N,\alpha)} = R^{N/2-\alpha}\frac{J_{\frac{N}{2}+\alpha}(R\,|x-y|)}{|x-y|^{\frac{N}{2}+\alpha}}.$$

But from (3.11), we see that $\widehat{g}(y) = H_R^\alpha(x-y)$; this last fact follows from the equality in (3.13). ∎

Proof of Theorem 3.1. We first observe from (3.9), (3.10), and (3.13) that there is a constant $K(\alpha, R)$ and an $\eta > 0$ such that for fixed R,

$$(3.20) \qquad |H_R^\alpha(x)| \le K(\alpha, R)/(1+|x|)^{N+\eta} \text{ for } x \in \mathbf{R}^N,$$

where $K(\alpha, R)$ is a constant depending on α and R. Consequently, the series

$$\sum_{m \in \Lambda_N} H_R^\alpha(x + 2\pi m) = H_R^{*\alpha}(x)$$

is absolutely convergent, and furthermore

$$(3.21) \qquad \lim_{R_1 \to \infty} \sum_{|m| \le R_1} H_R^\alpha(x + 2\pi m) = H_R^{*\alpha}(x)$$

uniformly for x in a bounded domain.

Set $S^j(x) = \sigma_j^\diamond(f, x)$, which is the trigonometric polynomial defined in (2.6). Then by (3.20), $S^j(y)H_R^\alpha(x-y) \in L^1(\mathbf{R}^N)$ with respect to y, and we obtain from Lemma 3.3 and (3.21) that for x in a bounded domain,

$$\begin{aligned}
\sigma_R^\alpha(S^j, x) &= \int_{\mathbf{R}^N} S^j(y)H_R^\alpha(x-y)dy \\
&= \lim_{R_1 \to \infty} \sum_{|m| \le R_1} \int_{T_N} S^j(y+2\pi m)H_R^\alpha(x-y-2\pi m)dy \\
&= \lim_{R_1 \to \infty} \int_{T_N} S^j(y)\Big(\sum_{|m| \le R_1} H_R^\alpha(x-y-2\pi m)\Big)dy \\
&= \int_{T_N} S^j(y)H_R^{*\alpha}(x-y)dy.
\end{aligned}$$

By Theorem 2.2, $S^j \to f$ in $L^1(T_N)$. Also, $H_R^{*\alpha} \in C(T_N)$. So from this last computation we can see by passing to the limit as $j \to \infty$, that

$$(3.22) \qquad \sigma_R^\alpha(f, x) = \int_{T_N} f(y)H_R^{*\alpha}(x-y)dy.$$

But $f(y)$ is defined in $\mathbf{R}^N$ by periodicity of period 2π in each variable. So we see that

$$(3.23) \qquad \int_{B(0,R_1+1)\setminus B(0,R_1)} |f(y)|\, dy = O(R_1^{N-1}) \text{ as } R_1 \to \infty.$$

This fact in conjunction with (3.20), implies that $f(y)H_R^\alpha(x-y) \in L^1(\mathbf{R}^N)$ with respect to y.

Hence, using (3.22), we can reverse the previous calculation and obtain

$$(3.24) \qquad \sigma_R^\alpha(f,x_0) = \int_{\mathbf{R}^N} f(y)H_R^\alpha(x_0-y)dy = \int_{\mathbf{R}^N} f(x+x_0)H_R^\alpha(x)dx.$$

Since the theorem is obviously true if $f(x)$ is a constant function, we can prove the theorem, with no loss in generality, if we assume that $f(x_0) = 0$. Therefore, from the hypothesis of the theorem,

$$\int_{B(0,r)} |f(x+x_0)|\, dx = o(r^N) \text{ as } r \to 0.$$

So using (3.15) and comparing (3.24) with (3.14), we see that locally the same proof will apply here as it was applied in the proof of Lemma 3.2. Consequently, to complete the proof of the theorem, we must to show that for fixed $\delta > 0$,

$$(3.25) \qquad \lim_{R\to\infty} \int_{\mathbf{R}^N\setminus B(0,\delta)} f(x+x_0)H_R^\alpha(x)dx = 0.$$

Using (3.13) in conjunction with the estimate in (3.12), we see that

$$\left| \int_{\mathbf{R}^N\setminus B(0,\delta)} \frac{f(x+x_0)}{\lambda(N,\alpha)} H_R^\alpha(x)dx \right| \leq R^{(N-1)/2-\alpha} \int_{\mathbf{R}^N\setminus B(0,\delta)} \frac{|f(x+x_0)|}{|x|^{\alpha+(N+1)/2}} dx$$

where $\lambda(N,\alpha) = c(N,\alpha)K_{\frac{N}{2}+\alpha}$ is a constant. Since $\alpha > (N-1)/2$, we see from (3.23) that the integral on the right-hand side of this last inequality is finite. Also we see that $(N-1)/2 - \alpha$ is strictly negative. Consequently, the right-hand side of this last inequality is $o(1)$ as $R \to \infty$.

We conclude that the limit in (3.25) is indeed valid, and we complete the proof of Theorem 3.1. ∎

$\alpha = (N-1)/2$ is called the critical index for Bochner-Riesz summability. What is very interesting about Theorem 3.1 is that it fails at the critical index for $N \geq 2$, even if $f = 0$ in a neighborhood of x_0. Bochner has shown, in particular that with $0 < \delta < 1$,

$$(3.26) \qquad \exists\, f \in L^1(T_N), N \geq 2, \text{with } f = 0 \text{ in } B(0,\delta)$$

such that
$$\limsup_{R\to\infty} \left| \sigma_R^{(N-1)/2}(f,0) \right| = \infty.$$

To see this ingenious counter-example, we refer the reader to [Boc, p. 193] or [Sh1, pp. 57-64].

It is clear from the Riemann-Lebesgue Lemma and the form of the Dirichlet kernel given in (2.1) that Bochner's counter-example itself does not hold when $N = 1$.

We close this section with the following corollary of Theorem 3.1:

Corollary 3.4. *Suppose $f \in L^1(T_N)$. Then for $\alpha > (N-1)/2$,*

$$\lim_{R \to \infty} \sigma_R^\alpha(x) = f(x) \qquad \text{for a.e. } x \in T_N.$$

Proof of Corollary 3.4. Since almost every $x \in T_N$ is in the Lebesgue set of f (see page 22), Corollary 3.4 follows immediately from Theorem 3.1. ∎

Exercises.

1. Show that Bochner's counter-example does indeed fail in dimension $N = 1$.

2. Find the third and fourth rows in the determinant corresponding to $\mathcal{J}_N(r, \theta_1, \dots, \theta_{N-2}, \phi)$ when $N = 5$.

3. By direct calculation, show that the following formula is true when $j = 3$:

$$\int_0^\pi (\sin \theta)^j \, d\theta = \Gamma(\frac{j+1}{2}) \Gamma(\frac{1}{2}) / \Gamma(\frac{j+2}{2}).$$

4. Given that $G(r)$ satisfies the conditions in (3.15) and that $\alpha > (N-1)/2$, $\delta > 0$ prove that

$$R^{N/2-(\alpha+\frac{1}{2})} \int_\delta^\infty r^{-(\frac{N}{2}+\alpha+\frac{1}{2})} dG(r)/dr \, dr = o(1) \text{ as } R \to \infty.$$

4. Abel Summability of Fourier Series

The Abel summability of Fourier series was defined in (1.3) of this chapter, and in this section, we shall prove three theorems regarding this method of summation. The first theorem we establish is an N-dimensional version of a well-known theorem in one dimension originally due to Fatou [Zy1, p. 100].

Theorem 4.1. *Let $f \in L^1(T_N)$, and for $t > 0$, set*

$$A_t(f, x) = \sum_{m \in \Lambda_N} \widehat{f}(m) e^{im \cdot x - |m| t}.$$

Also, set

$$\beta^-(x) = \limsup_{r \to 0} \frac{\int_{B(x,r)} f(y)dy}{|B(x,r)|} \quad and \quad \beta_-(x) = \liminf_{r \to 0} \frac{\int_{B(x,r)} f(y)dy}{|B(x,r)|} .$$

Then

$$\beta_-(x) \le \liminf_{t \to 0} A_t(f,x) \le \limsup_{t \to 0} A_t(f,x) \le \beta^-(x).$$

Of course, this theorem implies that in case $\beta^-(x) = \beta_-(x)$, then the Fourier series of f is Abel summable at x to this common value.

Proof of Theorem 4.1. To prove Theorem 4.1, we proceed in a manner similar to the proof given in Theorem 3.1. First, let $g \epsilon L^1(\mathbf{R}^N)$, and set

$$(4.1) \qquad \mathcal{A}_t(g,x) = \int_{\mathbf{R}^N} \widehat{g}(y) e^{iy \cdot x - |y|t} dy \qquad \text{for } t > 0,$$

where $\widehat{g}(y)$ is the Fourier transform of g and is defined above Lemma 3.2. Then, for $t > 0$, by Fubini's theorem,

$$(4.2) \qquad \mathcal{A}_t(g,x) = (2\pi)^{-N} \int_{\mathbf{R}^N} g(u) [\int_{\mathbf{R}^N} e^{iy \cdot (x-u) - |y|t} dy] du.$$

But, for $N \ge 2$,

$$\int_{\mathbf{R}^N} e^{iy \cdot (x-u) - |y|t} dy = |S_{N-2}| \int_0^\infty e^{-rt} r^{N-1} \int_0^\pi e^{i|x-u|r \cos \theta} (\sin \theta)^{N-2} d\theta.$$

Consequently,

$$(4.3) \qquad \int_{\mathbf{R}^N} e^{iy \cdot (x-u) - |y|t} dy = \omega_{N-2} \int_0^\infty e^{-rt} r^{N-1} \frac{J_{(N-2)/2}(r|x-u|)}{(r|x-u|)^{(N-2)/2}} dr,$$

where we have made use of the integral identity in (1.5) in Appendix A and $\omega_{N-2} = (2\pi)^{N/2}$ is the constant defined below (1.11) in Appendix A.

For $N = 1$, the equality in (4.3) continues to hold with $\omega_1 = (2\pi)^{1/2}$. This follows from a direct calculation that uses the well-known fact that

$$\cos t = (\pi/2)^{1/2} t^{1/2} J_{-1/2}(t) \quad \text{for } t > 0.$$

Next, we use the integral identity (1.7) in Appendix A and conclude from the equality in (4.3) that

$$\int_{\mathbf{R}^N} e^{iy \cdot (x-u) - |y|t} dy = b_N t [t^2 + |x - u|^2]^{-(N+1)/2}$$

where $b_N = (2)^{N/2} \Gamma(\frac{N+1}{2}) \omega_{N-2}(\pi)^{-\frac{1}{2}}$.

This last equality, in conjunction with (4.2), establishes the useful fact that for $t > 0$,

$$(4.4) \qquad \mathcal{A}_t(g,x) = (2\pi)^{-N} b_N \int_{\mathbf{R}^N} g(y) t [t^2 + |x - y|^2]^{-(N+1)/2} dy.$$

Next, we observe that the analog of Lemma 3.2 holds for $\mathcal{A}_t(g,x)$.

Also, we see that the analog of Lemma 3.3 holds, namely, if $S(x)$ is a trigonometric polynomial, then

$$(4.4') \qquad \mathcal{A}_t(S, x) = (2\pi)^{-N} b_N \int_{\mathbf{R}^N} S(y) t [t^2 + |x - y|^2]^{-(N+1)/2} dy.$$

To show that this is indeed the case, we need to only establish, as in the proof of Lemma 3.3, that

$$e^{iu \cdot x} e^{-|u|t} / b_N = (2\pi)^{-N} \int_{\mathbf{R}^N} e^{iu \cdot y} t [t^2 + |x - y|^2]^{-(N+1)/2} dy$$

for $u \in \mathbf{R}^N$ and $t > 0$. This equality will follow from the fact that the Fourier transform of $e^{iu \cdot x} e^{-|u|t} / b_N$ is

$$(2\pi)^{-N} t [t^2 + |x - y|^2]^{-(N+1)/2},$$

which is the statement three lines above (4.4) when u and y are interchanged.

Using the same technique that we used in the proof of Theorem 3.1 (i.e., see (3.25) through (3.27) in §3), to pass from Fourier integrals to Fourier series, we obtain from $(4.4')$ that for $f \in L^1(T_N)$,

$$(4.5) \qquad A_t(f, x) = (2\pi)^{-N} b_N \int_{\mathbf{R}^N} f(x + y) t [t^2 + |y|^2]^{-(N+1)/2} dy.$$

To prove Theorem 4.1, it is sufficient to just establish the last inequality stated in the conclusion, namely,

$$(4.6) \qquad \limsup_{t \to 0} A_t(f, x) \leq \beta^-(x).$$

For then the first inequality follows from a consideration of $-f$.

If $\beta^-(x) = \infty$, (4.6) is established. So we need only consider the two cases: (i) $\beta^-(x)$ is finite, or (ii) $\beta^-(x) = -\infty$ in establishing (4.6). It is clear that the inequality in (4.6) will follow in both these cases if we show that the following holds for $\gamma \epsilon \mathbf{R}$:

$$(4.7) \qquad \beta^-(x) < \gamma \Longrightarrow \limsup_{t \to 0} A_t(f, x) \leq \gamma.$$

We now establish (4.7). To do this, first of all, we observe from (4.5) that $f(y)$ identically one implies that

$$(4.8) \qquad (2\pi)^{-N} b_N t \int_{\mathbf{R}^N} [t^2 + |y|^2]^{-(N+1)/2} dy = 1 \quad \text{for} \quad t > 0.$$

Next, we set

$$(4.9) \qquad f_{[r]}(x) = \frac{\int_{B(0,r)} f(x + y) dy}{|B(0, r)|},$$

and use the hypothesis in (4.7) choose $\delta > 0$ so that

$$(4.10) \qquad f_{[r]}(x) < \gamma \quad \text{for} \quad 0 < r < \delta.$$

Observing that $f(x+y)|y|^{-(N+1)}$ $\epsilon L^1(\mathbf{R}^N \setminus B(0,\delta))$ with respect to y (because for fixed x, $f(x+y)\epsilon L^1(T_N)$ and is periodic of period 2π in each variable), we see that

$$\lim_{t \to 0} t \int_{\mathbf{R}^N \setminus B(0,\delta)} f(x+y)[t^2+|y|^2]^{-(N+1)/2} dy = 0.$$

Consequently, we obtain from (4.5) that
(4.11)
$$\limsup_{t \to 0} A_t(f,x) \leq (2\pi)^{-N} b_N \limsup_{t \to 0} \int_{B(0,\delta)} tf(x+y)[t^2+|y|^2]^{-(N+1)/2} dy.$$

From (4.9), we next observe that the integral on the right-hand side of this last inequality can be written as

$$t \int_0^{\delta} [t^2+r^2]^{-(N+1)/2} \frac{d[|B(0,r)| \, f_{[r]}(x)]}{dr} dr.$$

So we conclude from (4.9) and (4.10), after performing an integration by parts on this last integral, that
(4.12)
$$\limsup_{t \to 0} A_t(f,x) \leq \gamma(2\pi)^{-N} b_N \limsup_{t \to 0} (N+1)t \int_0^{\delta} r[t^2+r^2]^{-\frac{N+3}{2}} |B(0,r)| \, dr.$$

Likewise, after integrating by parts, we see from the identity in (4.8) that

$$(2\pi)^{-N} b_N \lim_{t \to 0} (N+1)t \int_0^{\delta} r[t^2+r^2]^{-\frac{N+3}{2}} |B(0,r)| \, dr = 1.$$

This last equality together with the inequality in (4.12) establishes the implication in (4.7) and concludes the proof to Theorem 4.1. ∎

The next theorem that we establish involves the concept of nontangential Abel summability. With $x_0 \epsilon \mathbf{R}^N$ and $\gamma > 0$, let $\mathcal{C}_\gamma(x_0)$ stand for the cone in $\mathbf{R}_+^{N+1}$ with vertex $(x_0,0)$ given as follows:

(4.13) $$\mathcal{C}_\gamma(x_0) = \{(x,t) : t > 0 \text{ and } \frac{t}{|x-x_0|} \geq \gamma \} \, .$$

We say that the Fourier series of f, namely $S[f]$, is nontangentially Abel summable at x_0 to the limit l if for every $\gamma > 0$,

$$\lim_{(x,t) \to ((x_0,0)} A_t(f,x) = l$$

where (x,t) tends to $(x_0,0)$ within the cone $\mathcal{C}_\gamma(x_0)$.

The nontangential Abel summability theorem that we shall present here is an improvement (for $N \geq 2$) over the usual one presented in books related to this subject (e.g., see [SW, p. 62]). In order to do this, we introduce the σ-set of f where $f \in L^1(T_N)$. We say x_0 is in the σ-set of f provided the following holds: $\forall \varepsilon > 0$, $\exists \delta > 0$ such that $|x - x_0| < \delta$ and $r < \delta$ implies that

$$(4.14) \qquad \left| \int_{B(x,r)} [f(y) - f(x_0)] dy \right| < \varepsilon (|x - x_0| + r)^N.$$

We prove the following theorem (see [Sh4]):

Theorem 4.2. *Let $f \epsilon L^1(T_N)$, and suppose that $x_0 \in \sigma\text{-set of } f$. Then $S[f]$ is nontangentially Abel summable at x_0 to $f(x_0)$.*

For $N = 1$, this result is the same as the result given in [Zy1, p. 61] which is evidently due to Fatou and states that if $F = \int f$ and F has a finite derivative equal to $f(x_0)$ (henceforth referred to as the Fatou condition at x_0), then nontangential Abel summability occurs at x_0. It is not difficult to show that for $N = 1$, $x_0 \in \sigma\text{-set of } f$ if and only if the Fatou condition holds for f at x_0.

For $N \geq 2$, this result about $x_0 \in \sigma\text{-set of } f$ has not appeared previously in any book and is due to the author (see [Sh 4]). The usual theorem proved is that if $x_0 \in$ Lebesgue set of f, nontangential Abel summability occurs at x_0, [SW, p. 62]. After we prove the above theorem, we shall show $x_0 \in$ Lebesgue set of f implies that $x_0 \in \sigma\text{-set of } f$. Also, we shall give an example of an $f \in L^\infty(T_2)$ such that x_0 is not in the Lebesgue set of f, but x_0 is in the $\sigma\text{-set of } f$.

Proof of Theorem 4.2. To prove the theorem, it is easy to see from the start that we can assume that $x_0 = 0$. Therefore, to prove the theorem, we assume that $\gamma > 0$ and that $\{(x_n, t_n)\}_{n=1}^\infty \subset \mathcal{C}_\gamma(0)$ with $x_n \to 0$ and $t_n \to 0$. The proof will be complete when we show that

$$(4.15) \qquad \lim_{n \to \infty} A_{t_n}(f, x_n) = f(0).$$

Given $\varepsilon > 0$, it is clear that the limit in (4.15) will follow if we show that

$$(4.16) \qquad \limsup_{n \to \infty} \left| \frac{A_{t_n}(f, x_n) - f(0)}{(2\pi)^{-N} b_N} \right| \leq 2(N+1)\varepsilon \left(\frac{1}{\gamma} + 1 \right)^N.$$

It follows from (4.5) and (4.8) in the proof of Theorem 4.1 that

$$\frac{A_{t_n}(f, x_n) - f(0)}{(2\pi)^{-N} b_N} = \int_{\mathbf{R}^N} [f(x_n + y) - f(0)] t_n [t_n^2 + |y|^2]^{-(N+1)/2} dy$$

where $b_N = (2)^{N/2} \Gamma(\frac{N+1}{2}) \omega_{N-2}(\pi)^{-\frac{1}{2}}$ and $\omega_{N-2} = (2\pi)^{N/2}$. Hence, the inequality in (4.16) will follow if we show that
(4.17)

$$\limsup_{n \to \infty} \left| \int_{\mathbf{R}^N} [f(x_n + y) - f(0)] t_n [t_n^2 + |y|^2]^{-(N+1)/2} dy \right| \leq 2(N+1)\varepsilon \left(\frac{1}{\gamma} + 1 \right)^N.$$

Next, we set
(4.18)
$$F_n(r) = \int_{B(0,r)} [f(x_n + y) - f(0)] dy = \int_0^r d\rho \int_{S(0,\rho)} [f(x_n + y) - f(0)] dS(y),$$

where $S(0, \rho) = \partial B(0, r)$. Using the fact that $0 \in \sigma$-set of f, we invoke (4.14) and choose a $\delta > 0$, so that

(4.19) $|F_n(r)| < \varepsilon(|x_n| + r)^N$ for $|x_n| < \delta$ and $r < \delta$.

Also, we observe that

$$\int_{\mathbf{R}^N \backslash B(0,\delta)} |f(x_n + y) - f(0)| / |y|^{N+1} dy \text{ is uniformly bounded in } n.$$

Furthermore, for $|y| \geq \delta$, $[t_n^2 + |y|^2]^{-(N+1)/2} \leq |y|^{-(N+1)}$. Hence, it follows that the inequality in (4.17) will be established if we show that
(4.20)
$$\limsup_{n \to \infty} \left| \int_{B(0,\delta)} [f(x_n + y) - f(0)] t_n [t_n^2 + |y|^2]^{-\frac{N+1}{2}} dy \right| \leq 2(N+1)\varepsilon(\frac{1}{\gamma} + 1)^N.$$

From (4.18), we see that $F_n(r)$ is absolutely continuous on the interval $(0, \delta)$ with $\frac{dF_n(r)}{dr}$ existing almost everywhere in $(0, \delta)$ and also in $L^1(0, \delta)$. Therefore, the integral in (4.20) is equal to

$$t_n \int_0^\delta \frac{dF_n(r)}{dr}(t_n^2 + r^2)^{-(N+1)/2} dr.$$

We conclude after integrating by parts, that the inequality in (4.20) will be established if we show

(4.21) $\limsup_{n \to \infty} \left| t_n \int_0^\delta r F_n(r)(t_n^2 + r^2)^{-(N+3)/2} dr \right| \leq 2(\gamma^{-1} + 1)^N \varepsilon.$

Next, we observe that $\int_0^\delta = \int_0^{t_n} + \int_{t_n}^\delta$. We shall deal with each of these cases separately and show

(4.22) $\limsup_{n \to \infty} \left| t_n \int_0^{t_n} r F_n(r)(t_n^2 + r^2)^{-(N+3)/2} dr \right| \leq (\gamma^{-1} + 1)^N \varepsilon$

and

(4.23) $\limsup_{n \to \infty} \left| t_n \int_{t_n}^\delta r F_n(r)(t_n^2 + r^2)^{-(N+3)/2} dr \right| \leq (\gamma^{-1} + 1)^N \varepsilon.$

Once the inequalities in (4.22) and (4.23) are established, then the inequality in (4.21) follows. So, to complete the proof of the theorem, it remains to show that the inequalities in (4.22) and (4.23) are valid.

We proceed with the situation in (4.22). For this case, $t_n < \delta$, and $0 < r < t_n$. Also, from (4.13) with $x_0 = 0$, $|x_n| \leq \gamma^{-1} t_n$. For this case Consequently, we see from (4.19), that

$$|F_n(r)| \leq \varepsilon(|(x_n| + r)^N \leq \varepsilon(\gamma^{-1} t_n + t_n)^N.$$

Therefore,

$$
\left| \int_0^{t_n} r F_n(r)(t_n^2 + r^2)^{-(N+3)/2} dr \right| \leq \varepsilon(\gamma^{-1}+1)^N t_n^N \int_0^{t_n} r(t_n^2 + r^2)^{-\frac{N+3}{2}} dr
$$

$$
\leq \varepsilon(\gamma^{-1}+1)^N t_n^N \int_0^{t_n} r t_n^{-(N+3)} dr
$$

$$
\leq \varepsilon(\gamma^{-1}+1)^N t_n^{-1}/2,
$$

and we conclude that the inequality in (4.22) does indeed hold.

So to complete the proof of the theorem, it remains to show that the inequality in (4.23) is valid. For this case, $t_n < r < \delta$, and from (4.13) with $x_0 = 0$, we also see that $|x_n| \leq \gamma^{-1} t_n \leq \gamma^{-1} r$. Consequently, we infer from (4.19) that

$$
|F_n(r)| \leq \varepsilon(|(x_n| + r)^N \leq \varepsilon(\gamma^{-1}r + r)^N.
$$

Therefore,

$$
\left| \int_{t_n}^\delta r F_n(r)(t_n^2 + r^2)^{-(N+3)/2} dr \right| \leq \varepsilon(\gamma^{-1}+1)^N \int_{t_n}^\delta r^{N+1}(t_n^2 + r^2)^{-\frac{N+3}{2}} dr
$$

$$
\leq \varepsilon(\gamma^{-1}+1)^N \int_{t_n}^\delta r^{N+1} r^{-(N+3)} dr
$$

$$
\leq \varepsilon(\gamma^{-1}+1)^N t_n^{-1},
$$

and, we conclude that the inequality in (4.23) is indeed valid. The proof of the theorem is therefore complete. ∎

Next we show that $x_0 \in$ Lebesgue set of f implies that $x_0 \in \sigma$-set of f. We recall that x_0 is in the Lebesgue set of f means that

$$
\lim_{\rho \to 0} \rho^{-N} \int_{B(x_0, \rho)} |f(y) - f(x_0)| \, dy = 0.
$$

Hence, given $\varepsilon > 0$, there exists $\delta > 0$ such that $\rho < \delta$ implies that

$$
\int_{B(x_0, \rho)} |f(y) - f(x_0)| \, dy < \varepsilon \rho^N.
$$

Since $B(x, r) \subset B(x_0, |x - x_0| + r)$, we see from this last observation that given $\varepsilon > 0$, $|x - x_0| < \delta/2$ and $r < \delta/2$ implies that

$$
\left| \int_{B(x,r)} [f(y) - f(x_0)] dy \right| \leq \int_{B(x_0, |x - x_0| + r)} |f(y) - f(x_0)| \, dy
$$

$$
< \varepsilon(|x - x_0| + r)^N.
$$

Hence, from (4.14), we infer that x_0 is indeed in the σ-set of f.

Next, we give an example of an $f \in L^\infty(T_2)$ such that 0 is not in the Lebesgue set of f, but 0 is in the σ-set of f .

To exhibit our example, we first consider the function $h_n(s)$ defined on the interval $(n+1)^{-1} \leq s \leq n^{-1}$ $\forall n \geq 1$. In order to this, we first introduce the five points $\{\xi_j^n\}_{j=0}^4$ that subdivide the interval $[(n+1)^{-1}, n^{-1}]$ into four equal intervals, namely,

$$\xi_j^n = \frac{1}{n+1} + \frac{j}{4n(n+1)} \qquad j = 0,1,2,3,4.$$

For example, when $n = 1$, we have the following subdivision where the first dot after $\frac{1}{2}$ is $\frac{5}{8}$, the next $\frac{3}{4}$, and the next $\frac{7}{8}$:

$$\frac{1}{2} \underline{\hspace{1cm} \cdot \hspace{1cm} \cdot \hspace{1cm} \cdot \hspace{0.5cm}} 1$$

So $\xi_0^1 = \frac{1}{2}$, $\xi_1^1 = \frac{5}{8}$, $\xi_2^1 = \frac{3}{4}$, $\xi_3^1 = \frac{7}{8}$, $\xi_4^1 = 1$.

For the general situation, we have

$$\frac{1}{n+1} \underline{\hspace{1cm} \cdot \hspace{1cm} \cdot \hspace{1cm} \cdot \hspace{0.5cm}} \frac{1}{n}$$

where the first dot after $\frac{1}{n+1}$ is $\xi_1^n = \frac{1}{n+1} + \frac{1}{4n(n+1)}$, the next dot $\xi_2^n = \frac{1}{n+1} + \frac{2}{4n(n+1)}$, and the next $\xi_3^n = \frac{1}{n+1} + \frac{3}{4n(n+1)}$.

We then define $h_n(s)$ to be linear in each of the intervals $[\xi_0^n, \xi_1^n]$, $[\xi_1^n, \xi_3^n]$, and $[\xi_3^n, \xi_4^n]$ with $h_n(\xi_0^n) = h_n(\xi_2^n) = h_n(\xi_4^n) = 0$ and $h_n(\xi_1^n) = 1$ and $h_n(\xi_3^n) = -1$. In other words,

$$\begin{aligned} h_n(s) &= 4n(n+1)(s - \xi_0^n) && \text{for } \xi_0^n \leq s \leq \xi_1^n \\ &= 4n(n+1)(\xi_2^n - s) && \text{for } \xi_1^n \leq s \leq \xi_3^n \\ &= 4n(n+1)(s - \xi_4^n) && \text{for } \xi_3^n \leq s \leq \xi_4^n. \end{aligned}$$

In particular, with $\frac{1}{2} = .5$, $\frac{5}{8} = .625$, $\frac{3}{4} = .75$, and $\frac{7}{8} = .875$, we get the following picture for $h_1(s)$:

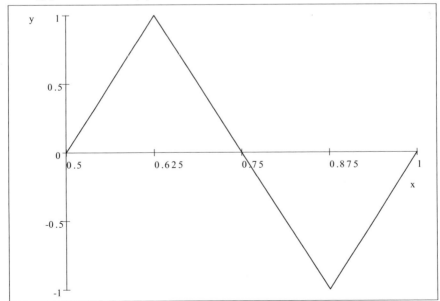

Next, we define $g(s)$ on the half-open interval $(0,1]$ as follows:

(4.24) $g(s) = h_n(s)$ for $s\epsilon[(n+1)^{-1}, n^{-1}]$ $n = 1, 2, ...,$

and then on **R** in the following manner:

$$g(s) \;=\; 0 \text{ for } s = 0 \text{ and } s \geq 1,$$
$$\;=\; -g(-s) \text{ for } s \leq 0.$$

It is clear that $g(s)$ is uniformly bounded in **R** and continuous everywhere except $s = 0$. However, if we define $G(s)=\int_0^s g(t)dt$ for $s \in$ **R**, we see that $G(0) = 0$, that $G(s)$ is an even function, and that

(4.25) $|G(s)| \leq [4n(n+1)]^{-1}$ for $s\epsilon[(n+1)^{-1}, n^{-1}]$ $n = 1, 2, ...$

Therefore,

$$\frac{|G(s)|}{s} \;\leq\; \frac{n+1}{4n(n+1)} \;\leq\; s \quad \text{for } s\epsilon[(n+1)^{-1}, n^{-1}] \quad n = 1, 2, ...,$$

and consequently, this last inequality plus the fact that G is an even function implies that

(4.26) $\dfrac{|G(s)|}{|s|} \leq |s|$ for $0 < |s| \leq 1.$

Since $G(0) = 0$, we obtain from (4.26) that the derivative of G exists at 0 with $\frac{dG}{ds}(0) = 0$. But then from the definition of G and the fact that $g(0) = 0$, we have that

(4.27) $\dfrac{dG}{ds}(s) = g(s) \qquad \forall s\epsilon\mathbf{R}.$

Also, we see that

$$\int_0^s |g(t)|\, dt \geq \sum_{k=n+1}^{\infty} \frac{1}{2k(k+1)} \quad \text{for } s\epsilon[(n+1)^{-1}, n^{-1}] \quad n = 1, 2, ...,$$

and

$$\sum_{k=n+1}^{\infty} \frac{1}{k(k+1)} \geq \sum_{k=n+2}^{\infty} \frac{1}{k^2} \geq \frac{1}{n+2} \geq \frac{s}{3} \text{ for } s\epsilon[(n+1)^{-1}, n^{-1}]$$

and $n = 1, 2, ...$

From these last two sets of inequalities, it follows that

(4.28) $s^{-1} \displaystyle\int_0^s |g(t)|\, dt \geq 6^{-1}$ for $0 < s \leq 1.$

We now define the function for our example, namely, $f(x_1, x_2)\ \epsilon L^\infty(T_2)$ as follows: for $x \in T_2,$

$$f(x_1, x_2) \;=\; g(x_1) \quad \text{for } (x_1^2 + x_2^2)^{\frac{1}{2}} < 2$$

(4.29)

$$\;=\; 0 \qquad\qquad \text{for } x \in T_2 \backslash B(0, 2).$$

We define $f(x_1, x_2)$ in the rest of $\mathbf{R}^2$ by the periodicity of period 2π in each variable.

Next, we let $Sq(0, r)$ be the square of side $2r$ centered at 0, and observe that $Sq(0, r) \subset B(0, 2r)$. Therefore, from (4.29), we have that for $0 < r < 1$,

$$\int_{B(0,2r)} |f(x)| \, dx \geq \int_{Sq(0,r)} |f(x)| \, dx$$

$$\geq \int_{-r}^{r} [\int_{-r}^{r} |f(x_1, x_2)| \, dx_1] x_2$$

$$\geq 4r \int_{0}^{r} |g(s)| \, ds.$$

Also, recalling that $g(0) = 0$, we have that $f(0) = 0$. Hence, we infer from this last set of inequalities and the inequality in (4.28) that

$$r^{-2} \int_{B(0,2r)} |f(x) - f(0)| \, dx \geq \frac{2}{3} \quad \text{for } 0 < r < 1.$$

Consequently,

$$\liminf_{r \to 0} r^{-2} \int_{B(0,2r)} |f(x) - f(0)| \, dx \geq \frac{2}{3},$$

and we conclude that 0 is not in the Lebesgue set of f.

To complete our example, it remains to show that 0 is in the σ-set of f. To accomplish this, we set

(4.30) $$F(x_1, x_2) = G(x_1) \quad \text{for} \quad (x_1, x_2) \epsilon \mathbf{R}^2,$$

and infer from (4.27) and (4.29) that F has a total derivative at each point of $B(0, 2)$, and furthermore, if $(x_1^2 + x_2^2)^{\frac{1}{2}} < 2$, then

(4.31) $$\frac{\partial F}{\partial x_1}(x_1, x_2) - f(x_1, x_2) \quad \text{and} \quad \frac{\partial F}{\partial x_2}(x_1, x_2) = 0.$$

We next invoke the version of Green's theorem given in [Sh8, p. 262] and obtain that for $(x_1^2 + x_2^2)^{\frac{1}{2}} < 1$ and $r < 1$,

(4.32) $$\int_{\partial B(x,r)} F(y_1, y_2) dy_2 = \int_{B(x,r)} [f(y) - f(0)] dy,$$

where we also have made use of the fact that $f(0) = 0$.

Now,

$$\int_{\partial B(x,r)} F(y_1, y_2) dy_2 = r \int_{0}^{2\pi} F(x_1 + r \cos\theta, x_2 + r \sin\theta) \cos\theta d\theta$$

$$= r \int_{0}^{2\pi} G(x_1 + r \cos\theta) \cos\theta d\theta.$$

So, from (4.32) and this last computation, we have that

$$(4.33) \qquad \left| \int_{B(x,r)} [f(y) - f(0)] dy \right| \le r \int_0^{2\pi} |G(x_1 + r\cos\theta)| \, d\theta.$$

Next, from the inequality in (4.26), we see that

$$(4.34) \qquad |G(x_1 + r\cos\theta)| \le |x_1 + r\cos\theta|^2 \quad \text{for} \quad |x_1 + r\cos\theta| < 1.$$

Consequently, given $\varepsilon > 0$ with $\varepsilon < 1$, we choose $\delta = \frac{\varepsilon}{2\pi}$. Then, from (4.33) and (4.34), we obtain that

$$\text{for } |x| < \delta \text{ and } r < \delta, \quad \left| \int_{B(x,r)} [f(y) - f(0)] dy \right| < \varepsilon(|x| + r)^2,$$

and we conclude from (4.14) that 0 is indeed in the σ-set of f. Therefore, 0 is not in the Lebesgue set of f, but it is in the σ-set of f, and our example is complete.

Next, with

$$(4.35) \qquad A_t(f, x) = \sum_{m \in \Lambda_N} \widehat{f}(m) e^{im \cdot x - |m|t} \quad \text{where } t > 0,$$

we shall prove the following theorem that we shall need in Chapter 2.

Theorem 4.3. *Let $f \in L^1(T_N)$ and suppose $A_t(f, x)$ is defined as in (4.35). Then*

$$(4.36) \qquad \lim_{t \to 0} \int_{T_N} |A_t(f, x) - f(x)| \, dx = 0.$$

To prove the theorem, we will first need the following lemma, which is sometimes known as the Poisson summation formula.

Lemma 4.4. *Set*

$$(4.37) \qquad P(x, t) = \sum_{m \in \Lambda_N} e^{im \cdot x - |m|t} \quad \text{for } t > 0.$$

Then

$$(4.38) \qquad P(x, t) = b_N t \sum_{m \in \Lambda_N} [t^2 + |x + 2\pi m|^2]^{-(N+1)/2},$$

for $t > 0$ where $b_N = 2^N \Gamma(\frac{N+1}{2}) \pi^{(N-1)/2}$.

Proof of Lemma 4.4. We shall call $P(x,t)$, which is defined in (4.37), the Poisson kernel and shall set

$$(4.39) \qquad P^*(x,t) = b_N t \sum_{m \in \Lambda_N} [t^2 + |x + 2\pi m|^2]^{-(N+1)/2}.$$

The proof of the lemma will be complete when we succeed in showing that

$$(4.40) \qquad P(x,t) = P^*(x,t).$$

In order to show that the equality in (4.40) is true, we observe from (4.37) and (4.39), for $t > 0$, that both functions are continuous and also periodic of period 2π in each variable. So from Corollary 2.3 (i.e., the completeness of the trigonometric system), to establish the equality in (4.40), it is sufficient to show

$$(4.41) \quad (2\pi)^{-N} \int_{T_N} P^*(y,t) e^{-im^* \cdot y} dy = e^{-|m^*|t} \quad \text{for } t > 0 \text{ and } m^* \in \Lambda_N.$$

From (4.39), we see that

$$(4.42) \qquad \int_{T_N} P^*(y,t) e^{-im^* \cdot y} dy = b_N t \int_{\mathbf{R}^N} [t^2 + |y|^2]^{-(N+1)/2} e^{-im^* \cdot y} dy.$$

On the other hand, from the formula four lines above (4.4), we see that the Fourier transform of $e^{-|u|t}$ is

$$(2\pi)^{-N} b_N t [t^2 + |y|^2]^{-(N+1)/2}.$$

Therefore, from Lemma 3.2 coupled with the Lebesgue dominated convergence theorem, we see that

$$(2\pi)^{-N} b_N t \int_{\mathbf{R}^N} [t^2 + |y|^2]^{-(N+1)/2} e^{-im^* \cdot y} dy = e^{-|m^*|t}.$$

So (4.42) joined with this last equality shows that the equality in (4.41) is indeed true and completes the proof of the lemma. ∎

Proof of Theorem 4.3. We first observe from (4.37) and (4.38) that the Poisson kernel, $P(x,t)$, has the following properties:

$$(4.43) \qquad \begin{aligned} &(i) \ P(x,t) \geq 0 \ \text{ for } x \in T_N \ \text{ and } t > 0; \\[4pt] &(ii) \ (2\pi)^{-N} \int_{T_N} P(x,t) dx = 1 \text{ for } t > 0. \end{aligned}$$

Also, we see that

$$|x + 2\pi m| \geq |2\pi m| - |x| \qquad \text{for } x \in T_N \ \text{ and } |m| \geq 1.$$

Consequently, it follows from the equality in (4.38) that for $0 < \delta < 1$,

$$(4.44) \qquad \sup_{x \in T_N - B(0,\delta)} |P(x,t)| \to 0 \quad \text{as } t \to 0.$$

Continuing with the proof of the theorem, we observe from (4.35) and (4.37) that

$$A_t(f, x) = (2\pi)^{-N} \int_{T_N} f(x - y) P(y, t) dy \text{ for } t > 0.$$

Hence, it follows from (4.43)(i) and(ii) that

$(2\pi)^N \int_{T_N} |A_t(f, x) - f(x)| \, dx$

$$\leq \int_{T_N} P(y, t) dy \int_{T_N} |f(x - y) - f(x)| \, dx.$$

Let $\varepsilon > 0$ be given. To complete the proof of the theorem, we conclude from this last inequality that it is sufficient to show that

$$(4.45) \qquad \limsup_{t \to 0} \int_{T_N} P(y, t) dy \int_{T_N} |f(x - y) - f(x)| \, dx \leq (2\pi)^N \varepsilon.$$

Since $f \in L^1(T_N)$ and we have a periodic of period 2π in each variable, it follows that there is a δ with $0 < \delta < 1$ such that

$$(4.46) \qquad \int_{T_N} |f(x - y) - f(x)| \, dx \leq \varepsilon \text{ for } |y| \leq \delta.$$

Also, we see that

$$\int_{T_N} |f(x - y) - f(x)| \, dx \leq 2 \|f\|_{L^1(T_N)} \text{ for } y \in \mathbf{R}^N.$$

So it follows from (4.46) and this last inequality that the iterated integrals on the left-hand side of the inequality in (4.45) are majorized by

$$2 \|f\|_{L^1(T_N)} \int_{T_N \setminus B(0, \delta)} P(y, t) dy$$

$$+ \varepsilon \int_{B(0, \delta)} P(y, t) dy.$$

We conclude from (4.43)(i) and (ii) and from (4.44) that the *lim sup* of this last sum as $t \to 0$ is less than or equal to $(2\pi)^N \varepsilon$. Hence, the inequality in (4.45) is indeed true, and the proof of the theorem is complete. ∎

Exercises.

1. Prove that if $g \in L^1(\mathbf{R}^N)$ and if g is continuous at x_0, then

$$\lim_{t \to 0} A_t(g, x_0) = g(x_0)$$

where $A_t(g, x_0)$ is defined in (4.1).

2. Using the result established in Exercise 1 and the identity three lines above (4.4), prove that

$$e^{-|u|t} / b_N = (2\pi)^{-N} \int_{\mathbf{R}^N} e^{iu \cdot y} t[t^2 + |y|^2]^{-(N+1)/2} dy$$

for $u \in \mathbf{R}^N$ and $t > 0$ where b_N is defined three lines above (4.4).

3. Using the identity in (4.8), prove that

$$(2\pi)^{-N} b_N \lim_{t \to 0} (N+1) t \int_0^\delta r [t^2 + r^2]^{-\frac{N+3}{2}} |B(0,r)| \, dr = 1.$$

4. Given $f \in L^1(T_1)$. Prove that f satisfies the Fatou condition at 0 if and only $0 \in \sigma$-set of f.

5. Given $f \in C(T_N)$. Using the properties of the Poisson kernel enumerated in (4.43) and (4.44), prove that

$$\lim_{t \to 0} A_t(f, x) = f(x) \text{ uniformly for } x \in T_N.$$

6. Using the mean-value theorem for harmonic functions (see Appendix C), prove the maximum principle for harmonic functions, i.e., if $\Omega \subset \mathbf{R}^N$ is an open connected set with $x_0 \in \Omega$, $v(x_0) = M$, and $v(x) \leq M \ \forall x \in \Omega$, then

$$v(x) \text{ harmonic in } \Omega \Rightarrow v(x) = M \quad \forall x \in \Omega.$$

7. Solve the following boundary-value problem: Given $f \in C(T_N)$ with $\int_{T_N} f \, dx = 0$, prove there exists a unique $v(x, t)$ where $x \in \mathbf{R}^N$ and $t > 0$ such that

$$\begin{aligned}
&(i) \ v(x, t) \text{ is harmonic in } \mathbf{R}_+^{N+1}, \\
&(ii) \text{ for } t > 0, \ v(x, t) \text{ is periodic of period } 2\pi \\
&\qquad \text{in the } x_j\text{-variable for } j = 1, ..., N, \\
&(iii) \lim_{t \to \infty} v(x, t) = 0 \quad \text{uniformly for } x \in T_N, \\
&(iv) \lim_{t \to 0} v(x, t) = f(x) \quad \text{uniformly for } x \in T_N.
\end{aligned}$$

5. Gauss-Weierstrass Summability of Fourier Series

Because of its importance in dealing with problems involving solutions to the heat equation, in this section, we present the Gauss-Weierstrass (G-W for short) method of summability. Given $f \epsilon L^1(T_N)$, for $t > 0$, we set

$$W_t(f, x) = \sum_{m \epsilon \Lambda_N} \widehat{f}(m) e^{im \cdot x - |m|^2 t}$$

and say S[f] is G-W summable at x_0 to $f(x_0)$ if

$$\lim_{t \to 0} W_t(f, x_0) = f(x_0).$$

We shall be primarily interested in nontangential G-W summability. In particular, we say S[f] is nontangentially G-W summable at x_0 to $f(x_0)$ if for every $\gamma > 0$,

$$(5.1) \qquad \lim_{(x,t) \to (x_0, 0)} W_t(f, x) = f(x_0)$$

where (x, t) tends to $(x_0, 0)$ within the cone $\mathcal{C}_\gamma(x_0)$. (The cone $\mathcal{C}_\gamma(x_0)$ is defined in (4.13) above.)

We shall prove the following theorem regarding nontangential G-W summability of S[f] and the σ-set of f. (The σ-set of f is defined in (4.14) above.)

Theorem 5.1. *Let $f \epsilon L^1(T_N)$, $N \geq 1$, and suppose that $x_0 \in \sigma$-set of f. Then $S[f]$ is nontangentially G-W summable at x_0 to $f(x_0)$.*

Proof of Theorem 5.1.   To prove the theorem, we first observe from (3.6) above that for $t > 0$,

$$(5.2) \qquad \int_{\mathbf{R}^N} e^{iy \cdot (x-u) - |y|^2 t} dy = (\frac{\pi}{t})^{N/2} e^{-|x-u|^2/4t}.$$

Also, we observe from (3.7) that

$$(5.3) \qquad (4\pi t)^{-N/2} \int_{\mathbf{R}^N} e^{-|y|^2/4t} dy = 1 \text{ for } t > 0.$$

Next, let $g \epsilon L^1(\mathbf{R}^N)$, and set

$$(5.4) \qquad \mathcal{W}_t(g, x) = \int_{\mathbf{R}^N} \widehat{g}(y) e^{iy \cdot x - |y|^2 t} dy \qquad \text{for } t > 0,$$

where $\widehat{g}(y)$ is the Fourier transform of g and is previously defined above in Lemma 3.2. Then, for $t > 0$, by Fubini's theorem,

$$\mathcal{W}_t(g, x) = (2\pi)^{-N} \int_{\mathbf{R}^N} g(u)[\int_{\mathbf{R}^N} e^{iy \cdot (x-u) - |y|^2 t} dy] du.$$

Consequently, we see from (5.2) that

$$(5.5) \qquad \mathcal{W}_t(g, x) = (4\pi t)^{-N/2} \int_{\mathbf{R}^N} g(y) e^{-|x-y|^2/4t} dy.$$

Using the same technique to pass from Fourier integrals to Fourier series that we used in the proof of Theorem 4.1 (i.e., see (4.4′) and (4.5) in §4), we see from (5.5) that for f satisfying the conditions in the hypothesis of the theorem,

$$(5.6) \qquad W_t(f, x) = (4\pi t)^{-N/2} \int_{\mathbf{R}^N} f(x+y) e^{-|y|^2/4t} dy.$$

To prove the theorem, we proceed in a manner similar to that used to previously prove Theorem 4.2. It is easy to see from the start that we can assume that $x_0 = 0$. Therefore, to prove the theorem, we assume that $\gamma > 0$ and that $\{(x_n, t_n)\}_{n=1}^{\infty} \subset \mathcal{C}_\gamma(0)$ with $x_n \to 0$ and $t_n \to 0$. The proof will be complete when we show that

$$(5.7) \qquad \lim_{n \to \infty} W_{t_n}(f, x_n) = f(0).$$

Given $\varepsilon > 0$, it is clear that the limit in (5.7) will follow if we show that

$$(5.8) \qquad \limsup_{n \to \infty} \left| \frac{W_{t_n}(f, x_n) - f(0)}{(4\pi)^{-N/2}} \right| \leq 2^{-1} \varepsilon (\frac{1}{\gamma} + 1)^N \eta_N,$$

where $\eta_N = \int_0^\infty r^{N+1} e^{-r^2/4} dr$.

It follows from (5.3) and (5.6) that

$$\frac{W_{t_n}(f, x_n) - f(0)}{(4\pi)^{-N/2}} = t_n^{-N/2} \int_{\mathbf{R}^N} [f(x_n + y) - f(0)] e^{-|y|^2/4t_n} dy.$$

Hence, the inequality in (5.8) will follow if we show that

$$(5.9) \quad \limsup_{n \to \infty} \left| t_n^{-N/2} \int_{\mathbf{R}^N} [f(x_n + y) - f(0)] e^{-|y|^2/4t_n} dy \right| \le 2^{-1} \varepsilon (\frac{1}{\gamma} + 1)^N \eta_N.$$

Next, we set

(5.10)

$$F_n(r) = \int_{B(0,r)} [f(x_n + y) - f(0)] dy = \int_0^r d\rho \int_{S(0,\rho)} [f(x_n + y) - f(0)] dS(y),$$

where $S(0, \rho) = \partial B(0, r)$. Using the fact that $0 \in \sigma$-set of f, we invoke (4.14) and choose a $\delta > 0$, so that

$$(5.11) \qquad |F_n(r)| < \varepsilon (|x_n| + r)^N \quad \text{for} \quad |x_n| < \delta \text{ and } r < \delta.$$

Also, we observe that

$$t_n^{-N/2} \int_{\mathbf{R}^N \setminus B(0,\delta)} |f(x_n + y) - f(0)| \, e^{-|y|^2/8t_n} dy$$

is uniformly bounded in n. Hence, since $t_n \to 0$ as $n \to \infty$,

$$\limsup_{n \to \infty} \left| t_n^{-N/2} \int_{\mathbf{R}^N \setminus B(0,\delta)} [f(x_n + y) - f(0)] e^{-|y|^2/4t_n} dy \right| = 0,$$

and we conclude that the inequality in (5.9) will follow if we show that

(5.12)

$$\limsup_{n \to \infty} \left| t_n^{-N/2} \int_{B(0,\delta)} [f(x_n + y) - f(0)] e^{-|y|^2/4t_n} dy \right| \le 2^{-1} \varepsilon (\frac{1}{\gamma} + 1)^N \eta_N.$$

From (5.10), we see that $F_n(r)$ is absolutely continuous on the interval $(0, \delta)$ with $\frac{dF_n(r)}{dr} \in L^1(0, \delta)$. Consequently, we see that the expression inside the absolute value sign on the left-hand side of the inequality in (5.12) is equal to

$$t_n^{-N/2} \int_0^\delta \frac{dF_n(r)}{dr} e^{-r^2/4t_n} dr.$$

We integrate by parts and see that the inequality in (5.12) will be established if we show that

$$(5.13) \qquad \limsup_{n \to \infty} \left| t_n^{-N/2} \int_0^\delta F_n(r) \frac{r}{t_n} e^{-r^2/4t_n} dr \right| \le \varepsilon (\frac{1}{\gamma} + 1)^N \eta_N.$$

Next, we observe that $\int_0^\delta = \int_0^{t_n} + \int_{t_n}^\delta$. We shall deal with each of these cases separately and show

$$(5.14) \qquad \limsup_{n \to \infty} \left| t_n^{-N/2} \int_0^{t_n} F_n(r) \frac{r}{t_n} e^{-r^2/4t_n} dr \right| \le 0$$

and

$$(5.15) \qquad \limsup_{n \to \infty} \left| t_n^{-N/2} \int_{t_n}^{\delta} F_n(r) \frac{r}{t_n} e^{-r^2/4t_n} dr \right| \leq \varepsilon (\frac{1}{\gamma} + 1)^N \eta_N.$$

Once the inequalities in (5.14) and (5.15) are established, then the inequality in (5.13) follows. So to complete the proof of the theorem, it remains to show that the inequalities in (5.14) and (5.15) are valid.

We proceed with the situation in (5.14). For this case, $t_n < \delta$, and $0 < r < t_n$. Also, $(x_n, t_n) \in C_\gamma(0)$. Therefore, from (4.13), $|x_n| \leq \gamma^{-1} t_n$ where $t_n \to 0$. Hence, we infer from (5.11), for this case,

$$|F_n(r)| \leq \varepsilon(|(x_n)| + r)^N \leq \varepsilon(\gamma^{-1} t_n + t_n)^N.$$

Therefore,

$$\left| t_n^{-N/2} \int_0^{t_n} F_n(r) \frac{r}{t_n} e^{-r^2/4t_n} dr \right| \leq \varepsilon(\gamma^{-1} + 1)^N t_n^{N/2} \left| \int_0^{t_n} \frac{r}{t_n} dr \right|$$

$$\leq \varepsilon(\gamma^{-1} + 1)^N t_n^{N/2} t_n 2^{-1}.$$

It is clear from this last inequality that the inequality in (5.14) is indeed true.

So to complete the proof of the theorem, it remains to show that the inequality in (5.15) is valid. For this case, $t_n < r < \delta$, and from (4.13), we also see that $|x_n| \leq \gamma^{-1} t_n \leq \gamma^{-1} r$. Hence, we infer from (5.11) that

$$|F_n(r)| \leq \varepsilon(|(x_n)| + r)^N \leq \varepsilon(\gamma^{-1} r + r)^N,$$

and obtain from the definition of η_N below (5.8) that

$$\left| t_n^{-N/2} \int_{t_n}^{\delta} F_n(r) \frac{r}{t_n} e^{-r^2/4t_n} dr \right| \leq \varepsilon \frac{(\gamma^{-1} + 1)^N}{t_n t_n^{N/2}} \int_{t_n}^{\delta} r^{N+1} e^{-r^2/4t_n} dr$$

$$\leq \varepsilon(\gamma^{-1} + 1)^N \int_{t_n^{\frac{1}{2}}}^{\infty} r^{N+1} e^{-r^2/4} dr$$

$$\leq \varepsilon(\gamma^{-1} + 1)^N \eta_N.$$

So the inequality in (5.15) is indeed valid, and the proof of the theorem is complete. ∎

In the sequel, we shall also need the following theorem regarding the G-W summability of the Fourier series of Borel measures on T_N.

Theorem 5.2. *Let μ_1 and μ_2 be nonnegative finite Borel measures on T_N, $N \geq 1$, and set $\mu = \mu_1 - \mu_2$. Also, define*

$$\widehat{\mu}(m) = (2\pi)^{-N} \int_{T_N} e^{-im \cdot x} d\mu(x).$$

Suppose $\mu_k[B(0, r_0)] = 0$ for $k = 1, 2$ and $0 < r_0 < 1$. Then if $0 < r_1 < r_0$,

$$\lim_{t \to 0} \sum_{m \epsilon \Lambda_N} \widehat{\mu}(m) e^{im \cdot x - |m|^2 t} = 0 \quad \text{uniformly for } x \in B(0, r_1).$$

Define $S[d\mu] = \sum_{m \epsilon \Lambda_N} \widehat{\mu}(m) e^{im \cdot x}$. Then the theorem says that if the total variation of μ in $B(0, r_0)$ is zero and if $0 < r_1 < r_0$, the series $S[d\mu]$ is uniformly G-W summable to zero in $B(0, r_1)$.

Proof of Theorem 5.2. For $t > 0$, set

(5.16) $$W_t(d\mu, x) = \sum_{m \epsilon \Lambda_N} \widehat{\mu}(m) e^{im \cdot x - |m|^2 t},$$

and extend each μ_k by periodicity of period 2π in each variable to all of $\mathbf{R}^N$, i.e., for $E \subset T_N$, $\mu_k(E + 2\pi m) = \mu_k(E) \; \forall m \in \Lambda_N$ and for $k = 1, 2$. Then with $\mu = \mu_1 - \mu_2$, use the same technique that we used to obtain the formula for $W_t(f, x)$ in (5.6); we can show that

$$\begin{aligned} W_t(d\mu, x) &= (4\pi t)^{-N/2} \int_{\mathbf{R}^N} e^{-|x-y|^2/4t} d\mu(y) \\ &= (4\pi t)^{-N/2} \int_{\mathbf{R}^N \backslash B(0, r_0)} e^{-|x-y|^2/4t} d\mu(y). \end{aligned}$$

For $x \in B(0, r_1)$ and $|y| \geq r_0$, it follows that $|x| \leq r_1 |y| / r_0$. Set $\delta = (1 - \frac{r_1}{r_0})^2$. Then $\delta > 0$, and we see from the above that

(5.17) $$|W_t(d\mu, x)| \leq (4\pi t)^{-N/2} \int_{\mathbf{R}^N \backslash B(0, r_0)} e^{-\delta |y|^2/4t} d\mu_3(y),$$

for $x \in B(0, r_1)$, where $\mu_3 = \mu_1 + \mu_2$.

It is clear that

$$\lim_{t \to 0} (4\pi t)^{-N/2} \int_{B(0, 2) \backslash B(0, r_0)} e^{-\delta |y|^2/4t} d\mu_3(y) = 0.$$

So we see from (5.17) that to establish the theorem, it is sufficient to show

(5.18) $$\limsup_{t \to 0} t^{-N/2} \int_{\mathbf{R}^N \backslash B(0, 2)} e^{-\delta |y|^2/4t} d\mu_3(y) \leq 0.$$

Since the number of integral lattice points in the annulus,

$$\{y : R \leq |y| < R + 1\} \quad \text{is} \quad O(R^{N-1}),$$

we infer that there is a positive constant c such that

$$\int_{B(0, R+1) \backslash B(0, R)} d\mu_3(y) \leq cR^{N-1} \quad \text{for } R \geq 1.$$

Consequently,

$$t^{-N/2} \int_{\mathbf{R}^N \setminus B(0,2)} e^{-\delta|y|^2/4t} d\mu_3(y) \; \le \; ct^{-N/2} \sum_{j=2}^{\infty} e^{-\delta|j|^2/4t} j^{N-1}$$

$$\le \; c \sum_{j=2}^{\infty} e^{-\delta|j|^2/4t} \left(\frac{j^2}{t}\right)^{N/2}.$$

Since $e^{-\delta s/4} s^{N/2}$ is a decreasing function for $s > s_0$, we see from this last set of inequalities and the integral test for dealing with series that there is a $t_0 > 0$ such that

$$t^{-N/2} \int_{\mathbf{R}^N \setminus B(0,2)} e^{-\delta|y|^2/4t} d\mu_3(y) \le c \int_0^{\infty} e^{-\delta|s|^2/4t} (s^2/t)^{N/2} ds$$

for $0 < t < t_0$. Hence,

$$t^{-N/2} \int_{\mathbf{R}^N \setminus B(0,2)} e^{-\delta|y|^2/4t} d\mu_3(y) \le ct^{1/2} \int_0^{\infty} e^{-\delta|s|^2/4} s^N ds$$

for $0 < t < t_0$. The lim sup inequality in (5.18) follows immediately from this last inequality, and the proof of the theorem is complete. ∎

In Chapter 3, in the section on the sets of uniqueness, we will also need the following theorem concerning the G-W summability of the Fourier series of a Borel measure on T_N.

Theorem 5.3. *Let μ be a nonnegative finite Borel measure on T_N, $N \ge 1$, and define*

$$\widehat{\mu}(m) = (2\pi)^{-N} \int_{T_N} e^{-im \cdot x} d\mu(x).$$

Suppose $\mu[B(0, r_0)] = 0$ where $0 < r_0 < 1$. Suppose, also, that j is a positive integer and that for $t > 0$, $W_t(d\mu, x)$ is defined by (5.16). Then if $0 < r_1 < r_0$,

$$\lim_{t \to 0} \Delta^j W_t(d\mu, x) = 0 \quad \text{uniformly for } x \in B(0, r_1),$$

where Δ^j stands for the j-th iterated Laplace operator.

Proof of Theorem 5.3. Extend μ by periodicity of period 2π in each variable to all of $\mathbf{R}^N$, i.e., for $E \subset T_N$, $\mu(E + 2\pi m) = \mu(E) \; \forall m \in \Lambda_N$. Then, as in the proof of Theorem 5.2, we can show that

$$W_t(d\mu, x) = (4\pi t)^{-N/2} \int_{\mathbf{R}^N \setminus B(0, r_0)} e^{-|x-y|^2/4t} d\mu(y).$$

Now, for $x \in B(0, r_1)$ and $|y| \geq r_0$, $t^{-N/2} e^{-|x-y|^2/4t}$ satisfies the heat equation $\Delta u = \partial u / \partial t$, where Δ is with respect to x. So, with $t > 0$,

$$(5.19) \qquad t^{-N/2} \Delta^j e^{-|x-y|^2/4t} = \partial^j t^{-N/2} e^{-|x-y|^2/4t} / \partial t^j.$$

Consequently,

$$(5.20) \qquad \Delta^j W_t(d\mu, x) = \int_{\mathbf{R}^N \backslash B(0, r_0)} \partial^j [(4\pi t)^{-N/2} e^{-|x-y|^2/4t}] / \partial t^j \, d\mu(y),$$

for $x \in B(0, r_1)$.

It is easy to see that the right-hand side of the equality in (5.19) is a finite linear combination of terms of the form

$$(|x - y|^{2k} t^{-n}) t^{-N/2} e^{-|x-y|^2/4t}$$

where $k = 0, ..., j$ and $n = 1, ..., 2j$.

Also, for $0 < |x| \leq r_1$ and $r_1 < r_0 \leq |y|$, we observe there is a constant c such that

$$|x - y|^{2k} \leq c |x - y|^{2j} \quad \text{for} \quad k = 0, ..., j - 1.$$

Likewise for $0 < t \leq 1$,

$$t^{-n} \leq t^{-2j} \quad \text{for} \quad n = 1, ..., 2j - 1.$$

We conclude from (5.20) that to establish the theorem, it is sufficient to show

$$(5.21) \qquad \lim_{t \to 0} \int_{\mathbf{R}^N \backslash B(0, r_0)} |x - y|^{2j} t^{-2j} (t^{-N/2} e^{-|x-y|^2/4t}) \, d\mu(y) = 0$$

uniformly for $x \in B(0, r_1)$.

As in the proof of Theorem 5.2, we set $\delta = (1 - \frac{r_1}{r_0})^2$ and observe that the integral in (5.21) is majorized by

$$(5.22) \qquad 4^j t^{-2j} t^{-N/2} \int_{\mathbf{R}^N \backslash B(0, r_0)} |y|^{2j} e^{-\delta |y|^2/4t} \, d\mu(y)$$

uniformly for $|x| \leq r_1$ and $0 < t \leq 1$.

Also, as in the proof of Theorem 5.2, we see there is a constant c_1 such that

$$\int_{B(0, R+1) \backslash B(0, R)} d\mu(y) \leq c_1 R^{N-1} \quad \text{for} \quad R \geq 1.$$

Consequently, the expression in (5.22) is in turn majorized by a constant (independent of t) multiple of

$$\sum_{p=2}^{\infty} (\frac{p^2}{t})^{2j+N/2} e^{-\delta p^2/4t} + o(1),$$

for $|x| \leq r_1$ and $0 < t \leq 1$, where the $o(1)$ comes from the part dealing with the integral over $B(0, 2) \backslash B(0, r_0)$.

As in the proof of Theorem 5.2, it follows from the integral test for series that the limit of this last summation is zero as $t \to 0$. Therefore, the limit of the expression in (5.22) is zero as $t \to 0$. But then the limit in (5.21) is also zero uniformly for $x \in B(0, r_1)$. ∎

Exercises.

1. Using the identity in (3.6), prove that for $t > 0$,

$$\int_{\mathbf{R}^N} e^{iy \cdot (x-u) - |y|^2 t} dy = \left(\frac{\pi}{t}\right)^{N/2} e^{-|x-u|^2/4t}.$$

2. Given $f \in C(T_N)$, prove

$$\lim_{t \to 0} W_t(f, x) = f(x) \text{ uniformly for } x \in T_N,$$

where $W_t(f, x)$ is defined by (5.6).

3. Using the stong maximum principle for the heat equation [Ev, p. 55], solve the following periodic boundary-value problem: Given $f \in C(T_N)$ with $\int_{T_N} f dx = 0$, prove there exists a unique $v(x, t)$ where $x \in \mathbf{R}^N$ and $t > 0$ such that

(i) $\frac{\partial v}{\partial t}(x, t) = \Delta v(x, t)$ $\forall x \in \mathbf{R}^N$ and $\forall t > 0$,
(ii) for $t > 0$, $v(x, t)$ is periodic of period 2π
        in the x_j − variable for $j = 1, ..., N$,
(iii) $\lim_{t \to \infty} v(x, t) = 0$ uniformly for $x \in T_N$,
(iv) $\lim_{t \to 0} v(x, t) = f(x)$ uniformly for $x \in T_N$.

6. Further Results and Comments

1. There is another method of summability used by by A. Beurling called absolute Abel summability. In particular, let $f \in L^1(T_N)$, $N \geq 2$. Set

$$f(x, t) = \sum_{m \in \Lambda_N} \widehat{f}(m) e^{im \cdot x - |m| t}$$

for $t > 0$. Say f is absolutely Abel summable at the point x_0 provided

$$\int_0^1 \left| \frac{\partial f}{\partial t}(x_0, t) \right| dt < \infty.$$

Let $Z \subset T_N$ be closed in the torus topology. Say Z is of ordinary capacity zero provided that

$$\int_{T_N} \int_{T_N} H_0(x - y) d\mu(x) d\mu(y) = \infty$$

for every μ which is a nonnegative finite Borel measure on T_N with

$$\mu(T_N) = 1 \text{ and } \mu(T_N \backslash Z) = 0$$

where $H_0(x)$ is the function introduced in Lemma 1.4 of Chapter 3. Motivated by the work of Beurling in [Beu], the following two results connecting absolute Abel summability and ordinary capacity were established in [LS].

Theorem A. *Let $Z \subset T_N$ be a closed set in the torus topology, $N \geq 2$, and let $f \in L^2(T_N)$. Suppose that*

$$(i) \quad \sum_{m \in \Lambda_N} |m|^2 \left| \widehat{f}(m) \right|^2 \; < \; \infty,$$

$$(ii) \quad \int_0^1 \left| \frac{\partial f}{\partial t}(x_0, t) \right| dt \; = \; +\infty \quad \forall x \in Z$$

Then Z is of ordinary capacity zero.

Theorem B. *Let $Z \subset T_N$ be a closed set in the torus topology, $N \geq 2$, and suppose that Z is of ordinary capacity zero. Then there exists an $f \in L^2(T_N)$ with $\sum_{m \in \Lambda_N} |m|^2 \left| \widehat{f}(m) \right|^2 < \infty$ such that*

$$\int_0^1 \left| \frac{\partial f}{\partial t}(x_0, t) \right| dt = +\infty \quad \forall x \in Z.$$

2. Arne Beurling, who was one of the leading analysts during the post World War II period, served as a codebreaker for the Swedish government during World War II itself. In a feat of the first order of magnitude, he single-handedly in a two-week period broke the German code that was passing over Swedish teephone cables going from Berlin to Norway. One can read all about this plus a biography of Beurling in a book published by the American Mathematical Society entitled "Codebreaker" by B. Beckman [Bec].

3. To establish Bochner's ingeneous result at the critical index $\frac{N-1}{2}$ stated in (3.26), we proceed in the following manner: Set

$$\Psi_R(x) = \sum_{|m| \leq R} e^{im \cdot x} \left(1 - \frac{|m|^2}{R^2} \right)^{(N-1)/2},$$

and establish the result stated in (3.26). It is sufficient to show

(6.1) $\qquad \begin{aligned} &\exists x_0 \in T_N \backslash B(0, \delta) \text{ and } \{R_j\}_{j=1}^{\infty} \text{ with } R_j \to \infty \\ &\text{such that } \lim_{j \to \infty} \Psi_{R_j}(x_0) = \infty. \end{aligned}$

To see that this is indeed the case, consider the Banach space $\mathcal{B}$ consisting of all real-valued functions in $L^1(T_N)$ that vanish a.e. in $B(0, \delta)$. Then

$$\sigma_{R_j}^{(N-1)/2}(f, 0) = (2\pi)^{-N} \int_{T_N} f(x) \Psi_{R_j}(x) \, dx$$

gives rise to a set of bounded linear functionals F_j on $\mathcal{B}$, i.e., $F_j(f) = \sigma_{R_j}^{(N-1)/2}(f, 0)$. If the result stated in (3.26) is false, then $\sup_j |F_j(f)|$ is finite for all $f \in \mathcal{B}$. But then by the Banach-Seinhaus theorem, $\sup_j \|F_j\|$ is finite. However,

$$\|F_j\| = (2\pi)^{-N} \sup_{x \in T_N \backslash B(0\delta)} \left| \Psi_{R_j}(x) \right|.$$

So the finiteness of $sup_j \|F_j\|$ is a contradiction to the statement in (6.1). We conclude that to establish (3.26), it is sufficient to show that (6.1) holds.

Bochner shows that (6.1) holds via a sequence of lemmas that involve the notion of a countable set $S = \{s_1, s_2, ...\}$ having numbers that are linearly independent with respect to integer coefficients (i.e., if $\{c_1, ..., c_n\}$ is a set of integers with $c_1^2 + \cdots + c_n^2 \neq 0$, then $\sum_{j=1}^{n} c_j s_j \neq 0$). To view the statement and proof of these lemmas, we refer the reader to [Boc1, p. 193] or [Sh1, pp. 57-64].

CHAPTER 2

Conjugate Multiple Fourier Series

1. Introduction

In this chapter, we shall deal with conjugate multiple Fourier series where the conjugacy is defined by means of Calderon-Zygmund kernels, which are of spherical harmonic type.

Also, the results that we present in this chapter will take place in dimension $N \geq 2$. In order to place this theory in its proper perspective, we shall first review some aspects of conjugate Fourier series in one dimension, which is defined via the Hilbert transform and the kernel x^{-1}.

If $g \in L^1(\mathbf{R})$, then the Hilbert transform of g, $\tilde{g}$, is defined as follows:

$$\tilde{g}(x) = \lim_{\varepsilon \to 0} \pi^{-1} \int_\varepsilon^\infty [g(x-y) - g(x+y)]/y \, dy.$$

Now this limit exists almost everywhere [Ti2, p. 132], and if, in addition, $\tilde{g} \in L^1(\mathbf{R}) \cap L^2(\mathbf{R})$, then

$$\widehat{\tilde{g}}(x) = -i(sgn x)\hat{g}(x)$$

where $sgn\ x = 1$ if $x > 0$, -1 if $x < 0$, and 0 if $x = 0$.

Even if $\tilde{g} \notin L^1(\mathbf{R}) \cap L^2(\mathbf{R})$, we still obtain [Ti2, p. 147] that for $\alpha > 0$,

$$\lim_{R \to \infty} \int_{-R}^R -i\ (sgn\ y)\ \hat{g}(y)e^{ixy}(1 - \frac{y}{R})^\alpha = \tilde{g}(x) \quad \text{for a.e. } x.$$

To pass from Fourier integrals to Fourier series, we first see from [Zy1, p. 73] that

$$(1.1) \qquad \frac{1}{2}\cot\frac{x}{2} = \frac{1}{x} + \sum_{n=-\infty}^{\infty} {}' [\frac{1}{x+2\pi n} - \frac{1}{2\pi n}] \qquad \text{for } x \neq 2\pi n.$$

Next, we observe that if $f \in L^1(T_1)$, then $\tilde{f}$, the conjugate function of f, is defined to be

$$(1.2) \qquad \tilde{f}(x) = \lim_{\varepsilon \to 0} \pi^{-1} \int_\varepsilon^\pi [f(x-y) - f(x+y)]\frac{1}{2}\cot\frac{y}{2}dy.$$

It is well-known [Zy1, p. 131] that this limit exists almost everywhere, and that if $\tilde{f} \in L^p(T_1), 1 < p \leq \infty$, then

$$\widehat{\tilde{f}}(n) = -i(sgn\ n)\hat{f}(n).$$

However, even if $\tilde{f} \notin L^1(T_1)$, we still obtain that $\sum_{-\infty}^{\infty} -i(sgn\ n)\widehat{f}(n)e^{inx}$ is Abel summable to $\tilde{f}(x)$ for almost every x.

We note also that if the limit in (1.2) exists, then using (1.1), it can be shown that

$$\lim_{\varepsilon \to 0} \int_{\varepsilon}^{\pi} [f(x-y)-f(x+y)]\frac{1}{2}\cot\frac{y}{2}dy = \lim_{\varepsilon \to 0}\lim_{R\to\infty} \int_{\varepsilon}^{R} [f(x-y)-f(x+y)]\frac{1}{y}dy.$$

We shall proceed in an analogous manner to develop the theory of N-dimensional conjugate Fourier series, $N \geq 2$. In particular, we focus on the one-dimensional function

$$x^{-1} = (sgn\ x)\,|x|^{-1} \quad \text{and} \quad sgn\ 1 + sgn\ -1 = 0.$$

and generalize this function to N-space by means of the kernel

(1.3) $$K(x) = W(x/\,|x|)\,|x|^{-N} \qquad \text{for } x \neq 0,$$

where

(1.4) $$W(x/\,|x|) = Q_n(x)/\,|x|^n\,.$$

Here, $Q_n(x)$ is a homogeneous real polynomial of degree n, $n \geq 1$, and is also a harmonic function, i.e., $\Delta Q_n(x) = 0 \ \forall\, x \in \mathbf{R}^N$. In other words, $Q_n(x)$ is a spherical harmonic function of degree n as is discussed in §3 of Appendix A.

With $S_{N-1} = \partial B(0,1)$, the unit $(N-1)$-sphere in $\mathbf{R}^N$, we observe that

(1.5) $$\int_{S_{N-1}} W(\xi)\,dS(\xi) = 0,$$

where $dS(\xi)$ is the natural volume element on S_{N-1}.

To see how the equality in (1.5) actually occurs, we set $x = r\xi$ where $\xi \in S_{N-1}$, and we see from the homogeneity of Q_n that $Q_n(r\xi) = r^n Q_n(\xi)$. Using this fact in conjunction with the familiar divergence theorem and the observation that $div \cdot \nabla Q_n(x) = 0$ gives the equality in (1.5).

$K(x)$ in (1.3) is a generalization of the Hilbert kernel x^{-1} in one dimension and is called a Calderon-Zygmund kernel of spherical harmonic type. This generalization persists in the sense that if $g(x) \in L^p(\mathbf{R}^N)$, $1 \leq p < \infty$, then $\tilde{g}(x)$ is defined to be the Calderon-Zygmund transform of g where

(1.6) $$\widehat{g}(x) = \lim_{\varepsilon \to 0} \int_{\mathbf{R}^N \setminus B(0,\varepsilon)} g(x-y)K(y)dy.$$

It is shown in [CZ1] or [SW, Chapter VI] that this limit exists almost everywhere. We shall state this fact as Theorem A.

Theorem A. *If $g \in L^p(\mathbf{R}^N)$, $1 \leq p < \infty$, then $\widetilde{g}(x)$, which is defined by (1.6), exists almost everywhere in $\mathbf{R}^N$.*

For Theorem B, which is also proved in [CZ1] or [SW, Chapter VI], we state the following:

Theorem B. *If* $g \in L^p(\mathbf{R}^N)$, $1 < p < \infty$, *then* $\widetilde{g} \in L^p(\mathbf{R}^N)$. *Also,* $\exists A_p > 0$ *independent of g such that*

$$(1.7) \qquad \qquad \|\widetilde{g}\|_{L^p(\mathbf{R}^N)} \leq A_p \|g\|_{L^p(\mathbf{R}^N)}.$$

It is our intention here to show that theorems similar to Theorems A and B hold for every function $f \in L^p(T_N)$, $N \geq 2$, where

$$T_N = \{x : -\pi \leq x_j < \pi, \ j = 1, ..., N\}.$$

In order to do this, we introduce $K^*(x)$, the periodic analogue of $K(x)$, as follows:

$$(1.8) \qquad K^*(x) = K(x) + \lim_{R \to \infty} \sum_{1 \leq |m| \leq R} [K(x + 2\pi m) - K(2\pi m)]$$

for x not equal to 2π times an integral lattice point, i.e., $x \neq 2\pi m$, where K(x) is defined by (1.3) and meets (1.4) and (1.5). Since

$$\frac{Q_n(x + 2\pi m)}{|x + 2\pi m|^{n+N}} - \frac{Q_n(2\pi m)}{|2\pi m|^{n+N}} = \frac{Q_n(x + 2\pi m) - Q_n(2\pi m)}{|x + 2\pi m|^{n+N}}$$

$$+ Q_n(2\pi m)\left[\frac{1}{|x + 2\pi m|^{n+N}} - \frac{1}{|2\pi m|^{n+N}}\right],$$

it is easy to see from this last equality, since $Q_n(x)$ is a homogeneous polynomial of degree n, that for $x \in B(0, R_0)$, where $R_0 > 10$, there is a constant $c(R_0)$ such that

$$(1.9) \qquad |K(x + 2\pi m) - K(2\pi m)| \leq \frac{c(R_0)}{|m|^{1 + N}} \quad \text{for } |m| \geq 2R_0.$$

It follows from this last inequality that the series in (1.8) converges uniformly and absolutely for x in any bounded domain.

Next, let m_0 be any fixed integral lattice point. What we want to show is that $K^*(x + 2\pi m_0) = K^*(x)$ for $x \epsilon T_N \backslash \{0\}$. To see that this is indeed the case, we observe from (1.8), for $x \in T_N \backslash \{0\}$,

$$|K^*(x + 2\pi m_0) - K^*(x)| \leq$$

$$(1.10) \qquad \lim_{R \to \infty} \sum_{R - |m_0| \leq |m| \leq R} \{|K(x + 2\pi m_0 + 2\pi m) - K(2\pi m)| +$$

$$|K(x + 2\pi m) - K(2\pi m)|\}.$$

However, the closure of T_N is contained in the ball $B(0, N\pi)$. So we infer from (1.9) that the absolute value inside the summation sign in (1.10) is majorized by

(1.11) $$2c(N\pi + 2\pi |m_0|) \, |m|^{-(N+1)}$$

for $x \in T_N$ and for R sufficiently large. With $A(R, m_0)$ designating the spherical annulus at the bottom of the summation sign in (1.10), it follows that for $m \in A(R, m_0)$ and for R sufficiently large, the expression in (1.11) is in turn majorized by

$$2c(N\pi + 2\pi |m_0|)(R - |m_0|)^{-(N+1)}.$$

As is well-known, the number of integral lattice points in $A(R, m_0) = O(R^{N-1})$ as $R \to \infty$. Consequently, we see that the right-hand side of the inequality in (1.10) is majorized by

$$\lim_{R \to \infty} \{O(R^{N-1})2c(N\pi + 2\pi |m_0|)(R - |m_0|)^{-(N+1)}\} = 0.$$

Hence, $K^*(x + 2\pi m_0) = K^*(x)$ for $x \in T_N \setminus \{0\}$, and we conclude $K^*(x)$ is a periodic function and is the N-dimensional analogue of the series in (1.1) that is equal to $\frac{1}{2} \cot \frac{x}{2}$.

$K^*(x)$ is called a periodic Calderon-Zygmund kernel of spherical harmonic type.

For $f \epsilon L^1(T_N)$, we define

(1.12) $$\widetilde{f}(x) = \lim_{\varepsilon \to 0} (2\pi)^{-N} \int_{T_N \setminus B(0,\varepsilon)} f(x - y) K^*(y) dy$$

at every point where this limit exists. This limit is the N-dimensional analogue of the limit in (1.2), and as we shall show, exists almost everywhere. $\widetilde{f}(x)$ is called the periodic Calderon-Zygmund transform of spherical harmonic type.

Using the ideas in [CZ2], we shall next establish the following two theorems regarding $\widetilde{f}(x)$.

Theorem 1.1. *If $f \in L^1(T_N)$, then $\widetilde{f}(x)$, which is defined by the limit in (1.12), exists almost everywhere in T_N.*

Theorem 1.2. *If $f \in L^p(T_N)$, $1 < p < \infty$, then $\widetilde{f} \in L^p(T_N)$. Also, $\exists C_p^* > 0$ and independent of f, such that*

(1.13) $$\left\|\widetilde{f}\right\|_{L^p(T_N)} \leq C_p^* \|f\|_{L^p(T_N)}.$$

Proof of Theorem 1.1. We set

(1.14) $$B_N = \{x = (x_1, ..., x_N) : |x_j| \leq 2\pi \quad \text{for } j = 1, ..., N\},$$

and

$$g(x) = f(x) \quad \text{for } x\epsilon B_N$$
$$= 0 \qquad \text{for } x\epsilon \mathbf{R}^N \backslash B_N.$$

Also, we set

$$K_1^*(x) = \lim_{R\to\infty} \sum_{1\leq|m|\leq R} [K(x+2\pi m) - K(2\pi m)] \quad \text{for } x\epsilon T_N,$$

and observe from (1.3) and (1.9) that there exists $C_K > 0$ such that

$$(1.15) \qquad |K_1^*(x)| \leq C_K \quad \text{for } x\epsilon T_N \quad \text{and} \quad |K(x)| \leq C_K \quad \text{for } x\epsilon \mathbf{R}^N \backslash T_N.$$

It follows from the above that $g\epsilon L^1(\mathbf{R}^N)$ and that

$$(1.16) \qquad \int_{T_N\backslash B(0,\varepsilon)} f(x-y)K^*(y)dy = \int_{T_N\backslash B(0,\varepsilon)} g(x-y)K^*(y)dy$$

for $x\epsilon T_N$ and $0 < \varepsilon < 1$. Since $K^*(x) = K(x) + K_1^*(x)$, we see from (1.15) that the right-hand side of the equality in (1.16) is equal to

$$\int_{\mathbf{R}^N\backslash B(0,\varepsilon)} g(x-y)K(y)dy + \int_{T_N\backslash B(0,\varepsilon)} g(x-y)K_1^*(y)dy$$

$$- \int_{\mathbf{R}^N\backslash T_N} g(x-y)K(y)dy.$$

Hence, it follows from (1.6), (1.12), and Theorem A that $\widetilde{f}(x)$ exists almost everywhere in T_N and that

$$(1.17) \quad \widetilde{f}(x) = \widetilde{g}(x) + \int_{T_N} g(x-y)K_1^*(y)dy - \int_{\mathbf{R}^N\backslash T_N} g(x-y)K(y)dy$$

for a.e. $x \in T_N$. This gives the conclusion to the proof of Theorem 1.1. ∎

Proof of Theorem 1.2. We use the terminology and notation just developed in the proof of Theorem 1.1, and observe from the definition of B_N given in (1.14) that there is a constant $\gamma_{N,p}$ independent of f such that

$$(1.18) \qquad \int_{B_N} |f(y)| \, dy \leq \gamma_{N,p} \|f\|_{L^p(T_N)}.$$

Also, since $g \in L^p(\mathbf{R}^N)$, where g is defined below (1.14), it follows from Theorem B that

$$(1.18') \qquad \int_{\mathbf{R}^N} |\widetilde{g}(x)|^p \, dx \leq A_p^p \int_{\mathbf{R}^N} |g(x)|^p \, dx = 2^N A_p^p \int_{T_N} |f(x)|^p \, dx,$$

where we also have made use of the fact that f is a periodic function.

Likewise, we see from (1.15) and (1.17) above that

$$\left|\widetilde{f}(x)\right| \leq |\widetilde{g}(x)| + 2C_K \int_{\mathbf{R}^N} |g(y)| \, dy$$

for a.e. $x \in T_N$. But then from the definition of g, (1.14), and (1.18), we obtain from this last inequality that

$$
\begin{aligned}
\left|\widetilde{f}(x)\right| &\leq |\widetilde{g}(x)| + 2C_K \int_{B_N} |f(y)|\, dy \\
&\leq |\widetilde{g}(x)| + 2\gamma_{N,p} C_K \|f\|_{L^p(T_N)}
\end{aligned}
$$

for a.e. $x \in T_N$.

Applying Minkowski's theorem, in turn, to this last inequality and also using (1.18′) above, we conclude that

$$
\begin{aligned}
\{\int_{T_N} \left|\widetilde{f}(x)\right|^p dx\}^{\frac{1}{p}} &\leq \{\int_{T_N} |\widetilde{g}(x)|^p\, dx\}^{\frac{1}{p}} + 2\gamma_{N,p}(2\pi)^{\frac{N}{p}} C_K \|f\|_{L^p(T_N)} \\
&\leq 2^{\frac{N}{p}} A_p \|f\|_{L^p(T_N)} + 2\gamma_{N,p}(2\pi)^{\frac{N}{p}} C_K \|f\|_{L^p(T_N)}.
\end{aligned}
$$

This proves Theorem 1.2 where $C_p^* = 2^{\frac{N}{p}} A_p + 2\gamma_{N,p}(2\pi)^{\frac{N}{p}} C_K$. ∎

Next, we define the principal-valued Fourier coefficient of $K^*(x)$ to be

$$
(1.19) \qquad \lim_{\varepsilon \to 0} (2\pi)^{-N} \int_{T_N \backslash B(0,\varepsilon)} e^{-im \cdot x} K^*(x) dx = \widehat{K^*}(m),
$$

and see from (1.8) and the first inequality in (1.15) that

$$
(1.20) \quad (2\pi)^N \widehat{K^*}(m) = \lim_{\varepsilon \to 0} \int_{T_N \backslash B(0,\varepsilon)} e^{-im \cdot x} K(x) dx + \int_{T_N} e^{-im \cdot x} K_1^*(x) dx.
$$

For $m = 0$, it is clear from (1.4) and (1.15) that the limit in (1.20) exists. If $m \neq 0$, we observe there exists a constant $\gamma(m)$ such that

$$
\left| e^{-im \cdot x} - 1 \right| \leq \gamma(m) |x| \quad \text{for } x \epsilon T_N.
$$

So we see from (1.3) and (1.5) that the limit in (1.20) does indeed exist. In any case, $\widehat{K^*}(m)$ is well-defined for all integral lattice points m.

Observing that $\int_{T_N} e^{-im \cdot x} dx = 0$ for $m \neq 0$, and that

$$
B(0, 2\pi(R - N)) \backslash T_N \subset \bigcup_{1 \leq |m| \leq R} (T_N + 2\pi m)
$$

and

$$
\bigcup_{1 \leq |m| \leq R} (T_N + 2\pi m) \subset B(0, 2\pi(R + N)) \backslash T_N,
$$

we infer furthermore from (1.3), (1.20), and the definition of $K_1^*(x)$ that

$$
(1.21) \quad \widehat{K^*}(m) = \lim_{\varepsilon \to 0} \lim_{R \to \infty} (2\pi)^{-N} \int_{B(0,R) \backslash B(0,\varepsilon)} e^{-im \cdot x} K(x) dx \quad \text{for } m \neq 0.
$$

If $m = 0$, proceeding in a similar manner, the previous argument shows that

$$\widehat{K^*}(0) = -\lim_{R \to \infty} \sum_{1 \leq |m| \leq R} K(2\pi m).$$

For $N = 1, \widehat{K^*}(0) = 0$, for $N \geq 2$, $\widehat{K^*}(0)$ may or may not be zero, [CZ2, p. 258]. However, in a number of important cases, it is true that $\widehat{K^*}(0) = 0$, e.g., for $N = 2$ and

$$K(x) = x_1 x_2 / |x|^4 \quad \text{or} \quad K(x) = (x_1^2 - x_2^2)/|x|^4 \text{ or } K(x) = x_1/|x|^3.$$

This last named kernel is often referred to as the Riesz kernel.

From Theorem 1.2, we know that if $f \in L^p(T_N)$, $1 < p < \infty$, then $\widetilde{f} \in L^p(T_N)$. So making use of $\widehat{K^*}(m)$ as defined in (1.19), we can establish the following theorem:

Theorem 1.3. If $f \in L^p(T_N), 1 < p < \infty$, then $\widehat{\widetilde{f}}(m) = \widehat{K^*}(m)\widehat{f}(m)$ $\forall m \in \Lambda_N$.

Proof of Theorem 1.3. If $h(x) = e^{im \cdot x}$, then from (1.12) we see that

$$\begin{aligned}
\widetilde{h}(x) &= e^{im \cdot x} \lim_{\varepsilon \to 0} (2\pi)^{-N} \int_{T_N \setminus B(0,\varepsilon)} e^{-im \cdot y} K^*(y) dy \\
&= e^{im \cdot x} \widehat{K^*}(m).
\end{aligned}$$

Hence, from (1.19), we see that the theorem is true in this case. Consequently, the theorem is true for any finite linear combination of exponentials, i.e., for any trigonometric polynomial.

Given an $f \in L^p(T_N)$, let $\sigma_n^\Diamond(f, x)$ be the finite linear combinations of exponentials given in (2.6) of Chapter 1. Also, let $m_0 \in \Lambda_N$. By what we have just shown,

$$(1.22) \qquad \lim_{n \to \infty} (2\pi)^{-N} \int_{T_N} e^{-im_0 \cdot x} \widetilde{\sigma}_n^\Diamond(f, x) dx = \widehat{K^*}(m_0)\widehat{f}(m_0).$$

Also, by Theorem 1.2,

$$(1.23) \qquad \left\| e^{-im_0 \cdot x} [\widetilde{\sigma}_n^\Diamond(f, \cdot) - \widetilde{f}] \right\|_{L^p(T_N)} \leq C_p \left\| \sigma_n^\Diamond(f, \cdot) - f \right\|_{L^p(T_N)}.$$

From Theorem 2.2 of Chapter 1, we know that the right-hand side of the inequality in (1.23) goes to zero as $n \to \infty$. Therefore, the left-hand side of the inequality also goes to zero. But then it follows that the limit in (1.22) is equal to

$$\lim_{n \to \infty} (2\pi)^{-N} \int_{T_N} e^{-im_0 \cdot x} \widetilde{f}(x) \, dx,$$

and we obtain from (1.22) that $\widehat{\widetilde{f}}(m_0) = \widehat{K^*}(m_0)\widehat{f}(m_0)$. This concludes the proof of Theorem 1.3.   ∎

We shall henceforth refer to

$$(1.24) \qquad \widetilde{S}_K(f) = \sum_{m \in \Lambda_N} \widehat{K^*}(m)\widehat{f}(m)e^{im\cdot x}$$

as the K-conjugate series of f, where K is given by (1.3).

Before proceeding to the next section in this chapter, we remark that with $K(x)$ given by (1.3) and (1.4), $\widehat{K^*}(m)$ is computed in Corollary 3.2 of Appendix A where it is shown that

$$(1.25) \qquad \widehat{K^*}(m) = \kappa_{n,N}Q_n(m)/|m|^n \quad \text{for } m \neq 0,$$

with

$$(1.26) \qquad \kappa_{n,N} = (-i)^n 2^{-N}\pi^{-\frac{N}{2}}\Gamma(\frac{n}{2})[\Gamma(\frac{n+N}{2})]^{-1}.$$

Also, we establish the following corollary.

Corollary 1.4. *If* $f \in L^1(T_N)$, $\widehat{f}(0) = 0$, *and* $\widetilde{f}(x_0)$ *exists, then*

$$\lim_{R\to\infty}\lim_{\varepsilon\to 0}(2\pi)^{-N}\int_{B(0,R)\backslash B(0,\varepsilon)} f(x_0 - y)K(y)dy = \widetilde{f}(x_0).$$

Proof of Corollary 1.4. Without loss in generality, we assume that $x_0 = 0$. Then by (1.12) and (1.15), we have that

$$(1.27) \quad (2\pi)^N\widetilde{f}(0) = \lim_{\varepsilon\to 0}\int_{T_N\backslash B(0,\varepsilon)} f(-y)K(y)dy + \int_{T_N} f(-y)K_1^*(y)dy.$$

Using the fact that $\widehat{f}(0) = 0$, we see from the definition of $K_1^*(y)$ that

$$\int_{T_N} f(-y)K_1^*(y)dy = \lim_{R\to\infty}\sum_{1\leq|m|\leq R}\int_{T_N} f(-y)K(y + 2\pi m).$$

But f is a periodic function. So, we conclude from this last limit and (1.27) that to establish the corollary, it is sufficient to show

$$\lim_{R\to\infty}[\sum_{1\leq|m|\leq R}\int_{T_N+2\pi m} f(-y)K(y)dy - \int_{B(0,2\pi(R+N))\backslash T_N} f(-y)K(y)dy] = 0.$$

To do this, we set

$$\mathcal{T}_R = \bigcup_{1\leq|m|\leq R}(T_N + 2\pi m)$$

and observe that this last limit can be rewritten as

$$(1.28) \qquad \lim_{R\to\infty}[\int_{\mathcal{T}_R} f(-y)K(y)dy - \int_{B(0,2\pi(R+N))\backslash T_N} f(-y)K(y)dy] = 0.$$

As we have observed previously in establishing the limit in (1.21),

$$B(0, 2\pi(R - N))\backslash T_N \subset T_R \subset B(0, 2\pi(R + N))\backslash T_N.$$

So we see that that the limit in (1.28) will hold if we show that

(1.29) $$\lim_{R \to \infty} \int_{B(0, 2\pi(R+N))\backslash B(0, 2\pi(R-N))} |f(-y)K(y)| \, dy = 0.$$

But the number of integral lattice points contained in the spherical annulus $B(0, 2\pi(R + N))\backslash B(0, 2\pi(R - N))$ is $O(R^{N-1})$, and for y in this same annulus, $|K(y)| = O(R^{-N})$. Hence,

$$\int_{B(0, 2\pi(R+N))\backslash B(0, 2\pi(R-N))} |f(-y)K(y)| \, dy \leq O(R^{N-1}) \, \|f\|_{L^1(T_N)} \, O(R^{-N})$$

as $R \to \infty$, and the limit in (1.29) is indeed valid. This concludes the proof of Corollary 1.4.   ∎

Exercises.

1. Given that $W(x/|x|) = Q_n(x)/|x|^n$ for $x \neq 0$ where $Q_n(x)$ is a spherical harmonic polynomial of degree n, prove that

$$\int_{S_{N-1}} W(\xi) \, dS(\xi) = 0,$$

where $dS(\xi)$ is the natural volume element on S_{N-1}.

2. Let $Q_n(x)$ be a spherical harmonic polynomial of degree 2 in dimension $N = 3$. Show that if $x \in B(0, R_0)$, where $R_0 > 10$, then there exists $c(R_0) > 0$ such that

$$\left| \frac{Q_n(x + 2\pi m) - Q_n(2\pi m)}{|x + 2\pi m|^{n+N}} \right| \leq \frac{c(R_0)}{|m|^{1+N}} \quad \text{for } |m| \geq 2R_0,$$

where $m \in \Lambda_N$.

3. Let $\{Q_{j,n}(x)\}_{j=1}^k$ be a set of spherical harmonic polynomials of degree n. Say the set of functions is linear independent if $a_j \in \mathbf{R}$ for $j = 1, ..., k$ and

$$\sum_{j=1}^k a_j Q_{j,n}(x) = 0 \text{ for } |x| = 1 \Rightarrow a_j = 0 \text{ for } j = 1, ..., k.$$

Show that in dimension $N = 3$, there are five linearly independent spherical harmonic polynomials of degree 2.

4. Prove that if $K(x) = Q_n(x)/|x|^{n+N}$ for $x \neq 0$ where $Q_n(x)$ is a spherical harmonic polynomial of degree n, then for $m \in \Lambda_N$,

$$\lim_{\varepsilon \to 0} \int_{T_N \backslash B(0, \varepsilon)} e^{-im \cdot x} K(x) dx \text{ exists and is finite.}$$

5. Given that Theorem 1.3 is true for a finite linear combination of exponentials, prove that

$$\lim_{n\to\infty} (2\pi)^{-N} \int_{T_N} e^{-im_0\cdot x}\, \widetilde{\sigma}_n^{\Diamond}(f,x)dx = \widehat{K^*}(m_0)\widehat{f}(m_0)$$

where $\sigma_n^{\Diamond}(f,x)$ is given by (2.6) of Chapter 1.

2. Abel Summability of Conjugate Series

In this section, we shall prove a theorem regarding the Abel summability of conjugate multiple Fourier series. In particular, when $K(x)$ is a Calderon-Zygmund kernel of spherical harmonic type (i.e., $K(x)$ meets the conditions in (1.3), (1.4), and (1.5)) and $K^*(x)$ is its periodic analogue defined in (1.8), we set

$$(2.1) \qquad \widetilde{S}_K(f) = \sum_{m\in\Lambda_N} \widehat{K^*}(m)\widehat{f}(m)e^{im\cdot x}$$

where $f \in L^1(T_N)$ and $\widehat{K^*}(m)$ is the principal-valued Fourier coefficient of $K^*(x)$. Also, $\widehat{K^*}(m)$, for $m \neq 0$, takes the value given in (1.25) and (1.26). (See Corollary 3.2 of Appendix A.)

The theorem, [Sh9, p. 44], which we prove regarding the series in (2.1), and the function $f \in L^1(T_N)$, which by assumption is also periodic of period 2π in each variable, is the following.

Theorem 2.1. *Let $f \in L^1(T_N)$ and $S(f) = \sum_{m\in\Lambda_N} \widehat{f}(m)e^{im\cdot x}$ be its Fourier series. Furthermore, let $K(x)$ be a Calderon-Zygmund kernel of spherical harmonic type that meets the conditions in (1.3), (1.4), and (1.5). Set*

$$(2.2) \qquad \widetilde{A}_t(f,x) = \sum_{1\leq|m|<\infty} \widehat{f}(m)\widehat{K^*}(m)e^{im\cdot x-|m|t}.$$

Then if x is in the Lebesgue set of f, i.e.,

$$(2.3) \qquad \lim_{r\to 0} r^{-N} \int_{B(0,r)} |f(x+y) - f(x)|\, dy = 0,$$

the following limit obtains

$$(2.4) \qquad \lim_{t\to 0}[\widetilde{A}_t(f,x) - \lim_{R\to\infty} (2\pi)^{-N} \int_{B(0,R)\backslash B(0,t)} f(x-y)K(y)dy] = 0.$$

Before proving the theorem, we observe that $\widetilde{A}_t(f,x)$ above only involves $\widehat{f}(m)$ for $m \neq 0$. Also, $\int_{B(0,R)-B(0,t)} K(y)dy = 0$. Consequently, with no loss in generality, we can assume that $\widehat{f}(0) = 0$. But in this case, we see from (1.12), (1.15), and Corollary 1.4 that if perchance the limit involving the integral in (2.4) does exist (separately) as $R \to \infty$ and $t \to 0$ and is finite, then this limit actually is $\tilde{f}(x)$.

Next, we use (2.4) from Appendix A and set

$$(2.5) \qquad A_n^\nu(t) = \frac{\Gamma(n/2)}{2^{N/2}\Gamma[(N+n)/2]} \int_0^\infty e^{-st} J_{\nu+n}(s) s^{\nu+1} ds,$$

for $t > 0$ where $\nu = (N-2)/2$. Also, with

$$K(x) = Y_n(x/|x|)|x|^{-N} \quad \text{for } x \neq 0,$$

we set

$$(2.6) \qquad K_{\nu,t}(x) = A_n^\nu(t/|x|)Y_n(x/|x|)|x|^{-N}$$

and establish the following lemma.

Lemma 2.2. *Let $S(x)$ be the trigonometric polynomial $\sum_{1 \leq |m| \leq R_1} b_m e^{im \cdot x}$, i.e., $S(x) = \sum_{m \in \Lambda_N} b_m e^{im \cdot x}$ where $b_m = 0$ for $m = 0$ and for $|m| > R_1$. For $t > 0$, set*

$$(2.7) \qquad \tilde{A}_t(S,x) = \sum_{1 \leq |m| \leq R_1} b_m \widehat{K^*}(m) e^{im \cdot x - |m|t}.$$

Then

$$(2.8) \qquad \tilde{A}_t(S,x) = (2\pi)^{-N} \lim_{R \to \infty} \int_{B(0,R)} K_{\nu,t}(y) S(x-y) dy.$$

Proof of Lemma 2.2. Since $S(x)$ is a finite linear combination of exponentials, it is clear using (1.21) above (i.e., $\widehat{K}(m) = \widehat{K^*}(m)$ for $m \neq 0$) that to establish the lemma, it is sufficient to show

$$\widehat{K}(u)e^{iu \cdot x - |u|t} = (2\pi)^{-N} \lim_{R \to \infty} \int_{B(0,R)} K_{\nu,t}(y) e^{iu \cdot x} e^{-iu \cdot y} dy$$

for $u \neq 0$. In other words, to establish the lemma, it is sufficient to show

$$\widehat{K}(u)e^{-|u|t} = (2\pi)^{-N} \lim_{R \to \infty} \int_{B(0,R)} (-1)^n K_{\nu,t}(y) e^{iu \cdot y} dy,$$

or to replace u with x so that

$$(2.9) \qquad \widehat{K}(x)e^{-|x|t} = (2\pi)^{-N}(-1)^n \lim_{R \to \infty} \int_{B(0,R)} K_{\nu,t}(y) e^{ix \cdot y} dy$$

for $x \neq 0$.

In order to show that (2.9) is valid, we set

$$(2.10) \qquad g(x) = \widehat{K}(x)e^{-|x|t}$$

and see from Theorem 3.1 in Appendix A that

$$\widehat{K}(x) = \kappa_{n,N} Q_n(x)|x|^{-n} = \kappa_{n,N} Y_n(\xi)$$

where $x = r\xi$. So it is clear from (2.10) that $g \in L^1(\mathbf{R}^N)$ for $t > 0$.

With $y = |y|\,\eta$ where $\eta \in S_{N-1}$, we next show that

$$(2.11) \qquad \widehat{g}(y) = (2\pi)^{-N}(-1)^n K_{\nu,t}(y)$$

by means of a computation where $\nu = (N-2)/2$. In this computation,

$$I(y,t) = \int_{\mathbf{R}^N} \widehat{K}(x)e^{-|x|t}e^{-ix\cdot y}dx.$$

So by definition,

$$\widehat{g}(y) = (2\pi)^{-N}I(y,t).$$

We will also make use of (2.5) and (2.6) above and (3.21''), and (3.28) in Appendix A.

$$
\begin{aligned}
I(y,t) &= \int_{\mathbf{R}^N} \kappa_{n,N}Q_n(x)\,|x|^{-n}\,e^{-|x|t}e^{-ix\cdot y}dx \\
&= \kappa_{n,N}\int_0^\infty e^{-rt}r^{N-1}\int_{S_{N-1}} Y_n(\xi)e^{-ir|y|\xi\cdot\eta}dS(\xi)dr \\
&= \kappa_{n,N}\int_0^\infty e^{-rt}r^{N-1}(-i)^n 2^{\nu+1}\pi^{\nu+1}\frac{J_{\nu+n}(|y|\,r)}{(|y|\,r)^\nu}Y_n(\eta)dr \\
&= K(y)(-i)^n\kappa_{n,N}(2\pi)^{N/2}\int_0^\infty e^{-rt/|y|}J_{\nu+n}(r)r^{\nu+1}dr \\
&= K(y)(-i)^n\kappa_{n,N}2^N\pi^{N/2}\Gamma(N+n)/2)A_n^\nu(t/\,|y|)/\Gamma(n)/2) \\
&= K(y)A_n^\nu(t/\,|y|)(-1)^n \\
&= (-1)^n K_{\nu,t}(y),
\end{aligned}
$$

which establishes the equality in (2.11).

From Lemma 3.2 in Chapter 1 and (2.10) and (2.11) above, we see that

$$(2.12) \quad (2\pi)^{-N}(-1)^n \lim_{R\to\infty}\int_{B(0,R)} K_{\nu,t}(y)(1-|y|^2/R^2)^\alpha e^{ix\cdot y}dy = \widehat{K}(x)e^{-|x|t}$$

for $\alpha > (N-1)/2$, which is almost (2.9), which we sought as a result. What we have to do is eliminate the factor $(1-|y|^2/R^2)^\alpha$.

With $y = s\eta$ and $x = r\xi$ where $\eta, \xi \in S_{N-1}$ and (2.6), we obtain

$$K_{\nu,t}(y) = A_n^\nu(\frac{t}{s})Y_n(\eta)s^{-N}$$

We set

$$(2.13) \qquad h(s) = A_n^\nu(\frac{t}{s})s^{-1}\int_{S_{N-1}} Y_n(\eta)e^{irs\eta\cdot\xi}dS(\eta),$$

and see that the limit in (2.12) can be rewritten as

$$(2.14) \qquad (2\pi)^{-N}(-1)^n \lim_{R\to\infty}\int_0^R h(s)(1-\frac{s^2}{R^2})^\alpha ds = \widehat{K}(x)e^{-|x|t}.$$

Making use of the identification of $h(s)$ given in (2.13), we infer from (2.7) in Appendix A that $h(s) \in L^1(0, R)$, $\forall R > 0$. Hence, if we show that

$$(2.15) \qquad \lim_{R \to \infty} \int_{B(0,R) \setminus B(0,1)} K_{\nu,t}(y) e^{ix \cdot y} dy \quad \text{exists and is finite,}$$

where $x \neq 0$ and $t > 0$, then it will follow that

$$\lim_{R \to \infty} \int_0^R h(s) ds \quad \text{exists and is finite.}$$

Consequently, we obtain Theorem 1.1 in Appendix B and from (2.14) above that

$$(2\pi)^{-N}(-1)^n \lim_{R \to \infty} \int_0^R h(s) ds = \widehat{K}(x) e^{-|x|t}$$

for $x \neq 0$ and $t > 0$. But, we see from (2.13) that this last limit is the same as the limit in (2.9), and the proof of the lemma will then be complete.

So it remains to show that the statement in (2.15) is true. To accomplish this, we first observe from both (3.13) and Theorem 3.1 in Appendix A that

$$\lim_{R \to \infty} \int_{B(0,R) \setminus B(0,1)} K(y) e^{ix \cdot y} dy \quad \text{exists and is finite.}$$

Consequently, to show that the statement in (2.15) is valid, as we see from (2.6), it is sufficient to show the following:

$$(2.16) \qquad \lim_{R \to \infty} \int_{B(0,R) \setminus B(0,1)} [A_n^\nu(t/|y|) - 1] K(y) e^{ix \cdot y} dy \quad \text{exists and is finite,}$$

for $x \neq 0$ and $t > 0$. But, by Theorem 2.1 in Appendix A,

$$|A_n^\nu(t/|y|) - 1| \leq C(N, n)\left(\frac{t}{|y|}\right)^{1/2} \quad \text{for } 2t \leq |y| < \infty.$$

Since

$$|K(y)| \leq c |y|^{-N} \quad \text{for} \quad |y| \geq 1,$$

where c is a positive constant, it is clear that that the statement in (2.16) is indeed valid, and the proof of the lemma is complete. ∎

Continuing with the preliminaries involved in the proof of Theorem 2.1, we next establish the following lemma.

Lemma 2.3. *Let $f \in L^1(T_N)$ and $S(f) = \sum_{m \in \Lambda_N} \widehat{f}(m) e^{im \cdot x}$ be its Fourier series. For $t>0$, set*

$$\widetilde{A}_t(f, x) = \sum_{1 \leq |m| < \infty} \widehat{f}(m) \widehat{K^*}(m) e^{im \cdot x - |m|t}.$$

Then

$$(2.17) \qquad \widetilde{A}_t(f, x) = (2\pi)^{-N} \lim_{R \to \infty} \int_{B(0,R)} K_{\nu,t}(y) f(x - y) dy,$$

where $K_{\nu,t}(y)$ is defined in (2.6).

Proof of Lemma 2.3. Since $\widetilde{A}_t(f,x)$ does not make use of $\widehat{f}(0)$ and since

$$\int_{B(0,R)} K_{\nu,t}(y)dy = 0 \quad \text{for } R > 0,$$

without loss in generality, we can assume from the start that

$$(2.18) \qquad\qquad \widehat{f}(0) = 0.$$

Next, we set

$$(2.19) \quad K^*_{\nu,t}(x) = K_{\nu,t}(x) + \lim_{R\to\infty} \sum_{1\le|m|\le R} [K_{\nu,t}(x+2\pi m) - K_{\nu,t}(2\pi m)]$$

for $x \in T_N$, and observe from (2.6) above and from (2.7) in Appendix A that $K_{\nu,t}(x)$ is bounded in $T_N \backslash \{0\}$. We write $K_{\nu,t}(x+2\pi m) - K_{\nu,t}(2\pi m)$ as

$$[A^\nu_n(\frac{t}{|x+2\pi m|}) - A^\nu_n(\frac{t}{|2\pi m|})]K(x+2\pi m)$$

$$+ A^\nu_n(\frac{t}{|2\pi m|})[K(x+2\pi m) - K(2\pi m)],$$

and make use of (2.8) in Appendix A as well as Theorem 2.1 in the Appendix A. By means of the same type of argument used to show that the series in (1.8) defining $K^*(x)$ is uniformly and absolutely convergent for x in a bounded domain, we obtain that the series defining $K^*_{\nu,t}(x)$ in (2.19) is uniformly and absolutely convergent for x in a bounded domain.

Hence, the series in (2.19) is uniformly convergent for $x \in T_N$. Also, from (2.18) above,

$$\int_{T_N} f(x-y)K_{\nu,t}(2\pi m)dy = 0 \quad \text{for } m \neq 0.$$

Consequently, we see that

$$(2.20) \quad \int_{T_N} f(x-y)K^*_{\nu,t}(y)dy = \lim_{R\to\infty} \sum_{|m|\le R} \int_{T_N} f(x-y)K_{\nu,t}(y+2\pi m)dy.$$

Using the same argument that we used in the proof of Corollary 1.4 above, we observe that the limit on the right-hand side of the equality in (2.20) is the same as

$$\lim_{R\to\infty} \int_{B(0,R)} K_{\nu,t}(y)f(x-y)dy.$$

Therefore, we have that

$$(2.21) \qquad \int_{T_N} f(x-y)K^*_{\nu,t}(y)dy = \lim_{R\to\infty} \int_{B(0,R)} K_{\nu,t}(y)f(x-y)dy,$$

for $x \in T_N$.

To complete the proof of this lemma, we set $S^j(x) = \sigma_j^{\Diamond}(f, x)$, which is the trigonometric polynomial defined in (2.6) of Chapter 1. From what we have just shown in (2.21)

$$\int_{T_N} S^j(x-y)K_{\nu,t}^*(y)dy = \lim_{R\to\infty} \int_{B(0,R)} K_{\nu,t}(y)S^j(x-y)dy,$$

for $x \in T_N$. Therefore, from (2.8) in Lemma 2.2, we obtain

$$(2.22) \qquad \widetilde{A}_t(S^j, x) = (2\pi)^{-N} \int_{T_N} S^j(x-y)K_{\nu,t}^*(y)dy,$$

for $x \in T_N$.

Now, from the fact that $S^j(x) = \sigma_j^{\Diamond}(f, x)$ and from Theorem 2.2 in Chapter 1, it follows that

$$(2.23) \qquad \lim_{j\to\infty} \int_{T_N} \left| S^j(f,x) - f(x) \right| dx = 0.$$

Let us write

$$(2.24) \qquad \widetilde{A}_t(S^j, x) = \sum_{1 \le |m| < \infty} \widehat{S^j}(m)\widehat{K^*}(m)e^{im\cdot x - |m|t}.$$

From (2.23) and Corollary 3.2 in Appendix A, we obtain that

$$\exists\, C > 0 \text{ such that } \left| \widehat{S^j}(m)\widehat{K^*}(m) \right| \le C \text{ for } m \in \Lambda_N \text{ and } \forall j.$$

Also, from (2.23), $\lim_{j\to\infty} \widehat{S^j}(m) = \widehat{f}(m)$ for every $m \in \Lambda_N$. Since for fixed $t > 0$, $\sum_{m\in\Lambda_N} Ce^{-|m|t} < \infty$, we conclude from (2.24) that

$$(2.25) \qquad \lim_{j\to\infty} \widetilde{A}_t(S^j, x) = \widetilde{A}_t(f, x) \quad \text{for } x \in T_N.$$

On the other hand, as we have observed, $K_{\nu,t}^*(y)$ is bounded in $T_N\backslash\{0\}$. Hence, we obtain from (2.23) that

$$(2.26) \qquad \lim_{j\to\infty} \int_{T_N} S^j(x-y)K_{\nu,t}^*(y)dy = \int_{T_N} f(x-y)K_{\nu,t}^*(y)dy,$$

for $x \in T_N$. Statement (2.17) in the lemma now follows immediately from (2.22), (2.25), and (2.26), and the proof of the lemma is complete. ■

Proof of Theorem 2.1. With no loss in generality, we can suppose from the start that the x that occurs in the statement of the theorem is equal to 0. Likewise without loss in generality, we can assume $f(0) = 0$. So assumption (2.3) of the theorem is replaced with the assumption that

$$(2.27) \qquad \lim_{t\to 0} t^{-N} \int_{B(0,t)} |f(y)|\, dy = 0.$$

Consequently, we see that the proof of the theorem will be complete when we succeed in showing that

$$(2.28) \qquad \lim_{t \to 0}[\widetilde{A}_t(f,0) - \lim_{R \to \infty}(2\pi)^{-N}\int_{B(0,R)\backslash B(0,t)} f(-y)K(y)dy] = 0.$$

To validate the limit in (2.28), we first show that

$$(2.29) \qquad \lim_{t \to 0}\int_{B(0,t)} f(-y)K_{\nu,t}(y)dy = 0.$$

To accomplish this, we observe from (2.6) above and from (2.7) in Appendix A that

$$\exists \, C > 0 \text{ such that } |K_{\nu,t}(y)| \le Ct^{-N} \quad \text{for } t > 0 \text{ and } y \ne 0.$$

Hence,

$$\left|\int_{B(0,t)} f(-y)K_{\nu,t}(y)dy\right| \le Ct^{-N}\int_{B(0,t)} |f(-y)| \, dy,$$

and we conclude from the limit in (2.27) that the limit in (2.29) is indeed true.

From the equality in (2.17) of Lemma 2.3 when $x = 0$ joined with the limit in (2.29), we see that

$$\lim_{t \to 0}[\widetilde{A}_t(f,0) - \lim_{R \to \infty}(2\pi)^{-N}\int_{B(0,R)\backslash B(0,t)} f(-y)K_{\nu,t}(y)dy] = 0.$$

Therefore, (2.28) will be established if we show that
(2.30)

$$\lim_{t \to 0}\lim_{R \to \infty}[\int_{B(0,R)\backslash B(0,t)} f(-y)K_{\nu,t}(y)dy - \int_{B(0,R)\backslash B(0,t)} f(-y)K(y)dy] = 0.$$

Using (2.6) above, we see that the limit in (2.30) is the same as the following:

$$(2.31) \qquad \lim_{t \to 0}\lim_{R \to \infty}\int_{B(0,R)\backslash B(0,t)} f(-y)[A_n^\nu(t/\,|y|) - 1]K(y)dy = 0.$$

So the proof of the theorem will be complete if we show that the limit in (2.31) is valid. In order to do this, let $\varepsilon > 0$ be given. (2.27) enables us to choose $\delta > 0$, so that

$$(2.32) \qquad \int_{B(0,t)} |f(-y)| \, dy \le \varepsilon t^N \quad \text{for } 0 < t \le \delta.$$

Next, we see from (2.9) in Appendix A, that

$$(2.33)$$
$$\left|\int_{B(0,R)\backslash B(0,t)} f(-y)[A_n^\nu(t/\,|y|) - 1]K(y)dy\right|$$
$$\le C^*(N,n)t^{\frac{1}{2}}\int_{B(0,R)\backslash B(0,t)} |f(-y)K(y)|\,|y|^{-\frac{1}{2}} \, dy$$

where $C^*(N,n)$ is a positive constant depending only on N and n.

Now,

$$\exists\, C_K > 0 \text{ such that } |K(y)| \le C_K\,|y|^{-N} \text{ for } y \ne 0,$$

where C_K is a constant depending only on K. Therefore, we can majorize the integral on the right-hand side of the inequality in (2.33) and obtain

(2.34)
$$t^{\frac{1}{2}} \int_{B(0,R)\setminus B(0,t)} |f(-y)K(y)|\,|y|^{-\frac{1}{2}}\,dy$$

$$\le C_K t^{\frac{1}{2}} \int_{B(0,R)\setminus B(0,t)} |f(-y)|\,|y|^{-(N+\frac{1}{2})}\,dy.$$

From (3.26) in Chapter 1, we see that

$$\int_{B(0,R)\setminus B(0,R-1)} |f(-y)|\,dy = O(R^{N-1}) \quad \text{as } R \to \infty.$$

From this fact, it is easy to obtain that

$$\lim_{R\to\infty} \int_{B(0,R)\setminus B(0,\delta)} |f(-y)|\,|y|^{-(N+\frac{1}{2})}\,dy < \infty,$$

where δ is given in (2.32). Consequently,

(2.35)
$$\lim_{t\to 0}\lim_{R\to\infty} t^{\frac{1}{2}} \int_{B(0,R)\setminus B(0,\delta)} |f(-y)|\,|y|^{-(N+\frac{1}{2})}\,dy = 0.$$

Next, we set

(2.36)
$$I(\delta,t) = \int_{B(0,\delta)\setminus B(0,t)} |f(-y)|\,|y|^{-(N+\frac{1}{2})}\,dy,$$

and

(2.37)
$$F(s) = \int_{B(0,s)} |f(-y)|\,dy$$
$$= \int_0^s r^{N-1}\left[\int_{S_{N-1}} |f(-r\eta)|\,dS(\eta)\right]dr.$$

Now from (2.36) and (2.37), we see that

$$
\begin{aligned}
I(\delta,t) &= \int_t^\delta s^{-(N+\frac{1}{2})}dF(s) \\
&= \left. s^{-(N+\frac{1}{2})}F(s)\right|_t^\delta + \left(N+\frac{1}{2}\right)\int_t^\delta s^{-(N+\frac{3}{2})}F(s)\,ds.
\end{aligned}
$$

Also, from (2.32), we have that

$$0 \le F(s) \le \varepsilon s^N \quad \text{for } t \le s \le \delta.$$

Consequently, we obtain from this last computation that

$$
\begin{aligned}
t^{\frac{1}{2}}\,|I(\delta,t)| &\le 2\varepsilon + \left(N+\frac{1}{2}\right)\varepsilon t^{\frac{1}{2}}\int_t^\delta s^{-\frac{3}{2}}ds \\
&\le (2N+3)\varepsilon.
\end{aligned}
$$

We conclude from (2.36) and this last inequality that

$$(2.38) \qquad \lim_{t \to 0} t^{\frac{1}{2}} \int_{B(0,\delta) \setminus B(0,t)} |f(-y)| \, |y|^{-(N+\frac{1}{2})} \, dy \leq (2N+3)\varepsilon.$$

Combining the limits in (2.35) and (2.38), we see that

$$\lim_{t \to 0} \lim_{R \to \infty} t^{\frac{1}{2}} \int_{B(0,R) \setminus B(0,t)} |f(-y)| \, |y|^{-(N+\frac{1}{2})} \, dy \leq (2N+3)\varepsilon.$$

As a consequence of this last result, we see from the inequalities in (2.33) and (2.34) that

$$\lim_{t \to 0} \lim_{R \to \infty} \left| \int_{B(0,R) \setminus B(0,t)} f(-y)[A_n^\nu(t/\,|y|) - 1]K(y)dy \right|$$

$$\leq C^*(N,n)C_K(2N+3)\varepsilon.$$

But ε is an arbitrary positive number. Hence,

$$\lim_{t \to 0} \lim_{R \to \infty} \left| \int_{B(0,R) \setminus B(0,t)} f(-y)[A_n^\nu(t/\,|y|) - 1]K(y)dy \right| = 0.$$

This limit validates the equality in (2.31) and completes the proof of the theorem. ∎

Exercises.

1. With $A_n^\nu(t)$ defined by (2.5), use Theorem 2.1 in Appendix A to prove that

$$\sum_{1 \leq |m| \leq R} [A_n^\nu(x + 2\pi m) - A_n^\nu(2\pi m)]$$

is uniformly and absolutely convergent for x in a bounded domain as $R \to \infty$.

2. Given $f \in L^1(T_N)$, prove that

$$\int_{T_N} f(y)K_{\nu,t}^*(y)dy = \lim_{R \to \infty} \sum_{|m| \leq R} \int_{T_N} f(y)K_{\nu,t}(y + 2\pi m)dy$$

where $K_{\nu,t}(y+2\pi m)$ and $K_{\nu,t}^*(y)$ are defined by (2.6) and (2.19), respectively.

3. Spherical Convergence of Conjugate Series

In this section, we shall prove two theorems regarding the spherical convergence of the series

$$\widetilde{S}_K(f) = \sum_{m \in \Lambda_N \setminus \{0\}} \widehat{K^*}(m)\widehat{f}(m)e^{im \cdot x}$$

where $\widehat{K^*}(m)$ is described in (1.25) above. The first theorem we shall prove is the following:

Theorem 3.1. *Suppose $f \in L^1(T_N)$, and K is a Calderon-Zygmund kernel of spherical harmonic type that meets the conditions in (1.3), (1.4), and (1.5). Suppose also that*

$$(3.1) \qquad \lim_{R \to \infty} \sum_{1 \le |m| \le R} \widehat{K^*}(m)\widehat{f}(m)e^{im \cdot x_0} = \alpha,$$

where α is a finite real number. Then

$$(3.2) \qquad \lim_{\varepsilon \to 0} \lim_{R \to \infty} (2\pi)^{-N} \int_{B(0,R) \setminus B(0,\varepsilon)} f(x_0 - x)K(x)dx = \alpha.$$

Proof of Theorem 3.1. Before actually starting the proof of this theorem, we observe that if x_0 is in the Lebesgue set of f, then there is nothing to prove because the theorem is then an immediate corollary of Theorem 2.1 above joined with Theorem 1.2 in Appendix B.

In order to prove the theorem, we observe that without loss in generality, we can assume from the start that $x_0 = 0$ and that $\widehat{f}(0) = 0$. Also, since the theorem is true for trigonometric polynomials (i.e., in this case, every point is in the Lebesgue set), we can also take $\alpha = 0$. So in view of (1.25), (3.1) becomes

$$(3.3) \qquad \lim_{R \to \infty} \sum_{1 \le |m| \le R} Q_n(\frac{m}{|m|})\widehat{f}(m) = 0,$$

where $K(x) = Q_n(x)/|x|^{N+n}$ and $n \ge 1$.

What we have to show is that the limit in (3.3) implies that

$$(3.4) \qquad \lim_{\varepsilon \to 0} \lim_{R \to \infty} (2\pi)^{-N} \int_{B(0,R) \setminus B(0,\varepsilon)} f(-x)K(x)dx = 0.$$

In order to do this, we set

$$(3.5) \qquad \chi_{n,\nu}(t) = \int_t^\infty J_{\nu+n}(r)/r^{\nu+1}dr$$

where $J_{\nu+n}(r)$ is the familiar Bessel function of the first kind of order $\nu + n$ and $\nu = (N-2)/2$.

For $R > 0$, we set

$$(3.6) \qquad S(R) = \sum_{0 < |m| \le R} Q_n(\frac{m}{|m|})\widehat{f}(m),$$

and observe that for $R > 1$,

$$\sum_{1 \le |m| \le R} Q_n(\frac{m}{|m|})\widehat{f}(m)\chi_{n,\nu}(|m|\,t) = \int_0^R \chi_{n,\nu}(rt)dS(r)$$

$$= t^{-\nu}\int_1^R S(r)J_{\nu+n}(rt)r^{-(\nu+1)}dr$$

$$+ S(R)\chi_{n,\nu}(Rt).$$

From the estimates (2.1) and (2.2) in Appendix A, it is clear that $\chi_{n,\nu}(t)$ is bounded on $(0, \infty)$. So it follows from (3.3), (3.6), and this last computation that for $t > 0$,

(3.7)
$$\lim_{R\to\infty} \sum_{1\le|m|\le R} Q_n\left(\frac{m}{|m|}\right)\widehat{f}(m)\chi_{n,\nu}(|m|\,t)$$
$$= t^{-\nu} \int_1^\infty S(r)J_{\nu+n}(rt)r^{-(\nu+1)}dr.$$

Next, we set for $s > 0$,

(3.8)
$$f(x, s) = \sum_{m\in\Lambda_N} \widehat{f}(m)e^{im\cdot x}e^{-|m|s}.$$

It follows from Theorem 4.3 in Chapter 1 that

(3.9)
$$\lim_{s\to 0} \int_{B(0,R)\backslash B(0,\varepsilon)} |f(x, s) - f(x)| = 0 \quad \text{for } 0 < \varepsilon < R < \infty.$$

From Appendix A (3.13) and two lines below (3.21′), we see with $y = m$, that

$$\int_{B(0,R)-B(0,\varepsilon)} e^{-im\cdot x}K(x)dx$$
$$= (-i)^n 2^{\nu+1}\pi^{\nu+1}Q_n\left(\frac{m}{|m|}\right)\int_\varepsilon^R \frac{J_{\nu+n}(|y|\,r)}{(|y|\,r)^\nu}r^{-1}dr.$$
$$= (-i)^n(2\pi)^{N/2}Q_n\left(\frac{m}{|m|}\right)[\chi_{n,\nu}(|m|\,\varepsilon) - \chi_{n,\nu}(|m|\,R)].$$

Observing that the series in (3.8) is absolutely convergent, we obtain from this last computation that

(3.10)
$$(2\pi)^{-N/2}\int_{B(0,R)\backslash B(0,\varepsilon)} f(-x, s)K(x)dx$$

$$= (-i)^n \sum_{m\in\Lambda_N} \widehat{f}(m)Q_n\left(\frac{m}{|m|}\right)e^{-|m|s}[\chi_{n,\nu}(|m|\,\varepsilon) - \chi_{n,\nu}(|m|\,R)].$$

Next, we set

(3.11)
$$g(t) = \lim_{R\to\infty} \sum_{1\le|m|\le R} Q_n\left(\frac{m}{|m|}\right)\widehat{f}(m)\chi_{n,\nu}(|m|\,t),$$

and observe from (3.7) that $g(t)$ is well-defined and finite for $t > 0$. Passing to the limit as $s \to 0$ in (3.10), we obtain from (3.9) and (3.11) that

(3.12)
$$(2\pi)^{-N/2}\int_{B(0,R)\backslash B(0,\varepsilon)} f(-x)K(x)dx = g(\varepsilon) - g(R).$$

Now from (3.7), (3.11), and the fact that $S(R) \to 0$ as $R \to \infty$, we have

$$|g(t)| \le ct^{-(\nu+\frac{1}{2})} \quad \text{for } t > 0,$$

where c is a constant. We conclude from (3.12) and this last inequality that

$$\lim_{R\to\infty} (2\pi)^{-N/2} \int_{B(0,R)\backslash B(0,\varepsilon)} f(-x)K(x)dx = g(\varepsilon).$$

It follows from (3.7), (3.11), and this last equality that the double limit in (3.4) will be valid if we show

$$(3.13) \qquad \lim_{t\to 0} t^{-\nu} \int_1^\infty S(r)J_{\nu+n}(rt)r^{-(\nu+1)}dr = 0.$$

To establish the limit in (3.13), we recall that

$$S(r) = o(1) \text{ as } r \to \infty.$$

Next, we note from the inequality in (2.1) of Appendix A, since $n \geq 1$, that

$$|J_{\nu+n}(s)| \leq cs^{\nu+1} \text{ for } 0 < s \leq 1,$$

where c is a constant. Consequently,

$$(3.14) \qquad \left| t^{-\nu} \int_1^{1/t} S(r)J_{\nu+n}(rt)r^{-(\nu+1)}dr \right| \leq t \int_1^{1/t} o(1)dr = o(1),$$

as $t \to 0$.

Also, from (2.2) in Appendix A, we obtain that

$$\left| t^{-\nu} \int_{1/t}^\infty S(r)J_{\nu+n}(rt)r^{-(\nu+1)}dr \right| \leq t^{-(\nu+\frac{1}{2})} \int_{1/t}^\infty o(1)r^{-(\nu+3/2)}dr.$$

Hence,

$$\lim_{t\to 0} \left| t^{-\nu} \int_{1/t}^\infty S(r)J_{\nu+n}(rt)r^{-(\nu+1)}dr \right| = 0.$$

We conclude from (3.14) and this last limit that the limit in (3.13) is indeed valid. Consequently, the double limit in (3.4) holds, and the proof of the theorem is complete. ∎

Next, we establish a necessary and sufficient condition for the convergence of the conjugate Fourier series at x_0 given that the following Tauberian condition prevails:

$$(3.15) \qquad \widehat{f}(m)\widehat{K^*}(m)e^{im\cdot x_0} + \widehat{f}(-m)\widehat{K^*}(-m)e^{-im\cdot x_0} \geq -A/|m|^N,$$

$\forall m \neq 0$, where A is a positive constant.

In particular, motivated by Hardy and Littlewood [HL], we show that the following theorem [Sh7] is valid.

Theorem 3.2.   *Suppose $f \in L^1(T_N)$, and K is a Calderon-Zygmund kernel of spherical harmonic type that meets the conditions in (1.3), (1.4),* *and (1.5). Suppose also that the condition (3.15) holds at the point x_0. Then a necessary and sufficient conditon that*

$$\tag{3.16} \lim_{R\to\infty} \sum_{1\le |m|\le R} \widehat{K^*}(m)\widehat{f}(m)e^{im\cdot x_0} = \alpha,$$

where α is a finite real number is that

$$\tag{3.17} \lim_{\varepsilon\to 0}\lim_{R\to\infty}(2\pi)^{-N}\int_{B(0,R)\setminus B(0,\varepsilon)} f(x_0 - x)K(x)dx = \alpha.$$

The necessary condition of the above theorem is an immediate corollary of Theorem 3.1. Hence, we need only establish the sufficient condition of the above theorem. In order to do this, we will first need the following Tauberian lemma.

Lemma 3.3. *Suppose that for $m\ne 0$, a_m is real-valued and that $a_m = O(|m|^j)$ as $|m| \to \infty$ for some nonnegative integer j. Suppose, furthermore, that there exists a positive constant A such that $a_m \ge -A/|m|^N$ $\forall m \ne 0$. For $t > 0$, set*

$$I(t) = \sum_{1\le |m|} a_m e^{-|m|t},$$

and suppose also that $\lim_{t\to 0}I(t) = \alpha$, where α is finite-valued. Then

$$\lim_{R\to\infty} \sum_{1\le |m|\le R} a_m = \alpha.$$

Proof of Lemma 3.3. To establish this lemma with no loss in generality, we can assume from the start that $A = 1$. Consequently, on observing that

$$\frac{d^2 I(t)}{dt^2} = \sum_{1\le |m|} |m|^2\, a_m e^{-|m|t},$$

we obtain the following:

$$(i)\ I(t) \text{ is in } C^\infty(0,\infty);$$

$$\tag{3.18} (ii)\ \lim_{t\to 0}I(t) = \alpha;$$

$$(iii)\ d^2 I(t)/dt^2 \ge -\sum_{1\le |m|} e^{-|m|t}/\ |m|^{N-2} \quad \text{for}\ \ t > 0.$$

Next, we observe from the Poisson summation formula in Lemma 4.4 of Chapter 1 that for $t > 0$,

$$\tag{3.19} \sum_{m\in\Lambda} e^{-|m|t} = b_N t^{-N} + b_N t \sum_{m\ne 0}[t^2 + |2\pi m|^2]^{-(N+1)/2},$$

where b_N is a positive constant. It follows from (3.19) and L'Hospital's rule that

(3.20) $$\lim_{t \to 0} \sum_{1 \leq |m|} \frac{e^{-|m|t}}{|m|^{N-2}} / t^{-2} \text{ exists and is finite.}$$

From (3.18)(iii) and (3.20), we consequently have that there is a positive constant b'_N such that

(3.21) $$\frac{d^2 I(t)}{dt^2} \geq -b'_N t^{-2} \quad \text{for } 0 < t < 1.$$

But then it follows from (3.18)(i) and (ii), (3.21), and Theorem 2.1 in Appendix B that

(3.22) $$\lim_{t \to 0} t \frac{dI(t)}{dt} = 0.$$

Since $A = 1$, we have by hypothesis that

(3.23) $$|m| a_m + |m|^{-(N-1)} \geq 0 \quad \text{for } m \neq 0.$$

Also, we have from (3.19) and L'Hospital's rule that there is a positive constant β_N such that

(3.24) $$\lim_{t \to 0} t \sum_{1 \leq |m|} \frac{e^{-|m|t}}{|m|^{N-1}} = \beta_N.$$

For R>0, set

$$(i) \ S_1(R) = \sum_{0 < |m| \leq R} |m| a_m + |m|^{-(N-1)},$$

$$(ii) S_2(R) = \sum_{0 < |m| \leq R} |m|^{-(N-1)}.$$

Then we observe from (3.24) that

(3.25) $$\lim_{t \to 0} t \int_0^\infty e^{-rt} dS_2(r) = \beta_N.$$

Consequently, it follows from Theorem 2.2 in Appendix B and (3.25) that

(3.26) $$R^{-1} \sum_{0 < |m| \leq R} |m|^{-(N-1)} \to \beta_N \quad \text{as } R \to \infty.$$

Also since $\frac{dI(t)}{dt} = - \sum_{0 < |m|} |m| a_m e^{-|m|t}$, we see from (3.22) and (3.25) that

$$\lim_{t \to 0} t \int_0^\infty e^{-rt} dS_1(r) = \beta_N.$$

Hence, it follows from (3.23), Theorem 2.2 in Appendix B, and this last limit that

$$R^{-1} \sum_{0 < |m| \leq R} [|m| a_m + |m|^{-(N-1)}] \to \beta_N \quad \text{as } R \to \infty.$$

We conclude from (3.26) and this last limit that

$$(3.27) \qquad R^{-1} \sum_{0<|m|\leq R} |m|\, a_m \to 0 \text{ as } R \to \infty.$$

Next, we set $S_3(R) = \sum_{0<|m|\leq R} |m|\, a_m$ and observe that

$$\int_0^R \frac{1-e^{-rt}}{r}\, dS_3(r) = \sum_{0<|m|\leq R} \frac{1-e^{-|m|t}}{|m|}\, |m|\, a_m.$$

Consequently, for $R > 1$,

$$\sum_{0<|m|\leq R} [1-e^{-|m|t}]a_m \;=\; R^{-1}S_3(R)[1-e^{-Rt}] - t\int_1^R S_3(r)e^{-rt}r^{-1}dr$$

$$+ \int_1^R S_3(r)[1-e^{-rt}]r^{-2}dr.$$

From (3.27), $S_3(r)r^{-1} = o(1)$ as $r \to \infty$. Hence, we infer from this last computation that

$$(3.28) \qquad \lim_{t\to 0} \sum_{0<|m|\leq t^{-1}} [1-e^{-|m|t}]a_m = 0.$$

Next, we observe that for $R > 1$,

$$\sum_{R<|m|} e^{-|m|t}a_m \;=\; -S_3(R)R^{-1}e^{-Rt} + t\int_R^\infty S_3(r)e^{-rt}r^{-1}dr$$

$$+ \int_R^\infty S_3(r)e^{-rt}r^{-2}dr.$$

We conclude from this computation and (3.27) that

$$\lim_{t\to 0} \sum_{t^{-1}<|m|} e^{-|m|t}a_m = 0.$$

Since by hypothesis, $\lim_{t\to 0}\sum_{0<|m|} e^{-|m|t}a_m = \alpha$, we obtain from this last limit that

$$\lim_{t\to 0} \sum_{0<|m|\leq t^{-1}} e^{-|m|t}a_m = \alpha,$$

and hence, from (3.28) that

$$\lim_{t\to 0} \sum_{0<|m|\leq t^{-1}} a_m = \alpha.$$

This concludes the proof of the lemma.   ∎

In order to prove Theorem 3.2, we will need two more lemmas.

Lemma 3.4. *Let $f(x)$ and $K(x)$ be as in the hypothesis of Theorem 3.2. Suppose also that the double limit in (3.17) holds. Then*

$$(3.29) \qquad \lim_{r \to 0} r^{-N} \int_{B(0,r)} f(x_0 - x) |x|^N K(x) dx = 0.$$

Proof of Lemma 3.4. We let $S(0,r)$ represent the (N-1)-sphere with center 0 and radius r and we let $dS(x)$ represent its natuural (N-1)-dimensional volume element. Then we define almost everywhere for $r > 0$,

$$h(r) = \int_{S(0,r)} f(x_0 - x) K(x) dS(x).$$

Then h meets the condition in the hypothesis of [Zy1, Lemma 7.23, p. 104]. Consequently,

$$\int_0^r h(s) s^N ds = o(r^N) \text{ as } r \to 0.$$

This establishes the limit in (3.29). ∎

We will also need the following lemma.

Lemma 3.5. *Let $f(x)$ and $K(x)$ be as in the hypothesis of Theorem 3.2. Suppose that the limit in (3.29) also holds. Let $\widetilde{A}_t(f, x)$ represent the expression in (2.2). Then*

$$\lim_{t \to 0}[\widetilde{A}_t(f, x_0) - \lim_{R \to \infty} (2\pi)^{-N} \int_{B(0,R) \backslash B(0,t)} f(x_0 - y) K(y) dy] = 0.$$

Proof of Lemma 3.5. With no loss in generality, we can assume $x_0 = 0$. Next, for $r > 0$, we set

$$(3.30) \qquad g(r) = (2\pi)^{-N} \int_{B(0,r)} f(-x) |x|^N K(x) dx$$

and observe by assumption that

$$(3.31) \qquad g(r) = o(r^N) \text{ as } r \to 0.$$

Now from (2.17), we have that

$$\widetilde{A}_t(f, 0) = (2\pi)^{-N} \lim_{R \to \infty} \int_{B(0,R)} K_{\nu,t}(y) f(-y) dy$$

where $K_{\nu,t}(y) = A_n^\nu(t/ |y|) K(y)$ and $\nu = (N - 2)/2$. Consequently, we have from (3.30) that

$$(3.32) \qquad \widetilde{A}_t(f, 0) = \lim_{R \to \infty} \int_0^R A_n^\nu(t/r) r^{-N} dg(r),$$

where $A_n^\nu(t)$ is defined in (2.5).

Next, we set $A_n^\nu(t)' = dA_n^\nu(t)/dt$ and see from (2.5) that

$$A_n^\nu(t)' = -c_n \int_0^\infty e^{-st} J_{\nu+n}(s) s^{\nu+2} ds,$$

where c_n is a positive constant.

From the estimates in Appendix A, $|J_{\nu+n}(s)| \leq c_n^* s^\nu$ for $s > 0$. Hence, it follows from this last equality that there is a positive constant c_n^{**} such that

$$(3.33) \qquad \left| A_n^\nu(t)' \right| \leq c_n^{**} t^{-(N+1)} \quad \text{for} \ \ t > 0.$$

Also, integrating by parts and using (3.31) and (2.7) in Appendix A gives the following formula:

$$\int_0^t A_n^\nu(t/r) r^{-N} dg(r) = A_n^\nu(1) t^{-N} g(t) + \int_0^t t \beta_n^\nu(t/r) r^{-N-2} g(r) dr$$

$$+ N \int_0^t A_n^\nu(t/r) r^{-N-1} g(r) dr$$

for $t > 0$ where $\beta_n^\nu(t) = A_n^\nu(t)'$. Hence, using (3.31) and (3.33) in conjunction with (2.7) in Appendix A, we obtain from this last formula that

$$\lim_{t \to 0} \int_0^t A_n^\nu(t/r) r^{-N} dg(r) = 0.$$

We conclude from (3.32) that the proof of the lemma will be complete once we show that

$$(3.34) \qquad \lim_{t \to 0} \lim_{R \to \infty} \int_t^R [A_n^\nu(t/r) - 1] r^{-N} dg(r) = 0.$$

To establish that this last double limit is valid, we use the estimate in Theorem 2.1 of Appendix A and see that

$$(3.35) \qquad |A_n^\nu(t/r) - 1| \leq C^*(N,n)(t/r)^{\frac{1}{2}} \quad \text{for} \ \ 0 < t \leq r < \infty,$$

where $C^*(N,n)$ is a positive constant. We shall show that the double limit in (3.34) holds by proving that, given $\eta > 0$,

$$(3.36) \qquad \limsup_{t \to 0} \lim_{R \to \infty} \left| \int_t^R [A_n^\nu(t/r) - 1] r^{-N} dg(r) \right| \leq C^{**}(N,n)\eta,$$

where $C^{**}(N,n)$ is a constant that depends only on N and n.

To accomplish this, using (3.31), we choose $\delta > 0$ so that

$$(3.37) \qquad |g(r)| \leq \eta r^N \quad \text{for} \ \ 0 < r \leq \delta.$$

Then from (3.35), we have that

$$\lim_{R \to \infty} \left| \int_\delta^R [A_n^\nu(t/r) - 1] r^{-N} dg(r) \right|$$

$$\leq C^*(N,n) t^{\frac{1}{2}} \lim_{R \to \infty} \int_{B(0,R) \setminus B(0,\delta)} |f(-x) K(x)| \, |x|^{-\frac{1}{2}} \, dx.$$

We see that the limit of the integral on the right-hand side of this last inequality is finite. Consequently, the left-hand side of this last inequality is

$O(t^{\frac{1}{2}})$ as $t \to 0$. We conclude that (3.36) will follow once we show

$$(3.38) \qquad \limsup_{t \to 0} \left| \int_t^\delta [A_n^\nu(t/r) - 1] r^{-N} dg(r) \right| \leq C^{**}(N, n) \eta.$$

Observing from (3.37) that

$$\limsup_{t \to 0} \left| [A_n^\nu(t/\delta) - 1] \delta^{-N} g(\delta) \right| \leq |A_n^\nu(0) - 1| \, \eta,$$

we see after integrating by parts that the estimate in (3.38) will follow if we show

$$\limsup_{t \to 0} \left| \int_t^\delta g(r) r^{-N} \{ N[A_n^\nu(t/r) - 1] r^{-1} + t \beta_n^\nu (t/r) r^{-2} \} dr \right| \leq C_*(N, n) \eta$$

where $\beta_n^\nu (t) = A_n^\nu(t)'$.

This last inequality, however, follows easily from (3.35), (3.37), and the uniform boundness in (2.10) of Appendix A. ∎

Proof of the Sufficiency Condition of Theorem 3.2 If we can show

$$(3.39) \qquad \lim_{t \to 0} \sum_{0 < |m|} \widehat{K^*}(m) \widehat{f}(m) e^{im \cdot x_0} e^{-|m|t} = \alpha,$$

the proof will be complete.

$$\sum_{0 < |m|} \widehat{K^*}(m) \widehat{f}(m) e^{im \cdot x_0} e^{-|m|t}$$

$$= 2^{-1} \sum_{0 < |m|} [\widehat{f}(m) \widehat{K^*}(m) e^{im \cdot x_0} + \widehat{f}(-m) \widehat{K^*}(-m) e^{-im \cdot x_0}] e^{-|m|t},$$

and also

$$\sum_{0 < |m| \leq R} \widehat{K^*}(m) \widehat{f}(m) e^{im \cdot x_0}$$

$$= 2^{-1} \sum_{0 < |m| \leq R} [\widehat{f}(m) \widehat{K^*}(m) e^{im \cdot x_0} + \widehat{f}(-m) \widehat{K^*}(-m) e^{-im \cdot x_0}].$$

Since the left-hand side of the inequality in (3.15) is real-valued, it follows from Lemma 3.3, (3.39), and these last two equalities that

$$\lim_{R \to \infty} \sum_{0 < |m| \leq R} \widehat{K^*}(m) \widehat{f}(m) e^{im \cdot x_0} = \alpha,$$

and the sufficiency condition of the theorem would be established.

To show that the limit in (3.39) is indeed valid, we proceed as follows. We have by assumption that the double limit in (3.17) holds. Hence, by Lemma 3.4, we have that the limit in (3.29) holds. Consequently, from Lemma 3.5, we have that that the double limit in (3.17), which is α, is the same as the limit of $\widetilde{A}_t(f, x_0)$, i.e.,

$$\lim_{t \to 0} \widetilde{A}_t(f, x_0) = \alpha.$$

Since $\widetilde{A}_t(f, x)$ represents the expression in (2.2), this last limit is the same as the one in (3.39). ∎

Exercises.

   1. Prove that in dimension $N \geq 2$,

$$\lim_{t \to 0} t^2 \sum_{1 \leq |m|} \frac{e^{-|m|t}}{|m|^{N-2}} \quad \text{exists and is finite.}$$

2. Prove that if $S_3(R) = \sum_{0 < |m| \leq R} |m| \, a_m$ where $R > 1$, then

$$\sum_{R < |m|} e^{-|m|t} a_m \quad = \quad -S_3(R) R^{-1} e^{-Rt} + t \int_R^\infty S_3(r) e^{-rt} r^{-1} dr$$

$$+ \int_R^\infty S_3(r) e^{-rt} r^{-2} dr.$$

3. With $A_n^\nu(t)$ defined in (2.5) and $g(r) = o\left(r^N\right)$ as $r \to 0$, prove that

$$\lim_{t \to 0} \int_0^t A_n^\nu(t/r) r^{-N} dg(r) = 0,$$

where $g(r)$ is defined in (3.30).

4. The C^α-Condition

   Given $f \in C(T_N)$, i.e., $f \in C(R^N)$ and is periodic of period 2π in each variable, we say $f \in C^\alpha(T_N)$, $0 < \alpha < 1$, provided there is a constant C_* such that

(4.1) $$|f(x + y) - f(x)| \leq C_* |y|^\alpha \quad \forall x, y \in R^N.$$

   If f meets (4.1), we will also say that f is in $Lip \, \alpha$ on T_N, $0 < \alpha < 1$. It turns out that $f \in C^\alpha(T_N)$ implies that $\widetilde{f} \in C^\alpha(T_N)$ where $\widetilde{f}(x)$ is the periodic Calderon-Zygmund transform of f of spherical harmonic type defined in (1.12) above. This basic fact is also highly useful in demonstrating that certain partial differential equations have classical solutions, as will be shown in §5.

   We establish that $\widetilde{f} \in C^\alpha(T_N)$ in the following theorem, which uses the presentation given in [CZ2, pp. 262-265].

Theorem 4.1. *Let $f \in C^\alpha(T_N)$, $0 < \alpha < 1$, and suppose $K(x)$ is a Calderon-Zygmund kernel of spherical harmonic type that meets the conditions in (1.3), (1.4), and (1.5). Set*

$$(4.2) \qquad \widetilde{f}(x) = \lim_{\varepsilon \to 0}(2\pi)^{-N} \int_{T_N \setminus B(0,\varepsilon)} f(x-y)K^*(y)dy$$

where $K^*(x)$ is defined in (1.8). Then the limit in (4.2) exists for every $x \in R^N$, and $\widetilde{f} \in C^\alpha(T_N)$.

Proof of Theorem 4.1. It will be apparent from the proof we give below that from the start we can assume

$$\lim_{\varepsilon \to 0} \int_{T_N \setminus B(0,\varepsilon)} K^*(y)dy = 0.$$

Therefore, for $x \in R^N$,

$$(4.3) \qquad (2\pi)^N \widetilde{f}(x) = \lim_{\varepsilon \to 0} \int_{T_N \setminus B(0,\varepsilon)} [f(x-y) - f(x)]K^*(y)dy.$$

Since by (1.8) and (1.15),

$$(4.4) \qquad K^*(y) = K(y) + K_1^*(y)$$

where $K_1^*(y)$ is uniformly bounded for $y \in T_N$ and $|y|^\alpha K(y) \in L^1(T_N)$, we have that the limit in (4.3) exists for every $x \in T_N$. Hence, the limit in (4.2) exists for every $x \in T_N$.

It remains to show that $\widetilde{f}(x) \in C^\alpha(T_N)$. As we have just observed, $\forall x \in T_N$, as a function of y, $[f(x-y) - f(x)]K^*(y) \in L^1(T_N)$. Also we see from (4.4) that there is a constant C_{1*}, independent of x, such that

$$|[f(x-y) - f(x)]K^*(y)| \leq C_{1*}(|y|^{\alpha-N} + |y|^\alpha) \qquad \forall x \in T_N.$$

As a consequence, for $h \in B(0, 1/8)$,

$$(4.5) \qquad \int_{B(0,3|h|)} |[f(x-y) - f(x)]K^*(y)| \, dy \leq C_{2*} |h|^\alpha,$$

$\forall x \in T_N$.

From (4.3) and (4.5), we see that

$$(4.6) \qquad (2\pi)^N \widetilde{f}(x) = \int_{T_N \setminus B(0,3|h|)} [f(x-y) - f(x)]K^*(y)dy + A_1(x,h)$$

where

$$(4.7) \qquad |A_1(x,h)| \leq C_{2*} |h|^\alpha \qquad \forall x \in T_N.$$

Next, we observe that

$$(2\pi)^N \widetilde{f}(x+h) = \int_{T_N} [f(x+h-y) - f(x+h)]K^*(y)dy$$

$$= \int_{T_N - h} [f(x-y) - f(x+h)]K^*(y+h)dy.$$

It is clear from the boundedness of both $K(x)$ and $K_1^*(x)$ in a neighborhood of the boundary of T_N, that that we can adjust this last integral and obtain that for $h \in B(0, 1/8)$,

$$(4.8) \quad (2\pi)^N \widetilde{f}(x+h) = \int_{T_N} [f(x-y) - f(x+h)]K^*(y+h)dy + A_2(x,h)$$

where

$$(4.9) \qquad |A_2(x,h)| \le C_{3*} |h|^\alpha \qquad \forall x \in T_N.$$

Using the C^α-condition on f in the integrand in (4.8) and observing that

$$\int_{B(0,3|h|)} |y+h|^{\alpha-N} \, dy \le \int_{B(0,4|h|)} |y|^{\alpha-N} \, dy,$$

we see from (4.4) and (4.8) that

$$(4.10) \quad \begin{aligned} (2\pi)^N \widetilde{f}(x+h) &= \int_{T_N \setminus B(0,3|h|)} [f(x-y) - f(x+h)]K^*(y+h)dy \\ &\quad + A_2(x,h) + A_3(x,h) \end{aligned}$$

where

$$(4.11) \qquad |A_3(x,h)| \le C_{4*} |h|^\alpha \qquad \forall x \in T_N.$$

Next, we claim that there exists a constant $C_{7*} > 0$ such that

$$(4.12) \qquad \left| \int_{T_N \setminus B(0,3|h|)} K^*(y+h)dy \right| \le C_{7*} \qquad \forall h \in B(0, 1/8).$$

To see that this last inequality is true, we observe from (4.4) that it is sufficient to establish (4.12) with $K^*(y+h)$ replaced by $K(y+h)$ and T_N replaced by $B(0,1)$.

So set $I(h)$ equal to the expression on the left-hand side of the inequality in (4.12) with $K^*(y+h)$ replaced by $K(y+h)$ and T_N replaced by $B(0,1)$, and observe that

$$(4.13) \qquad I(h) = \left| \int_{B(0,1) \setminus B(0,3|h|)} [K(y+h) - K(y)]dy \right|.$$

We need to show that $I(h)$ is uniformly bounded for $|h| \le 1/8$.

Now $K(y+h) - K(y)$ is given by

$$\frac{Q_n(y+h)}{|y+h|^{n+N}} - \frac{Q_n(y)}{|y|^{n+N}} = \frac{Q_n(y+h) - Q_n(y)}{|y+h|^{n+N}}$$
$$+ Q_n(y)\left[\frac{1}{|y+h|^{n+N}} - \frac{1}{|y|^{n+N}}\right].$$

It is clear from the fact that $Q_n(y)$ is a homogeneous polynomial of degree n that this last equality implies there exists a constant C_{5*} such that

$$(4.14) \quad |K(y+h) - K(y)| \le C_{5*} |h| \, |y|^{-(N+1)} \qquad \forall y \in B(0,1) - B(0, 3|h|)$$

and for $|h| \le 1/8$.

Since

$$|h| \int_{|3h|}^{\infty} t^{-2} dt = 1/3,$$

it follows from (4.14) in conjunction with the definition of $I(h)$ given in (4.13) that

$$I(h) \leq C_{6*} \quad \forall h \in B(0, 1/8).$$

This fact establishes the inequality in (4.12), and we obtain as a consequence that

$$|f(x) - f(x+h)| \left| \int_{T_N - B(0,3|h|)} K^*(y+h) dy \right|$$

$$\leq C_* C_{7*} |h|^\alpha$$

$\forall h \in B(0, 1/8)$ and $\forall x \in T_N$.

From this last inequality in conjunction with (4.6)-(4.11), we obtain that $(2\pi)^N [\widetilde{f}(x+h) - \widetilde{f}(x)] =$

$$(4.15) \quad \int_{T_N \setminus B(0,3|h|)} [f(x-y) - f(x)][K^*(y+h) - K^*(y)] dy + A_4(x,h)$$

where $|A_4(x,h)| \leq C_{8*} |h|^\alpha \quad \forall h \in B(0, 1/8) \quad$ and $\quad \forall x \in T_N$.

Next, we claim that there exists a constant C_{9*} such that

$$(4.16) \quad |K^*(y+h) - K^*(y)| \leq C_{9*} |h| |y|^{-(N+1)}$$

$\forall h \in B(0, 1/8)$ and $\forall y \in T_N \setminus B(0, 3|h|)$.

To establish the inequality in (4.16), we recall that $K^*(y) = K(y) + K_1^*(y)$ where $K_1^*(y)$ is given in a previous expression above (1.15). Hence,

$$|K_1^*(y+h) - K_1^*(y)| \leq \lim_{R \to \infty} \sum_{1 \leq |m| \leq R} |K(y+h+2\pi m) - K(y+2\pi m)|.$$

It is easy to see from a decomposition similar to that given above (4.14) that the absolute value inside the summation sign in this last inequality is majorized by a constant multiple of

$$|h| |2\pi m|^{-(N+1)} \quad \text{for} \quad |m| \geq 1,$$

when $h \in B(0, 1/8)$ and $y \in T_N - B(0, 3|h|)$. Consequently, there exists a constant C_{10*} such that

$$|K_1^*(y+h) - K_1^*(y)| \leq C_{10*} |h|.$$

Therefore, to establish the inequality in (4.16), it is sufficient to show

$$|K(y+h) - K(y)| \leq C_{11*} |h| |y|^{-(N+1)}$$

$\forall h \in B(0, 1/8)$ and $\forall y \in T_N \setminus B(0, 3|h|)$. But this is essntially the inequality already obtained in (4.14). Hence, claim (4.16) is valid.

We next return to the integral in (4.15) and observe from the claim in (4.16) that

$$\int_{T_N \setminus B(0,3|h|)} [f(x-y) - f(x)][K^*(y+h) - K^*(y)] dy$$

$$(4.17) \qquad \leq C_* C_{9*} \int_{T_N \setminus B(0,3|h|)} |h|^\alpha |h| |y|^{-(N+1)} dy \leq C_{12*} |h|^\alpha$$

$\forall h \in B(0, 1/8)$ and $\forall x \in T_N$, where once again we used the fact that

$$|h| \int_{3|h|}^\infty t^{-2} dt = 1/3.$$

From the equality in (4.15) in conjunction with the inequality in (4.17), we obtain

$$\left| (2\pi)^N [\widetilde{f}(x+h) - \widetilde{f}(x)] \right| \leq (C_{8*} + C_{12*}) |h|^\alpha,$$

$\forall h \in B(0, 1/8)$ and $\forall x \in T_N$. Hence, indeed $\widetilde{f} \in C^\alpha(T_N)$. ∎

Exercises.

1. With $h = (h_1, h_2)$ where $|h| \leq \frac{1}{8}$ and h_1, h_2 are nonnegative, prove that there exists a constant $C_* > 0$ such that

$$\left| \int_{T_2 \setminus B(-h,h)} K(y+h) \, dy - \int_{T_2 - h \setminus B(-h,h)} K(y+h) \, dy \right| \leq C_* |h|.$$

5. An Application of the C^α-Condition

In this section, we show that Theorem 4.1 is very useful in obtaining a classical solution to a periodic boundary value problem in partial differential equations. Our example is fairly elementary. For a more sophisticated example showing how Theorem 4.1 can be used, we refer the reader to §2 of Chapter 6 below, which deals with a classical solution to a boundary value problem for the stationary Navier-Stokes equations.

Let $f \in C^\alpha(T_N)$, $0 < \alpha < 1$. We recall this means that f is periodic of period 2π in each variable and satisfies

$$|f(x+y) - f(x)| \leq C_* |y|^\alpha \qquad \forall x, y \in R^N$$

where C_* is a positive constant.

Given $g \in C(T_N)$, we say u is a classical solution of the periodic boundary value problem

$$(5.1) \qquad\qquad -\Delta u(x) = g(x) \quad \forall x \in T_N$$

provided the following holds:
   (i) $u \in C^2(\mathbf{R}^N)$;
   (ii) $u(x)$ is periodic of period 2π in each variable;
   (iii) $-\Delta u(x) = g(x) \quad \forall x \in T_N$.

Here, Δu is the familiar Laplace operator applied to u, namely,

$$\Delta u = \sum_{j=1}^{N} \frac{\partial^2 u}{\partial x^2}.$$

By using specific Calderon-Zygmund kernels $K(x)$ in conjunction with Theorem 4.1, we establish the following result concerning the boundary value problem (5.1).

Theorem 5.1. *Let $f \in C^\alpha(T_N)$, $N \geq 2$, $0 < \alpha < 1$. Suppose that $\widehat{f}(0) = 0$. Then there exists $u \in C^{2+\alpha}(T_N)$, which is a classical solution of the periodic boundary value problem*

$$(5.2) \qquad\qquad -\Delta u(x) = f(x) \quad \forall x \in T_N.$$

$u \in C^{2+\alpha}(T_N)$ means that $u(x)$ meets conditions (i) and (ii) just below (5.1) and, in addition,

$$\partial^2 u(x)/\partial x_j \partial x_k \in C^\alpha(T_N) \quad \text{for} \ \ j,k = 1, ..., N.$$

To show that the condition $\widehat{f}(0) = 0$ is a necessary condition for the solution of the periodic boundary value problem (5.2), we observe that if also $v \in C^2(T_N)$, then it follows from Green's second identity that

$$\int_{\partial T_N} [u\frac{\partial v}{\partial n} - v\frac{\partial u}{\partial n}]ds = \int_{T_N} [u\Delta v - v\Delta u]dx.$$

Taking $v = 1$ in this last equality gives the value zero to the left-hand side. Hence, $\int_{T_N} \Delta u dx = 0$, which we use with (5.2), 2nd in turn implies $\widehat{f}(0) = 0$.

To prove Theorem 5.1, we will need the following lemma:

Lemma 5.2. *Suppose g, g_j, $g_{jk} \in C(T_N)$ for $j,k=1, ..., N$. Suppose also that*

$$\begin{aligned} \widehat{g_j}(m) &= im_j\widehat{g}(m) \ \forall m \in \Lambda_N , \quad j = 1, ..., N, \quad and \\ \widehat{g_{jk}}(m) &= -m_j m_k \widehat{g}(m) \ \forall m \in \Lambda_N , \quad j,k = 1, ..., N. \end{aligned}$$

Then $g \in C^2(T_N)$ and $\partial^2 g(x)/\partial x_j \partial x_k = g_{jk}(x)$ for $x \in T_N$.

Proof of Lemma 5.2. To establish the lemma, it is clearly sufficient to show

$$(5.3) \qquad\qquad \partial g(x)/\partial x_1 = g_1(x) \text{ for } x \in T_N,$$

for everything else will follow in a similar manner.

We let $\sigma_n^\Diamond(g, x)$ designate the *n-th* iterated Fejer sum of g as defined in (2.6) of Chapter 1. Also, let x^* be a fixed but arbitrary point in T_N. We are given that $\widehat{g_1}(m) = im_1\widehat{g}(m)$ $\forall m \in \Lambda_N$. Since $\sigma_n^\Diamond(g, x)$ is a trigonometric polynomial, it follows that

$$\partial \sigma_n^\Diamond(g, x)/\partial x_1 = \sigma_n^\Diamond(g_1, x) \quad \text{for } x \in \mathbf{R}^N,$$

and therefore that

$$\sigma_n^\Diamond(g, x_1^* + t, x_2^*, ..., x_N^*) - \sigma_n^\Diamond(g, x_1^*, x_2^*, ..., x_N^*)$$

$$= \int_{x_1^*}^{x_1^*+t} \sigma_n^\Diamond(g_1, s, x_2^*, ..., x_N^*) ds$$

for $t \in \mathbf{R}$.

By Theorem 2.1 in Chapter 1, $\lim_{n\to\infty} \sigma_n^\Diamond(g_1, x) = g_1(x)$ uniformly for $x \in \mathbf{R}^N$. So using this uniformity, we obtain from this last equality that

$$(5.4) \quad g(x_1^* + t, x_2^*, ..., x_N^*) - g(x_1^*, x_2^*, ..., x_N^*) = \int_{x_1^*}^{x_1^*+t} g_1(s, x_2^*, ..., x_N^*) ds$$

for $t \in \mathbf{R}$.

Next, we divide both sides of the equality in (5.4) by $t \neq 0$, pass to the limit as $t \to 0$, and conclude that

$$\partial g(x^*)/\partial x_1 = g_1(x^*).$$

Since x^* is an arbitrary point in T_N, (5.3) is established, and the proof of the lemma is complete. ∎

Proof of Theorem 5.1. Since $(N-1)x_1^2 - x_2^2 - \cdots - x_N^2$ is a homogeneous polynomial of degree 2 and also a harmonic function, it follows that

$$K(x) = \frac{(N-1)x_1^2 - x_2^2 - \cdots - x_N^2}{|x|^{N+2}}$$

is a Calderon-Zygmund kernel of spherical harmonic type, and therefore from Corollary 3.2 in Appendix A that

$$\widehat{K^*}(m) = \kappa \frac{(N-1)m_1^2 - m_2^2 - \cdots - m_N^2}{|m|^2} \quad \forall m \neq 0,$$

where κ is a positive constant. Hence, by Theorem 4.1,

$$\sum_{m \neq 0} \widehat{f}(m) \frac{(N-1)m_1^2 - m_2^2 - \cdots - m_N^2}{|m|^2} e^{im \cdot x}$$

is the Fourier series of a function in $C^\alpha(T_N)$. Since by assumption,

$$\sum_{m \neq 0} \widehat{f}(m) \frac{m_1^2 + m_2^2 + \cdots + m_N^2}{|m|^2} e^{im \cdot x}$$

is also the Fourier series of a function in $C^\alpha(T_N)$, we obtain by addition that the same holds for

$$\sum_{m\neq 0} \widehat{f}(m)\frac{m_1^2}{|m|^2}e^{im\cdot x}.$$

Similarly, we see that

$$\sum_{m\neq 0} \widehat{f}(m)\frac{m_j^2}{|m|^2}e^{im\cdot x}$$

is the Fourier series of a function in $C^\alpha(T_N)$ for $j = 1, ..., N$.

Likewise, since $x_j x_k$ is a spherical harmonic function for $j \neq k$, we proceed in a similar manner and obtain that

(5.5) $u_{jk} \in C^\alpha(T_N),$

where

(5.6) $$S[u_{jk}] = \sum_{m\neq 0} \widehat{f}(m)\frac{m_j m_k}{|m|^2}e^{im\cdot x}$$

for $j, k = 1, ..., N$.

Next, we invoke (iv) and (vii) of Lemma 1.4 in Chapter 3 and see that $H_0, H_1, ..., H_N \in L^1(T_N)$ where

(5.7)
$$(i)\ S[H_0] = \sum_{m\neq 0} \frac{1}{|m|^2}e^{im\cdot x}\ \text{and}$$

$$(ii)\ S[H_j] = \sum_{m\neq 0} \frac{im_j}{|m|^2}e^{im\cdot x}\ \text{ for }\ j = 1, ..., N.$$

We set

(5.8) $$u_j(x) = (2\pi)^{-N}\int_{T_N} f(x-y)H_j(y)dy\ \text{ for }\ j = 0, 1, ..., N,$$

and obtain, since $f \in C^\alpha(T_N)$, that

(5.9) $u_j \in C^\alpha(T_N)\quad\text{for}\quad j = 0, 1, ..., N.$

Also, from (i) in (5.7) in conjunction with (5.8), we obtain that

(5.10) $$S[u_0] = \sum_{m\neq 0} \frac{\widehat{f}(m)}{|m|^2}e^{im\cdot x},$$

and from (ii) in (5.7) that

(5.11) $$S[u_j] = \sum_{m\neq 0} \frac{im_j\widehat{f}(m)}{|m|^2}e^{im\cdot x}\ \text{ for }\ j = 1, ..., N.$$

If we temporarily think of u_0 as g in Lemma 5.2, we see from (5.6), (5.10), and (5.11) that all the conditions in the hypothesis of Lemma 5.2 are met. Consequently, $u_0 \in C^2(T_N)$, and

$$\partial^2 u_0(x)/\partial x_j \partial x_k = u_{jk}(x)\ for\ x \in T_N.$$

But then from (5.5), we have that $u_0 \in C^{2+\alpha}(T_N)$ and from (5.6) that

$$-\Delta u_0(x) = f(x) \quad \forall x \in T_N.$$

This fact establishes the theorem with $u(x) = u_0(x)$. ∎

Exercises.

1. With $N = 2$ and $H_1(x)$ the function delinated in (5.7)(ii), given $f \in C^\alpha(T_2)$, $0 < \alpha < 1$, proves that $v_1(x) \in C^{1+\alpha}(T_2)$, where

$$v_1(x) = (2\pi)^{-2} \int_{T_2} f(x - y) H_1(y) \, dy.$$

2. With $N = 2$, given $f \in C^\alpha(T_2)$, $0 < \alpha < 1$, and $\widehat{f}(0) = 0$, find a vector-valued function $\mathbf{v}(x) = (v_1(x), v_2(x))$ with $v_1, v_2 \in C^{1+\alpha}(T_2)$ such that

$$\nabla \cdot \mathbf{v}(x) = f(x) \quad \forall x \in T_2.$$

6. An Application of the L^p-Condition

In this section, we show that Theorem 1.2 is very useful in obtaining distribution solutions to a periodic boundary value problem in partial differential equations. The result obtained here will prove valuable in Chapter 6 when we deal with solutions to the stationary Navier-Stokes equations.

We recall that $\phi \in C^\infty(T_N)$ means that $\phi \in C^\infty(\mathbf{R}^N)$ and periodic of period 2π in each variable. With $1 \le p < \infty$, by $\|\phi\|_{W^{1,p}(T_N)}$, we mean the following:

$$\|\phi\|_{W^{1,p}(T_N)}^p = \int_{T_N} \left[|\phi|^p + \sum_{j=1}^N \left| \frac{\partial \phi}{\partial x_j} \right|^p \right] dx.$$

The space $W^{1,p}(T_N)$ is the space we obtain by using the method of Cauchy sequences to close $C^\infty(T_N)$ wlth respect to the $\|\cdot\|_{W^{1,p}(T_N)}$-norm.

Similarly, by $\|\phi\|_{W^{2,p}(T_N)}$, we mean

$$\|\phi\|_{W^{2,p}(T_N)}^p = \int_{T_N} \left[|\phi|^p + \sum_{j=1}^N \left| \frac{\partial \phi}{\partial x_j} \right|^p + \sum_{j,k=1}^N \left| \frac{\partial^2 \phi}{\partial x_j \partial x_k} \right|^p \right] dx.$$

The space $W^{2,p}(T_N)$ is the space we obtain by using the method of Cauchy sequences to close $C^\infty(T_N)$ with respect to the $\|\cdot\|_{W^{2,p}(T_N)}$-norm.

Given $f \in L^p(T_N)$ with $\int_{T_N} f \, dx = 0$, we say u is a distribution solution of the periodic boundary value problem

(6.1) $$-\Delta u = f$$

provided that $u \in L^1(T_N)$ and

$$(6.2) \qquad -\int_{T_N} u \Delta \phi \, dx = \int_{T_N} f \phi \, dx \quad \forall \phi \in C^\infty(T_N).$$

With $1 < p < \infty$ and using specific Calderon-Zygmund kernels in conjunction with Theorem 1.2, we intend to prove the following theorem regarding distribution solutions of the boundary value problem (6.1).

Theorem 6.1. Let $f \in L^p(T_N)$, $1 < p < \infty$, $N \geq 2$. Suppose that $\widehat{f}(0) = 0$. Set

$$u_0(x) = (2\pi)^{-N} \int_{T_N} f(x - y) H_0(y) dy$$

where H_0 is defined in (5.7). Then $u_0 \in W^{2,p}(T_N)$ and is a distribution solution of the periodic boundary value problem (6.1).

To prove Theorem 6.1, we will need the following lemma.

Lemma 6.2. Suppose $g \in L^p(T_N)$, $1 < p < \infty$, and $h \in L^1(T_N)$. Set

$$v(x) = \int_{T_N} h(x - y) g(y) \, dy.$$

Then $v \in L^p(T_N)$, and

$$(6.3) \qquad \|v\|_{L^p(T_N)} \leq \|h\|_{L^1(T_N)} \|g\|_{L^p(T_N)}.$$

Proof of Lemma 6.2. Let $p' = p/(p - 1)$. Then writing $|h(x - y)| = |h(x - y)|^{1/p} |h(x - y)|^{1/p'}$, we see that

$$
\begin{aligned}
|v(x)| &\leq \int_{T_N} |h(x - y)| \, |g(y)| \, dy \\
&\leq [\int_{T_N} |h(x - y)| \, |g(y)|^p \, dy]^{1/p} [\int_{T_N} |h(x - y)| \, dy]^{1/p'}.
\end{aligned}
$$

Consequently,

$$\int_{T_N} |v(x)|^p \, dx \leq \|h\|_{L^1(T_N)}^{p/p'} \|h\|_{L^1(T_N)} \|g\|_{L^p(T_N)}^p.$$

The inequality in (6.3) follows immediately from this last inequality. ∎

Proof of Theorem 6.1. Using $H_0, H_1, ..., H_N \in L^1(T_N)$ as defined in (5.7), we set

$$(6.4) \qquad u_j(x) = (2\pi)^{-N} \int_{T_N} f(x - y) H_j(y) dy \quad \text{for} \quad j = 0, 1, ..., N,$$

and obtain from Lemma 6.2, since $f \in L^p(T_N)$ and $H_j \in L^1(T_N)$, that

(6.5) $$u_j \in L^p(T_N) \quad \text{for} \quad j = 0, 1, ..., N.$$

Also, from (i) in (5.7) in conjunction with (6.4), we obtain that

(6.6) $$S[u_0] = \sum_{m \neq 0} \frac{\widehat{f}(m)}{|m|^2} e^{im \cdot x},$$

and from (ii) in (5.7) that

(6.7) $$S[u_j] = \sum_{m \neq 0} \frac{im_j \widehat{f}(m)}{|m|^2} e^{im \cdot x} \quad \text{for} \quad j = 1, ..., N.$$

Next, using the material in the first paragraph of the proof of Theorem 5.1 in conjunction with Theorem 1.2, we see that

$$\sum_{m \neq 0} \widehat{f}(m) \frac{m_j^2}{|m|^2} e^{im \cdot x}$$

is the Fourier series of a function in $L^p(T_N)$ for $j = 1, ..., N$.

Likewise, since $x_j x_k$ is a spherical harmonic function for $j \neq k$, we proceed in a similar manner and use Theorem 1.2 to obtain that

$$\sum_{m \neq 0} \widehat{f}(m) \frac{m_j m_k}{|m|^2} e^{im \cdot x}$$

is the Fourier series of a function in $L^p(T_N)$ for $j \neq k$ and $j, k = 1, ..., N$.

As a consequence of all this, we see that

(6.8)
$$(i) \; \exists u_{jk} \in L^p(T_N)$$
with
$$(ii) \; S[u_{jk}] = -\sum_{m \neq 0} \widehat{f}(m) \frac{m_j m_k}{|m|^2} e^{im \cdot x}$$

for $j, k = 1, ..., N$.

Next, we let $\sigma_n^\Diamond(u_0, x)$ designate the n-th iterated Fejer sum of u_0 as defined in (2.6) of Chapter 1. Since $\sigma_n^\Diamond(u_0, x)$ is a trigonometric polynomial, it follows from (6.6)-(6.8) that

(6.9)
$$(i) \; \frac{\partial \sigma_n^\Diamond(u_0, x)}{\partial x_j} = \sigma_n^\Diamond(u_j, x),$$

$$(ii) \; \frac{\partial^2 \sigma_n^\Diamond(u_0, x)}{\partial x_j \partial x_k} = \sigma_n^\Diamond(u_{jk}, x)$$

for $j, k = 1, ..., N$.

From Theorem 2.2 in Chapter 1, we obtain that

(6.10) $$\lim_{n \to \infty} \left\| \sigma_n^\Diamond(u_j, \cdot) - u_j \right\|_{L^p(T_N)} = 0$$

for $j = 0, 1, ..., N$, and

$$\lim_{n \to \infty} \left\| \sigma_n^\Diamond(u_{jk}, \cdot) - u_{jk} \right\|_{L^p(T_N)} = 0$$

for $j, k = 1, ..., N$.

We conclude from (6.9) and these last two limits that

$$(6.11) \qquad u_0 \in W^{2,p}(T_N).$$

Also since $\sigma_n^\Diamond(u_0, x)$ is a trigonometric polynomial, we see from (6.8)(ii) and (6.9)(ii) that

$$-\Delta \sigma_n^\Diamond(u_0, x) = \sigma_n^\Diamond(f, x).$$

On the other hand, it is clear that

$$\int_{T_N} \sigma_n^\Diamond(u_0, x) \Delta \phi \, dx = \int_{T_N} \Delta \sigma_n^\Diamond(u_0, x) \phi \, dx \quad \forall \phi \in C^\infty(T_N).$$

So

$$-\int_{T_N} \sigma_n^\Diamond(u_0, x) \Delta \phi \, dx = \int_{T_N} \sigma_n^\Diamond(f, x) \phi \, dx \quad \forall \phi \in C^\infty(T_N).$$

Passing to the limit as $n \to \infty$ on both sides of this last equation, we obtain that

$$-\int_{T_N} u_0 \Delta \phi \, dx = \int_{T_N} f \phi \, dx \quad \forall \phi \in C^\infty(T_N).$$

This gives (6.2) with u_0 replacing u. Hence, u_0 is a distribution solution of (6.1). From (6.11), we see also that $u_0 \in W^{2,p}(T_N)$.

This establishes the theorem. ∎

Exercises.

1. With $N = 2$ and $\sigma_n^\Diamond(u_0, x)$ as in the proof of Theorem 6.1, use Green's second identity to prove

$$\int_{T_N} \sigma_n^\Diamond(u_0, x) \Delta \phi \, dx = \int_{T_N} \Delta \sigma_n^\Diamond(u_0, x) \phi \, dx \quad \forall \phi \in C^\infty(T_N).$$

2. With $1 < p < \infty$, define $W^{4,p}(T_2)$ in a manner analogous to the definition of $W^{2,p}(T_2)$. Given $f \in L^p(T_2)$ with $\int_{T_N} f \, dx = 0$, we say v is a distribution solution of the periodic boundary value problem

$$(6.12) \qquad \Delta^2 v = f$$

provided that $v \in L^1(T_2)$ and

$$\int_{T_2} v \Delta^2 \phi \, dx = \int_{T_2} f \phi \, dx \quad \forall \phi \in C^\infty(T_2).$$

Find a distrbution solution v to the periodic boundary value problem (6.12) with $v \in W^{4,p}(T_2)$.

7. Further Results and Comments

1. Originally from Poland, Antoni Zygmund was a professor of mathematics at the University of Chicago from 1947 to 1980, where Alberto Calderon was one of his Ph.D. students. Zygmund is known for founding one of the leading schools in mathematical analysis called "The School of Antoni Zygmund." A listing of 179 names belonging to this school as well as a biography of Zygmund can be found in the book "A Century of Mathematics, Part III," published by the American Mathematical Society [Du].

2. Calderon-Zygmund kernels of spherical harmonic type of the form $Q_k(x)/|x|^{N+k}$ when $Q_k(x)$ is a homogeneous linear polynomial are generally called Riesz kernels and the corresponding principal-valued transforms are generally called Riesz transforms. Both the periodic analogue of this case and when $Q_k(x)$ is also a spherical harmonic polynomial of degree 2 were considered in a 1938 paper by J. Marcinkiewicz [Mar]. Evidently, in these two cases, he proved Theorem 1.2 stated above (see [CZ2, p. 262]).

Josef Marcinkiewicz was one of the four Ph.D. students that Zygmund had while he was a professor at the University of Wilno (see [Du, p. 345 and p. 349]).

3. Calderon and Zygmund, while working with local properties of solutions to elliptic partial differential equations, introduced another important concept involving pointwise L^1-total differentials, [CZ3]. To define this notion, let $B(0,1) \subset \mathbf{R}^N$ be the unit N-ball. Let $x_0 \in B(0,1)$. $u \in t_1^1(x_0)$ provided the following holds: $u \in L^1(B(0,1))$ and there exists a linear polynomial $P(x)$ such that

$$\lim_{\rho \to 0} \rho^{-N-1} \int_{B(0,\rho)} |u(x_0+x) - P(x)|\, dx = 0.$$

If $P(x) = \alpha_0 + \sum_{j=1}^{N} \alpha_j x_j$, we set $\alpha_j = u_{x_j}(x_0)$ for $j = 1, ..., N$.

Once this definition is given, an obvious question arises in dimension $N = 2$, namely the following: Suppose $u, v \in L^1(B(0,1))$ and $u, v \in t_1^1(x)$ $\forall x \in B(0,1)$. Suppose, futhermore, the analogue of the Cauchy-Riemann equations hold, i.e.,

$$\begin{aligned} u_{x_1}(x) &= v_{x_2}(x) \quad \forall x \in B(0,1), \\ u_{x_2}(x) &= -v_{x_1}(x) \quad \forall x \in B(0,1). \end{aligned}$$

Is it true that there exists $U, V \in C^\infty(B(0,1))$ with $f(z) = U(x) + iV(x)$ holomorphic in $B(0,1)$ where $z = x_1 + ix_2$ such that

$$u(x) = U(x) \text{ and } v(x) = V(x) \quad \text{a.e. in } B(0,1)?$$

This fundamental question was an open problem for approximately ten years. It was finally solved with the aid of multiple Fourier series in a manuscript that appeared in the Annals of Mathematics [Sh16]. We refer the reader to this reference for the details of the solution.

CHAPTER 3

Uniqueness of Multiple Trigonometric Series

1. Uniqueness for Abel Summability

In this section, we shall deal with the best possible results involving uniqueness questions that arise from the Abel summability of multiple trigonometric series. A corollary to the main result presented in this section will be a solution to a problem which was open for approximately *one hundred years*: Is the analog of Cantor's famous uniqueness theorem valid for double trigonometric series, i.e., is it true that

$$(1.1) \qquad \lim_{R \to \infty} \sum_{|m| \leq R} a_m e^{im \cdot x} = 0 \quad \forall x \in T_2 \Longrightarrow a_m = 0 \quad \forall m?$$

In other words, the analog of Cantor's theorem on a one-dimensional trigonometric series for the circular convergence of double trigonometric series was the open question. Cantor's work on one-dimensional trigonometric series started with a number of publications beginning in 1870 (see [Da, pp. 33-34]) and eventually led to his ground-breaking development in set theory.

The answer in the affirmative to the question stated above in (1.1) follows from the work of Shapiro [Sh2] in 1957 and Cooke [Co] in 1971. In between these times, many famous mathematicians including the logician and analyst Paul J. Cohen [Coh] had worked on trying to solve this problem on double series. The same question replacing circular convergence with square convergence is still open (see [AW3, p.24]). Also, the analogous problem on S_2 for spherical harmonics is still open.

In this section, we concentrate on the ideas in the reference [Sh2] mentioned above and, in particular, establish the following theorem (see [AW3, p.10]) and its corollary.

Theorem 1.1. *Given the trigonometric series $\sum_{m \in \Lambda_N} a_m e^{im \cdot x}$ where the a_m are arbitrary complex numbers and $N \geq 1$, suppose that*
  (i) $\sum_{R-1 < |m| \leq R} |a_m| = o(R)$ *as $R \to \infty$,*
  (ii) $\lim_{t \to 0} \sum_{m \in \Lambda_N} a_m e^{im \cdot x - |m|t} = 0$ *$\forall x \in T_N$.*
Then $a_m = 0$ $\forall m \in \Lambda_N$.

Corollary 1.2. *Given the trigonometric series $\sum_{m \in \Lambda_N} a_m e^{im \cdot x}$ where the a_m are arbitrary complex numbers and $N \geq 2$, suppose that*
  (i) $\sum_{R-1 < |m| \leq R} |a_m| = o(R)$ *as $R \to \infty$,*

$(ii)\lim_{t\to 0}\sum_{m\in\Lambda_N} a_m e^{im\cdot x-|m|t}=0 \quad \forall x\in T_N\backslash\{0\}.$
Then $a_m=0 \ \forall m\in\Lambda_N.$

Before proceeding, we show that the corollary is not true in one-dimension. Working with $N=1$, we see from the equalities in (4.37) and (4.38) of Chapter 1 that for $t>0$,

$$(1.2)\quad P(s,t)=\sum_{n=-\infty}^{\infty} e^{ins-|n|t}=b_1 t \sum_{n=-\infty}^{\infty} [t^2+(2\pi n+s)^2]^{-1} \quad \forall s\in T_1$$

where b_1 is a positive constant. It is clear that $\lim_{t\to 0} P(s,t)=0 \ \forall s\in T_1\backslash\{0\}$, and that $\lim_{t\to 0} P(0,t)=\infty$. Since the coefficients of the trigonometric series in the second term in (1.1) are $a_n=1$ for all n, we have $\sum_{R-1<|n|\le R}|a_n|=2=o(R)$ as $R\to\infty$. Consequently, Corollary 1.2 is false in dimension $N=1$.

Next, we observe from (1.2) that for $t>0$,

$$\frac{\partial P(s,t)}{\partial s} = \sum_{n=-\infty}^{\infty} ine^{ins-|n|t} \text{ and also,}$$

$$\frac{\partial P(s,t)}{\partial s} = -b_1 t \sum_{n=-\infty}^{\infty} 2(2\pi n+s)[t^2+(2\pi n+s)^2]^{-2}$$

$\forall s\in T_1.$ Hence,

$$(1.3)\qquad \lim_{t\to 0}\sum_{n=-\infty}^{\infty} ine^{ins-|n|t}=0 \quad \text{for all } s\in T_1.$$

To show that little "o" cannot be replaced by big "O" in condition (i) of Theorem 1.1, we set

$$(1.4)\qquad a_m=\begin{cases} 0 \text{ for } m_2^2+...+m_N^2\ne 0 \\ im_1 \text{ for } m_2^2+...+m_N^2=0 \end{cases}.$$

Then

$$\sum_{m\in\Lambda_N} a_m e^{im\cdot x-|m|t}=\sum_{n=-\infty}^{\infty} ine^{ins-|n|t},$$

where $x_1=s$, and we see from (1.3) that condition (ii) in Theorem 1.1 holds for all $x\in T_N$. On the other hand, from the definition in (1.4), it is clear that

$$2(R-1)<\sum_{R-1<|m|\le R}|a_m|\le 2R.$$

So our assertion that little "o" cannot be replaced by big "O" in condition (i) of Theorem 1.1 is substantiated, and Theorem 1.1 from this point of view is a best possible result.

Theorem 1.1 in dimension $N=1$ is due to Verblunsky and the proof in this dimension can be found in [Zy1, p. 352]. The proof of Theorem 1.1 in dimension $N\ge 2$ requires new ideas. These new ideas are to be found in

[Sh2], and we shall use them here in our proof of Theorem 1.1. About these new ideas, Professor A. Beurling's comment was " Very skillful, indeed!" Professor A. Zygmund said about Theorem 1.1 in dimension $N \geq 2$, "Results of this nature come about only once in a generation."

To prove Theorem 1.1, we need the following lemma:

Lemma 1.3. *Let $G \in L^1(T_N)$, and let $S[G] = \sum_{m \in \Lambda_N} \widehat{G}(m)e^{im \cdot x}$. For $t > 0$, set $G(x,t) = \sum_{m \in \Lambda_N} \widehat{G}(m)e^{im \cdot x - |m|t}$. Suppose $\lim_{t \to 0} G(x_0, t) = G(x_0)$ exists and is finite. Set $\gamma^* = \lim \sup_{t \to 0} - \sum_{m \in \Lambda_N} |m|^2 \widehat{G}(m)e^{im \cdot x - |m|t}$ and define γ_* similarly using $\lim \inf_{t \to 0}$. Then*

(a)$\Delta_* G(x_0) \leq \gamma^*$ and

(b)$\gamma_* \leq \Delta^* G(x_0)$.

In the above lemma, $\Delta_* G(x_0)$ refers to the lower generalized Laplacian of G at x_0 as defined in (2.2) of Appendix C, namely,

$$(1.5) \qquad \Delta_* G(x_0) = 2(N+2) \liminf_{r \to 0} \frac{G_{[r]}(x_0) - G(x_0)}{r^2}$$

where $G_{[r]}(x_0)$ is the volume mean of G at x_0 defined in (2.1) of Appendix C. (For the one-dimensional analogue of the above result, see [Zy1, p. 353].)

Proof of Lemma 1.3. To prove the lemma, it is sufficient to prove *(a)*, for *(b)* will then follow by considering $-G(x)$. With no loss in generality, we can assume that $x_0 = 0$ and $G(0) = 0$.

If $\Delta_* G(0) = -\infty$ or if $\gamma^* = +\infty$, *(a)* is already established. So we can assume $\Delta_* G(0) > -\infty$ and $\gamma^* < +\infty$.

Suppose *(a)* does not hold. Then there exists an $\eta \in \mathbf{R}$ such that $\Delta_* G(0) > \eta > \gamma^*$. Since we can find a periodic function $\lambda(x) \in C^\infty(T_N)$ with the property that $\lambda(0) = 0$ and $\Delta\lambda(0) = \eta$, we can assume $\eta = 0$. We prove the lemma by showing

$$(1.6) \qquad \Delta_* G(0) > 0 > \gamma^*$$

leads to a contradiction.

With $G_t(0,t) = dG(0,t)/dt$ and $G_{tt}(0,t) = dG_t(0,t)/dt$, and observing from (1.6) that $\gamma^* < 0$, we obtain that

$$\limsup_{t \to 0} -G_{tt}(0,t) = \gamma^* < 0.$$

Consequently, $G_{tt}(0,t) > 0$ for t sufficiently small. Therefore, for t sufficiently small, $G_t(0,t)$ is a strictly increasing function of t, i.e., there exists a $t_0 > 0$ such that $G_t(0,t)$ is a strictly increasing function of t in the interval $0 < t < t_0$. Also, $G(0,t)/t = G_t(0,s)$ where $0 < s < t$ by the mean-value theorem, since $G(0,t) = 0$. Therefore, $\limsup_{t \to 0} d[G(0,t)/t]/dt < 0$ is incompatible with the fact that $G_t(0,t)$ is a strictly increasing function of t in the interval

$0 < t < t_0$. Hence, $\limsup_{t \to 0} d[G(0,t) /t]/dt < 0$ is incompatible with the fact that $\gamma^* < 0$. We consequently conclude from (1.6) that if we can show

$$(1.7) \qquad \Delta_* G(0) > 0 \Longrightarrow \liminf_{t \to 0} -d[G(0,t)/t]/dt > 0,$$

we will have arrived at a contradiction.

We now establish (1.7). Observing from (2.1) in Appendix C that $|B(0,R)| \, G_{[r]}(0) = \int_{B(0,R)} G(y) dy$ and from (4.5) and Theorem 4.1 of Chapter 1 that

$$G(0,t) = (2\pi)^{-N} b_N \int_{\mathbf{R}^N} G(x) t [t^2 + |x|^2]^{-(N+1)/2} dx,$$

we see that

$$\begin{aligned} \frac{G(0,t)}{t} &= \alpha \int_0^\infty [t^2 + r^2]^{-(N+1)/2} dr^N G_{[r]}(0) \\ &= \alpha' \int_0^\infty [t^2 + r^2]^{-(N+3)/2} r^{N+1} G_{[r]}(0) dr \end{aligned}$$

where α and α' are positive constants. Consequently, since $G_{[r]}(0) = O(1)$ as $R \to \infty$,

$$(1.8) \qquad -d[G(0,t)/t]/dt = \beta t \int_0^\infty [t^2 + r^2]^{-(N+5)/2} r^{N+1} G_{[r]}(0) dr,$$

for $t > 0$ where β is a positive constant.

By the assumption in (1.7), there exists $\eta_0 > 0$ and $\delta > 0$ such that

$$G_{[r]}(0) > \eta_0 r^2 \quad \text{for } 0 < r < \delta.$$

We thus obtain from (1.8) that

$$\begin{aligned} \liminf_{t \to 0} -d[G(0,t)/t]/dt &\geq \liminf_{t \to 0} \beta t \int_0^\delta [t^2 + r^2]^{-(N+5)/2} \eta_0 r^{N+3} dr \\ &\geq \beta \eta_0 \int_0^\infty [1 + r^2]^{-(N+5)/2} r^{N+3} dr \\ &> 0. \end{aligned}$$

This establishes (1.7), and completes the proof of the lemma. ∎

Proof of Theorem 1.1. From a consideration of the series

$$\sum_{m \in \Lambda_N} (a_m + \bar{a}_{-m}) e^{im \cdot x} \quad \text{and} \quad \sum_{m \in \Lambda_N} i(a_m - \bar{a}_{-m}) e^{im \cdot x},$$

we see from the start that it is sufficient to prove the theorem under the additional assumption

$$(1.9) \qquad \bar{a}_m = a_{-m} \quad \forall m \in \Lambda_N.$$

Next, for $t > 0$, we define the following functions.

$$f(x,t) = \sum_{m \in \Lambda_N} a_m e^{im \cdot x - |m|t}$$

$$f_1(x,t) = -\sum_{|m|>0} a_m |m|^{-1} e^{im \cdot x - |m|t}$$

(1.10)

$$F(x,t) = -\sum_{|m|>0} a_m |m|^{-2} e^{im \cdot x - |m|t}$$

$$f^*(x) = \limsup_{t \to 0} f(x,t) \quad \text{and} \quad f_*(x) = \liminf_{t \to 0} f(x,t),$$

and observe that

(1.11) $\lim_{t \to 0} f_1(x,t)$ and $\lim_{t \to 0} F(x,t)$ exist and are finite

for $x \in T_N$.

To see this, fix $x \in T_N$. Then by hypothesis (ii) of the theorem, there exists a constant K depending on x such that $|f(x,t)| \leq K$ for $t > 0$. Hence by the mean-value theorem for $0 < t_1 < t_2$, there exists an s such that

$$|f_1(x,t_2) - f_1(x,t_1)| = |f(x,s) - a_0|(t_2 - t_1)$$

where $0 < t_1 < s < t_2$. Therefore, $f_1(x,t)$ satisfies the Cauchy criterion for convergence and the first part of (1.11) is established. Repeating this argument for $F(x,t)$ establishes the second part.

Using (1.11), we define the periodic function $F(x)$ by

(1.12) $$F(x) = \lim_{t \to 0} F(x,t) \quad \forall x \in \mathbf{R}^N.$$

As is easily seen, condition (i) in the hypothesis of the theorem implies that $\sum_{|m|>0} |a_m|^2 |m|^{-4} < \infty$. Consequently, we obtain from (1.10) and (1.12) that $F(x) \in L^2(T_N)$ and

(1.13) $$S[F] = -\sum_{|m|>0} a_m |m|^{-2} e^{im \cdot x}.$$

Also, we observe from (1.5) and (1.6) in Appendix A that

(1.14) $$|B(x,r)|^{-1} \int_{B(x,r)} e^{im \cdot y} dy = \mu_N e^{im \cdot x} J_{N/2}(|m|\,r)(|m|\,r)^{-N/2}$$

where μ_N is a constant depending on N but not on m. So with $F_{[r]}(x)$ designating the volume mean of F in a ball of radius r centered at x, we obtain from (1.13) that

(1.15) $$F_{[r]}(x) = -\mu_N \sum_{|m|>0} a_m |m|^{-2} e^{im \cdot x} J_{N/2}(|m|\,r)(|m|\,r)^{-N/2}.$$

Using the Bessel estimate in (2.2) of Appendix A and condition (i) in the hypothesis of the theorem, it is clear that the multiple series in (1.15) is absolutely convergent.

Next, we obtain from Lemma 1.3 and condition (ii) in the hypothesis of the theorem that

$$(1.16) \qquad \Delta^* F(x) \geq -a_0 \quad \text{and} \quad \Delta_* F(x) \leq -a_0 \qquad \forall x \in T_N.$$

If $F(x)$ is continuous in all of $\mathbf{R}^N$, then it would follow from the inequalities in (1.16) and Theorem 2.2 in Appendix C that $F(x) + a_0 |x|^2 / 2N$ is harmonic in $\mathbf{R}^N$. As we shall see, from this latter fact it is easy to obtain the conclusion of the theorem. So we see what we have to do is establish the continuity of F in $\mathbf{R}^N$.

In order to establish this continuity, we set

$$(1.17) \qquad \alpha_1(t) = \sup_{0 < r < t} \sup_{x \in T_N} \left| F_{[r]}(x) - F(x, r) \right|,$$

and we will show that

$$(1.18) \qquad \lim_{t \to 0} \alpha_1(t) = 0.$$

It follows from the series representations of $F(x, r)$ in (1.10) and of $F_{[r]}(x)$ in (1.15) that

$$\sup_{x \in T_N} \left| F_{[r]}(x) - F(x, r) \right|$$

$$(1.19) \qquad \leq \sum_{|m| > 0} |a_m| \, |m|^{-2} \left| e^{-|m|r} - \mu_N J_{N/2}(|m| r)(|m| r)^{-N/2} \right|.$$

We split the sum on the right-hand side of the inequality in (1.19) into two parts, A_r and B_r. A_r will designate the sum over the lattice points $m, 1 \leq |m| \leq r^{-1}$, and B_r will designate the sum over the lattice points $m, |m| \geq r^{-1}$. To establish (1.18), it is sufficient to show that $\lim_{r \to 0} A_r = 0$ and $\lim_{r \to 0} B_r = 0$.

Observing from the equality in (1.14) that

$$\lim_{r \to 0} \mu_N J_{N/2}(r)(r)^{-N/2} = 1,$$

we see from (1.1) in Appendix A that there is a constant K such that

$$\left| e^{-r} - \mu_N J_{N/2}(r)(r)^{-N/2} \right| \leq Kr \quad \text{for} \ \ 0 < r \leq 1.$$

Hence, we obtain from condition (i) in the hypothesis of the theorem and from (1.19) that

$$A_r \leq Kr \sum_{1 \leq |m| \leq r^{-1}} |a_m| \, |m|^{-1} = ro(r^{-1}) \ \ \text{as} \ r \to 0.$$

Therefore, $\lim_{r \to 0} A_r = 0$.

Using the fact that there exists a constant K such that $\left| J_{N/2}(r) \right| \leq Kr^{-1/2}$, we obtain from (1.19) and from condition (i) in the hypothesis of the theorem that for r small

$$B_r \leq o(1) \int_{r^{-1}-1}^{\infty} e^{-rs} s^{-1} ds + o(1) r^{-(N+1)/2} \int_{r^{-1}-1}^{\infty} s^{-(N+3)/2} ds$$

as $r \to 0$.

We conclude that $\lim_{r \to 0} B_r = 0$, and consequently that the limit in (1.18) is established.

The next fact that we establish concerning $F(x)$ is the following: set

$$(1.20) \qquad \alpha_2(t) = \sup_{0 < r < t} \sup_{|x-y| \le r, x, y \in \mathbf{R}^N} \left| F_{[r]}(x) - F_{[r]}(y) \right|,$$

then

$$(1.21) \qquad \lim_{t \to 0} \alpha_2(t) = 0.$$

From (1.15),

$$\mu_N^{-1} \sup_{|x-y| \le r, x, y \in \mathbf{R}^N} \left| F_{[r]}(x) - F_{[r]}(y) \right|$$

$$\le \sup_{|x-y| \le r, x, y \in \mathbf{R}^N} \sum_{1 \le |m| \le r^{-1}} \{ |a_m| \left| J_{N/2}(|m| r) \right| (|m| r)^{-N/2} |m|^{-2}$$

$$\left| e^{im \cdot x} - e^{im \cdot y} \right| \}$$

$$+ 2 \sum_{r^{-1} \le |m|} |a_m| \left| J_{N/2}(|m| r) \right| (|m| r)^{-N/2} |m|^{-2}$$

$$= A_r' + B_r'.$$

To establish the limit in (1.21), we have to show that A_r' and B_r' tend to zero as r goes to zero. The $\lim_{r \to 0} B_r' = 0$ was already shown when we established (1.18). For A_r', we observe from (2.1) in Appendix A that

$$2^{-1} A_r' \le \sum_{1 \le |m| \le r^{-1}} |a_m| \left| J_{N/2}(|m| r) \right| (|m| r)^{-N/2} |m|^{-2} |m| r$$

$$\le O(1) r \sum_{1 \le |m| \le r^{-1}} |a_m| |m|^{-1}$$

$$\le O(1) r o(r^{-1})$$

as $r \to 0$. Consequently, $\lim_{r \to 0} A_r' = 0$, and the limit in (1.21) is established.

Next, using the fact that $f(x, t)$ is continuous for $t > 0$ and periodic in the x-variables, we see that there exists a sequence

$$t_1 > t_2 > \cdots > t_n \cdots \to 0$$

such that

$$(1.22) \qquad \sup_{x \in \mathbf{R}^N} \sup_{t_n \le t \le t_{n+1}} |f(x, t) - f(x, t_n)| \le 1.$$

Let $B(x_0, r_0)$, $r_0 > 0$ be an arbitrary but fixed open ball in $\mathbf{R}^N$. We propose to show that $F(x)$ is continuous in $B(x_0, r_0)$. So, let

$$(1.23) \qquad E = \{ x \in B(x_0, r_0) : F \text{ is not continuous at } x \}.$$

We will now show that E is the empty set.

By condition (ii) in the theorem,

$$\lim_{n\to\infty} |f(x, t_n)| = 0 \quad \forall x \in B(x_0, r_0).$$

If $\overline{B}(y, s) \subset B(x_0, r_0), s > 0$, then by the Baire category theorem [Zy1, p. 29 (12.3)], there exists $\overline{B}(y', s') \subset \overline{B}(y, s), s' > 0$, and a $K > 0$, such that

$$|f(x, t_n)| \leq K \quad \forall x \in \overline{B}(y', s') \quad \text{and} \quad \forall n.$$

From (1.22), it therefore follows that

$$|f(x, t)| \leq K + 1 \quad \forall x \in \overline{B}(y', s') \text{ and for } 0 < t < t_1.$$

As a consequence, employing the same technique used to establish the limits in (1.11), we see that

$$\lim_{t\to 0} F(x, t) = F(x) \quad \text{uniformly for } x \in \overline{B}(y', s').$$

We conclude that E defined in (1.23) is a nondense (nowhere dense) set in $B(x_0, r_0)$.

Next, we note that E has no isolated points. If z_0 were an isolated point of E, then as we have observed earlier, there would exist $s_0 > 0$ such that $F(x) + a_0 |x|^2 / 2N$ would be harmonic in the punctured N-ball $B(z_0, 2s_0) \setminus \{z_0\}$. Therefore, by the mean-value theorem for harmonic functions, for $x \in B(z_0, r) \setminus \{z_0\}$, where $0 < r < s_0$,

$$F(z_0) + a_0 |z_0|^2 / 2N - F(x) - a_0 |x|^2 / 2N$$
$$= F(z_0) + a_0 |z_0|^2 / 2N - F_{[|z_0 - x|]}(x)$$
$$- a_0 \int_{B(x, |z_0 - x|)} |y|^2 \, dy / |B(x, |z_0 - x|)| 2N.$$

As a consequence, from (1.17) and (1.20),

$$\left| F(z_0) + a_0 |z_0|^2 / 2N - F(x) - a_0 |x|^2 / 2N \right|$$
$$\leq \left| F(z_0) - F_{[|z_0 - x|]}(x) \right|$$
$$+ |a_0| \left| \int_{B(x, |z_0 - x|)} (|z_0|^2 - |y|^2) dy \right| / |B(x, |z_0 - x|)| 2N$$
$$\leq \left| F(z_0) - F_{[|z_0 - x|]}(x) \right| + o(1)$$
$$\leq \left| F(z_0) - F(z_0, |z_0 - x|) \right| + \left| F(z_0, |z_0 - x|) - F_{[|z_0 - x|]}(z_0) \right|$$
$$+ \left| F_{[|z_0 - x|]}(z_0) - F_{[|z_0 - x|]}(x) \right| + o(1)$$
$$\leq o(1) + \alpha_1(r) + \alpha_2(r).$$

Using (1.18) and (1.21), we conclude that

$$\lim_{x\to z_0} [F(x) + a_0 |x|^2 / 2N] = F(z_0) + a_0 |z_0|^2 / 2.$$

Hence, E contains no isolated points.

Next, let $\overline{E}$ designate the closure of E. Since $|f(x, t)| \to 0$ for $x \in \overline{E}$ and since $\overline{E}$ is a perfect set, we can obtain once again from the Baire category theorem and (1.22) that if $\overline{E}$ is a nonempty set, there is a $z_0 \in E$ and an $s_0 > 0$ with $\overline{B}(z_0, 2s_0) \subset B(x_0, r_0)$ and a constant K such that

$$(1.24) \qquad |f(z,t)| \leq K + 1 \quad \text{for } z \in \overline{E} \cap \overline{B}(z_0, 2s_0).$$

Using the same techniques that were used to establish the limits in (1.11), we see from (1.24) that

$$(1.25) \qquad \lim_{t \to 0} F(z,t) = F(z) \quad \text{uniformly for } z \in \overline{E} \cap \overline{B}(z_0, 2s_0).$$

Consequently, $F(z)$ is a continuous function when z is restricted to the closed set $\overline{E} \cap \overline{B}(z_0, 2s_0)$. Therefore, given an $\varepsilon > 0$, choose s_1 so that $0 < s_1 < s_0$ and such that

$$(1.26) \qquad |F(z) - F(z_0)| \leq \varepsilon \quad \text{for } z \in \overline{E} \cap \overline{B}(z_0, s_1).$$

Next, using (1.17), (1.20), and (1.25), choose s_2 so that the following five items hold:
(1.27)

(i) $\alpha_1(s) < \varepsilon$ for $0 < s < s_2$;

(ii) $\alpha_2(s) < \varepsilon$ for $0 < s < s_2$;

(iii) $|a_0| (2|z_0| + 3s_2)s_2 < \varepsilon$;

(iv) $|F(z,s) - F(z)| < \varepsilon$ for $0 < s < s_2$ and $z \in \overline{E} \cap \overline{B}(z_0, 2s_0)$;

(v) $2s_2 < s_1$.

We propose to show that

$$(1.28) \qquad |F(x) - F(z_0)| < 5\varepsilon \quad \text{for } x \in B(z_0, s_2).$$

If $x \in B(z_0, s_2)$ and $x \in \overline{E}$, then (1.28) holds by virtue of (1.26) and (1.27)(v). We can therefore suppose that $x \in B(z_0, s_2)$ and $x \notin \overline{E}$. Let z' be the closest point in $\overline{E}$ (or one of the closest if more than one exists) to x. Then $|z' - x| = s_3 < s_2$ and $F(y) + a_0 |y|^2 /2N$ is a harmonic function in $B(x, s_3)$. Therefore,

$$F(x) = F_{[s_3]}(x) + a_0 \int_{B(x,s_3)} |y|^2 \, dy/ |B(x, s_3)| \, 2N - a_0 |x|^2 /2N.$$

Consequently,

$$\begin{aligned}
|F(x) - F(z_0)| &\leq |F(x) - F(z')| + |F(z') - F(z_0)| \\
&\leq |F_{[s_3]}(x) - F(z')| + |F(z') - F(z_0)| \\
&\quad + |a_0| \left| \int_{B(x,s_3)} (|y|^2 - |x|^2) dy \right| / |B(x, s_3)| \, 2N.
\end{aligned}$$

Hence, from (1.26) and (1.27)(iii), we obtain

$$(1.29) \qquad |F(x) - F(z_0)| \le |F_{[s_3]}(x) - F(z')| + 2\varepsilon.$$

We can estimate the first term on the right-hand side of this last inequality using (1.17), (1.20), and (1.27)(i), (ii), and (iv) as follows:

$$
\begin{aligned}
\left|F_{[s_3]}(x) - F(z')\right| &\le \left|F_{[s_3]}(x) - F_{[s_3]}(z')\right| + \left|F_{[s_3]}(z') - F(z', s_3)\right| \\
&\quad + \left|F(z', s_3) - F(z')\right| \\
&\le \alpha_2(s_3) + \alpha_1(s_3) + \varepsilon \\
&\le 3\varepsilon.
\end{aligned}
$$

This last part coupled with the inequality in (1.29) establishes the inequality in (1.28).

We conclude that F is indeed continuous at z_0. Therefore, E must be the empty set and F is a continuous function in the open ball $B(x_0, r_0)$. But $B(x_0, r_0)$ was an arbitrary ball contained in $\mathbf{R}^N$. Hence, F is a continuous function in all of $\mathbf{R}^N$. As we observed earlier, this fact implies that

$$(1.30) \qquad F(x) + a_0 |x^2| / 2N \text{ is a harmonic function in } \mathbf{R}^N.$$

From (1.10) and (1.12), we see that $F(x)$ is a periodic function of period 2π in each variable. Since $F(x)$ is also a continuous function, we have that it is bounded in all of $\mathbf{R}^N$. Hence, we have from (1.30) that the harmonic function $F(x) + a_0 |x^2| / 2N = O(|x^2|)$ as $|x| \to \infty$. We conclude from well-known properties of harmonic functions (Theorem 1.10 in Appendix C) that the harmonic function that we are dealing with is a polynomial of at most degree 2, i.e.,

$$F(x) + a_0 |x^2| / 2N = b_0 + \sum_{j=1}^N b_j x_j + \sum_{j=1}^N \sum_{k=1}^N b_{jk} x_j x_k,$$

where b_0, b_j, b_{jk} are constants for $j, k = 1, ..., N$.

But the only way possible that the polynomial

$$b_0 + \sum_{j=1}^N b_j x_j + \sum_{j=1}^N \sum_{k=1}^N b_{jk} x_j x_k - a_0 |x^2| / 2N$$

can be bounded in all of $\mathbf{R}^N$ is when it is identically constant. Therefore,

$$F(x) = b_0 \quad \forall x \in \mathbf{R}^N.$$

From (1.13), we see first that $b_0 = 0$ and next that $F(x) = 0 \quad \forall x \in \mathbf{R}^N$. This implies that

$$a_m = 0 \text{ for } |m| > 0.$$

From condition (ii) in the hypothesis of the theorem, we finally obtain that $a_0 = 0$, and the proof of the theorem is complete. ∎

Next, we define several functions that will be useful in proving Corollary 1.2. Set

(1.31)
$$\Phi(x) = (2\pi)^N [|S_{N-1}| (N-2)]^{-1} |x|^{-(N-2)} \quad \forall x \in T_N \backslash \{0\} \quad \text{and} \quad N \geq 3$$

$$= (2\pi) \log |x|^{-1} \quad \forall x \in T_2 \backslash \{0\} \quad \text{and} \quad N = 2,$$

and extend Φ to $\mathbf{R}^N \backslash \cup_{m \in \Lambda_N} \{2\pi m\}$ by periodicity of period 2π in each variable. In the above equation, $|S_{N-1}|$ designates the volume of the unit $(N-1)$-sphere, i.e., $|S_{N-1}| = 2\pi^{N/2}/\Gamma(N/2)$.

Also, for $t > 0$, set

(1.32)
$$H_0(x,t) = \sum_{|m|>0} |m|^{-2} e^{im \cdot x - |m|t},$$

and for $j = 1, ..., N$, set

(1.33)
$$H_j(x,t) = \sum_{|m|>0} im_j |m|^{-2} e^{im \cdot x - |m|t}.$$

We establish a lemma concerning the functions just introduced.

Lemma 1.4. *The following facts hold for the functions defined in (1.31), (1.32), and (1.33) where $N \geq 2$:*

(i) $\lim_{t \to 0} H_j(x,t) = H_j(x)$ *exists and is finite* $\forall x \in \mathbf{R}^N \backslash \cup_{m \in \Lambda_N} \{2\pi m\}$ *and* $j = 0, 1, ..., N$;

(ii) $\lim_{|x| \to 0}[H_0(x) - \Phi(x)]$ *exists and is finite;*

(iii) $\lim_{|x| \to 0}[H_j(x)] + (2\pi)^N x_j / |S_{N-1}| |x|^N]$ *exists and is finite for* $j = 1, ..., N$;

(iv) $H_j(x) \in L^1(T_N)$ *for* $j = 0, 1, ..., N$;

(v) $H_0(x) - |x|^2 /2N$ *is harmonic in* $\mathbf{R}^N \backslash \cup_{m \in \Lambda_N} \{2\pi m\}$;

(vi) $H_j(x)$ *is harmonic in* $\mathbf{R}^N \backslash \cup_{m \in \Lambda_N} \{2\pi m\}$ *for* $j = 1, ..., N$;

(vii) $\lim_{t \to 0} \int_{T_N} |H_j(x,t) - H_j(x)| \, dx = 0$ *for* $j = 0, 1, ..., N$.

Proof of Lemma 1.4. We first show that there exists a continuous periodic function $\psi(x)$ such that

$$\widehat{\Phi}(m) - \widehat{\psi}(m) = |m|^{-2} \quad \text{for} \quad m \neq 0$$

(1.34)

$$\widehat{\Phi}(0) - \widehat{\psi}(0) = 0.$$

Also,

(1.35)
$$\sum_{m \in \Lambda_N} \left| \widehat{\psi}(m) \right| < \infty.$$

Observing from Green's second identity [Ke, p. 215] that for $N \geq 3$ and $m \neq 0$,

$$\int_{T_N - B(0,\varepsilon)} |x|^{-(N-2)} e^{-im\cdot x} dx$$

$$= \int_{\partial[T_N \backslash B(0,\varepsilon)]} e^{-im\cdot x} |m|^{-2} \partial \; |x|^{-(N-2)} / \partial n \; dS + o(1)$$

with a similar calculation holding in dimension 2 for $\log |x|^{-1}$, we obtain from (1.31) that for $m \neq 0$,

$$\widehat{\Phi}(m) = |m|^{-2}$$

(1.36)

$$+ b_N |m|^{-2} \sum_{j=1}^{N} \int_{T_{N-1}} e^{-i(m\cdot x - m_j x_j)} \frac{\cos m_j \pi}{(|x|^2 - x_j^2 + \pi^2)^{N/2}} dx_1 \cdots dx_j^* \cdots dx_N,$$

where b_N is a constant depending on N but not on m, and dx_j^* stands for the deletion of dx_j.

For $m = 0$, define $\gamma_0 = \widehat{\Phi}(0)$. For $m \neq 0$, define γ_m to be the value of the second expression on the right-hand side of the equality in (1.36). As a consequence,

$$\widehat{\Phi}(m) - \gamma_m = |m|^{-2} \quad \text{for} \;\; m \neq 0$$
$$\widehat{\Phi}(0) - \gamma_0 = 0.$$

To show that $\sum_{m \in \Lambda_N} |\gamma_m| < \infty$ prevails, we observe that for $m_1 \neq 0$ and $j \neq 1$,

$$\int_{-\pi}^{\pi} e^{im_1 x_1} (|x|^2 - x_j^2 + \pi^2)^{-N/2} dx_1$$

$$= N m_1^{-1} \int_{-\pi}^{\pi} \frac{x_1 \sin m_1 x_1}{(|x|^2 - x_j^2 + \pi^2)^{(N+2)/2}} dx_1$$

and obtain consequently that there is a constant b'_N such that for $m \neq 0$,

$$(|m| + 1)^2 |\gamma_m|$$

$$\leq b'_N \sum_{j=1}^{N} (|m_1| + 1)^{-1} \cdots (|m_j| + 1)^{-1*} \cdots (|m_N| + 1)^{-1}.$$

Also, we observe that

$$[|m_1| + 1) \cdots (|m_N| + 1)]^{(N+1)/N}$$

$$\leq (|m| + 1)^2 (|m_1| + 1) \cdots (|m_j| + 1)^* \cdots (|m_N| + 1).$$

These last two inequalities together imply that

$$\sum_{m \in \Lambda_N} |\gamma_m| \leq N b'_N \sum_{m \in \Lambda_N} [|m_1| + 1) \cdots (|m_N| + 1)]^{-(N+1)/N} < \infty.$$

So we set

$$\psi(x) = \sum_{m \in \Lambda_N} \gamma_m e^{im\cdot x}$$

and obtain from this last inequality that $\psi(x)$ is a continuous periodic function. As a consequence

$$(1.37) \qquad \widehat{\psi}(m) = \gamma_m \qquad \forall m \in \Lambda_N,$$

and we infer from the above that the conditions set forth in (1.34) and (1.35) prevail.

Continuing with the proof of the lemma, we next define

$$(1.38) \qquad H_0(x) = \Phi(x) - \psi(x) \ \text{ for } x \in \mathbf{R}^N \backslash \cup_{m \in \Lambda_N} \{2\pi m\}$$

and observe first that $H_0(x) \in L^1(T_N)$ and next from (1.37), (1.36), and the paragraph below it that

$$\begin{aligned} \widehat{H_0}(m) &= \widehat{\Phi}(m) - \widehat{\psi}(m) = |m|^{-2} & \forall m \neq 0 \\ &= 0 & \text{for } m = 0. \end{aligned}$$

With $H_0(x,t)$ defined as in (1.32), we consequently obtain from Theorem 4.1 in Chapter 1 and the fact that $\Phi(x) - \psi(x)$ is continuous and periodic in $\mathbf{R}^N \backslash \cup_{m \in \Lambda_N} \{2\pi m\}$ that

$$\lim_{t \to 0} H_0(x,t) = H_0(x) \text{ exists and is finite } \text{ for } x \in \mathbf{R}^N \backslash \cup_{m \in \Lambda_N} \{2\pi m\}.$$

Also, from Theorem 4.3 in Chapter 1,

$$\lim_{t \to 0} \int_{T_N} |H_0(x,t) - H_0(x)| \, dx = 0.$$

Furthermore, from (1.38), $H_0(x) - \Phi(x) = \psi(x) \ \text{ for } x \in B(0,1) \backslash \{0\}$. But, ψ is a continuous function in all of $B(0,1)$. Therefore,

$$\lim_{|x| \to 0} [H_0(x) - \Phi(x)] \text{ exists and is finite},$$

and condition (ii) in Lemma 1.4 is established.

Next, using the Poisson kernel, $P(x,t) = \sum_{m \in \Lambda_N} e^{im \cdot x - |m|t} \ \text{ for } t > 0$, defined in (4.37) of Chapter 1, we see from (4.38) in Chapter 1 that

$$\lim_{t \to 0} \sum_{m \in \Lambda_N} e^{im \cdot x - |m|t} = 0 \ \text{ for } \ x \in \mathbf{R}^N \backslash \cup_{m \in \Lambda} \{2\pi m\}.$$

Consequently, it follows from Lemma 1.3 above that

$$\Delta_*[H_0(x) - |x|^2 / 2N] \leq 0 \leq \Delta^*[H_0(x) - |x|^2 / 2N]$$

for $x \in \mathbf{R}^N \backslash \cup_{m \in \Lambda_N} \{2\pi m\}$.

Since $H_0(x) - |x|^2 / 2N$ is continuous in $\mathbf{R}^N \backslash \cup_{m \in \Lambda_N} \{2\pi m\}$, we conclude from this last set of inequalities and Theorem 2.2 in Appendix C that

$$(1.39) \qquad H_0(x) - |x|^2 / 2N \text{ is harmonic in } \mathbf{R}^N \backslash \cup_{m \in \Lambda_N} \{2\pi m\}.$$

If we check back and see what we have already established in this proof, we see that we actually have established all that is stated in Lemma 1.4 pertaining to $H_0(x)$. So, to complete the proof of the lemma, it remains to show that the statement in Lemma 1.4 for $H_j(x)$, $j = 1, ..., N$ is also valid. To accomplish this, we proceed as follows.

Since $\Phi(x)$ is harmonic in $B(0,1)\backslash\{0\}$, we obtain from (1.38) and (1.39) that

$$(1.40) \qquad \psi(x) + |x|^2/2N \text{ is harmonic in } B(0,1)\backslash\{0\}.$$

But $\psi(x)$ is continuous in all of $\mathbf{R}^N$. Hence, from Lemma 1.5 in Appendix C, we have that $\psi(x) + |x|^2/2N$ is harmonic in all of $B(0,1)$, and consequently that

$$(1.41) \qquad \psi(x) \in C^\infty(B(0,1)).$$

But then it follows from (1.31), (1.38), and (1.41) that

$$\lim_{|x|\to 0}[\frac{\partial H_0(x)}{\partial x_j} + (2\pi)^N x_j/|S_{N-1}| |x|^N] \text{ exists and is finite for } j = 1, ..., N.$$

Consequently, $\frac{\partial H_0(x)}{\partial x_j} \in L^1(T_N)$. We define

$$H_j(x) = \partial H_0(x)/\partial x_j \qquad \forall x \in \mathbf{R}^N\backslash\cup_{m\in\Lambda_N}\{2\pi m\},$$

and we see from the above that (iii), (iv), and (vi) in the lemma are now established.

It remains to show that conditions (i) and (vii) are valid.

From Theorem 4.1 in Chapter 1, condition (i) in the lemma concerning $H_j(x)$ for $j = 1, ..., N$ will be established once we show

$$(1.42) \qquad \lim_{t\to 0} H_j(x,t) = \partial H_0(x)/\partial x_j \qquad \forall x \in \mathbf{R}^N\backslash\cup_{m\in\Lambda_N}\{2\pi m\}$$

for $j = 1, ..., N$.

It is clear from (1.33) that the equality in (1.42) will follow once we demonstrate that

$$(1.43) \qquad \begin{aligned} \widehat{\frac{\partial H_0}{\partial x_j}}(m) &= im_j/|m|^2 \quad \text{for } |m| > 0 \\[2ex] \widehat{\frac{\partial H_0}{\partial x_j}}(0) &= 0 \end{aligned}$$

for $j = 1, ..., N$. Also, from Theorem 4.3 in Chapter 1, condition (vii) in the lemma will follow once we establish (1.43).

So the proof of the lemma will be complete once we demonstrate that (1.43) is true.

(1.43) will be valid if the following two items obtain

$$(1.44) \qquad \begin{aligned} &(i)\ \textstyle\int_{T_N} \partial H_0(x)/\partial x_j dx = 0; \\[2ex] &(ii)\ \textstyle\int_{T_N} e^{-im\cdot x}\partial H_0(x)/\partial x_j = (2\pi)^N im_j/|m|^2 \quad \text{for } |m| > 0. \end{aligned}$$

To demonstrate that (i) of (1.44) is indeed true, we define $\mathbf{v}(x)$ to be the vector field in $T_N\backslash\{0\}$ whose j-th component is $H_0(x)$ and whose other

$(N-1)$ components are zero. Then by the divergence theorem, (i) of (1.44) is true provided

$$(1.45) \qquad \lim_{\varepsilon \to 0} \int_{\partial[T_N - B(0,\varepsilon)]} \mathbf{v}(x) \cdot \mathbf{n}(x) dS(x) = 0$$

where $\mathbf{n}(x)$ is the outward pointing unit normal. But $H_0(x)$ is a periodic function. Consequently, $\int_{\partial T_N} \mathbf{v}(x) \cdot \mathbf{n}(x) dS(x) = 0$. On the other hand,

$$\left| \int_{\partial B(0,\varepsilon)} \mathbf{v}(x) \cdot \mathbf{n}(x) dS(x) \right| \leq c\varepsilon^{N-1} \max_{|x|=\varepsilon} |H_0(x)|$$

where c is a positive constant. From (1.31) and condition (ii) of the lemma, we see that the right-hand side of this last inequality goes to zero as $\varepsilon \to 0$. So the limit in (1.45) is established.

To complete the proof of the lemma, we have to show that (ii) of (1.44) is true. To do this, we now let $\mathbf{w}(x)$ be the vector field defined in $T_N \backslash \{0\}$ whose j-th component is $e^{-im \cdot x} H_0(x)$ and whose other $(N-1)$ components are zero. The same reasoning as the above shows that

$$\lim_{\varepsilon \to 0} \int_{\partial[T_N \backslash B(0,\varepsilon)]} \mathbf{w}(x) \cdot \mathbf{n}(x) dS(x) = 0.$$

On the other hand, the divergence theorem gives
$\int_{\partial[T_N \backslash B(0,\varepsilon)]} \mathbf{w}(x) \cdot \mathbf{n}(x) dS(x)$

$$= \int_{T_N \backslash B(0,\varepsilon)} \partial[e^{-im \cdot x} H_0(x)]/\partial x_j dx.$$

We conclude that

$$\begin{aligned} \int_{T_N} e^{-im \cdot x} \partial H_0(x)/\partial x_j dx &= \int_{T_N} i m_j H_0(x) e^{-im \cdot x} dx \\ &= (2\pi)^N i m_j |m|^{-2} \qquad \text{for } |m| > 0 \end{aligned}$$

(ii) of (1.44) is established, and the proof of the lemma is complete. ∎

Proof of Corollary 1.2. To prove this corollary, we proceed exactly as in the proof of Theorem 1.1 and observe, as before, that without loss in generality, we can assume that

$$\bar{a}_m = a_{-m} \quad \forall m \in \Lambda_N.$$

Also, as before, we obtain that the function

$$(1.46) \qquad F(x) = \lim_{t \to 0} F(x,t) \ \forall x \in \mathbf{R}^N \backslash \cup_{m \in \Lambda_N} \{2\pi m\},$$

where $F(x,t)$ is given in (1.10), is well-defined.

From the proof given in Theorem 1.1, we furthermore obtain that

$$(1.47) \qquad F(x) + a_0 |x|^2 /2N \text{ is harmonic in } \mathbf{R}^N \backslash \cup_{m \in \Lambda_N} \{2\pi m\}.$$

In addition, from the comments above (1.13), we see that $F(x)$ is also in $L^2(T_N)$. Hence, it follows from Theorems 1.3 and 1.4 in Appendix C that

$$F(x) + a_0 |x|^2 /2N = b_0 \log r + \sum_{j=1}^{N} b_j x_j r^{-N} + u_0(x) \qquad \text{for } N = 2,$$

$$= b_0 r^{-(N-2)} + \sum_{j=1}^{N} b_j x_j r^{-N} + u_0(x) \quad \text{for } N \geq 3,$$

$\forall x \in B(0,1)-0$, where $u_0(x)$ is harmonic in $B(0,1)$ and the b_j are constants.

We next invoke Lemma 1.4 above and see from the set of equalities just established that there are constants $b_0^*, b_1^*, ..., b_N^*$, such that

$$\lim_{|x| \to 0} [F(x) + a_0 |x|^2 /2N - \sum_{j=0}^{N} b_j^* H_j(x)] = \beta,$$

where β is a finite number.

Using this last limit in conjunction with (v) and (vi) of Lemma 1.4, we obtain from the periodicity of $F(x)$, which is clear from (1.46), that the function

$$(1.48) \qquad V(x) = F(x) - \sum_{j=0}^{N} b_j^* H_j(x)$$

for $x \in \mathbf{R}^N \setminus \cup_{m \in \Lambda_N} \{2\pi m\}$, and

$$V(2\pi m) = \beta \text{ for } m \in \Lambda_N$$

is both periodic and continuous in $\mathbf{R}^N$. In addition, from (1.47), (1.48), and (v) and (vi) of Lemma 1.4, we have that

$$(1.49) \qquad V(x) + (a_0 + b_0^*) |x|^2 /2N \text{ is harmonic in } \mathbf{R}^N.$$

Since $V(x)$ is a periodic and continuous function, and hence bounded, (1.49) and Theorem 1.8 in Appendix C tell us that $V(x)$ is a polynomial of degree at most two. But the only way a continuous periodic function can be a polynomial of degree at most two is when it is identically a constant. Consequently, from (1.48),

$$(1.50) \qquad F(x) - \sum_{j=0}^{N} b_j^* H_j(x) = \beta$$

for $x \in \mathbf{R}^N \setminus \cup_{m \in \Lambda_N} \{2\pi m\}$.

We see from (1.13) that $\widehat{F}(m) = -a_m / |m|^2$ for $m \neq 0$. Therefore, Lemma 1.4 and (1.50), in turn, imply that

$$a_m + b_0^* = - \sum_{j=1}^{N} im_j b_j^* \quad \text{for } m \neq 0.$$

By condition (i) in the corollary, the left-hand side of this last equality is $o(|m|)$ as $|m| \to \infty$. Consequently, we obtain that

$$b_j^* = 0 \quad \text{for } j = 1, ..., N,$$

and therefore that

$$a_m = -b_0^* \quad \text{for } m \neq 0.$$

But, by condition (i) of the corollary,

$$\sum_{R-1 < |m| \leq R} |a_m| = o(R) \quad \text{as } R \to \infty.$$

However, the number of lattice points in the annulus $R - 1 < |m| \leq R$ is $O(R^{N-1})$ and not $o(R^{N-1})$ as $R \to \infty$. Hence, $b_0^* = 0$, and therefore,

$$a_m = 0 \quad \text{for } m \neq 0.$$

This, along with condition (ii) of the corollary, then implies that

$$a_0 = 0.$$

These last two equalities complete the proof of the corollary. ∎

Exercises.

1. Given $f(t) = \sum_{n=1}^{\infty} a_n e^{-nt}$ for $t > 0$ where $\sum_{n=1}^{\infty} |a_n| < \infty$ and $\sum_{n=1}^{\infty} a_n = 0$, prove

$$\limsup_{t \to 0} -\frac{d^2 f}{dt^2}(t) < 0 \Rightarrow \limsup_{t \to 0} d(\frac{f(t)}{t})/dt \geq 0.$$

2. Prove that condition (i) in the hypothesis of Theorem 1.1 implies that

$$\sum_{|m| > 0} |a_m|^2 |m|^{-4} < \infty.$$

3. Using (1.5) and (1.6) in Appendix A, prove that for $r > 0$,

$$|B(x,r)|^{-1} \int_{B(x,r)} e^{im \cdot y} dy = \mu_N e^{im \cdot x} J_{N/2}(|m| r)(|m| r)^{-N/2}$$

where μ_N is a constant depending on N but not on m.

4. Prove that, in dimension $N = 3$, the number of integral lattice points contained in the spherical annulus $B(0, R+1) \setminus B(0, R)$ is $O(R^2)$ and not $o(R^2)$ as $R \to \infty$.

2. Uniqueness for Circular Convergence

In this section, we focus on the problem that was open for approximately one hundred years before it was solved, namely, the analogue of Cantor's Theorem for double trigonometric series discussed at the beginning of §1. After we present this result, we will exhibit an analogous problem involving Cantor's theorem on the two-sphere S_2, which is still open.

The theorem we present here, which settled this one-hundred year old problem, is the following:

Theorem 2.1. *Given the trigonometric series $\sum_{m \in \Lambda_2} b_m e^{im \cdot x}$ where the b_m are arbitrary complex numbers and dimension $N = 2$, set*

$$S_R(x) = \sum_{|m| \leq R} b_m e^{im \cdot x},$$

and suppose that

(2.1) $$\lim_{R \to \infty} S_R(x) = 0 \quad \text{for } x \in T_2 \backslash \{0\}.$$

Then $b_m = 0 \quad \forall m \in \Lambda_2$.

Theorem 2.1 is proved using the clever 1971 result of Roger Cooke [Co], combined with Corollary 1.2 above, which was established in 1957 by Shapiro [Sh2]. We present Cooke's result in Theorem 2.2 below.

The analogue of Theorem 2.1 in dimension $N \geq 3$ was established in 1996 by Bourgain [Bou]. For a good presentation of Bourgain's result, we refer the reader to the expository article of Ash and Wang [AW1]. For another expository comment about Theorem 2.1, see [Ash, p. 94].

Theorem 2.2. *Given the trigonometric series $\sum_{m \in \Lambda_2} b_m e^{im \cdot x}$ where $\bar{b}_m = b_{-m}$ and dimension $N = 2$, set $B_n(x) = \sum_{|m|^2 = n} b_m e^{im \cdot x}$, and suppose*

(2.2) $$\lim_{n \to \infty} B_n(x) = 0 \quad \text{for a.e. } x \in T_2.$$

Then

(2.3) $$\lim_{n \to \infty} \sum_{|m|^2 = n} |b_m|^2 = 0.$$

To prove Theorem 2.2, we will need the following lemma.

Lemma 2.3. *With $B_n(x)$ defined as in Theorem 2.2, where $\bar{b}_m = b_{-m}$,*

(2.4) $$\left\{ \int_{T_2} |B_n(x)|^4 \, dx \right\}^{1/4} \leq \left\{ \int_{T_2} |B_n(x)|^2 \, dx \right\}^{1/2}.$$

Proof of Lemma 2.3.   To prove the lemma, we first observe that

(2.5)
$$|B_n(x)|^2 = \sum_{|m|^2=n} \sum_{|p|^2=n} b_m \bar{b}_p e^{i(m-p)\cdot x}$$

and

(2.6)
$$\int_{T_2} |B_n(x)|^2 \, dx = 4\pi^2 \sum_{|m|^2=n} |b_m|^2 .$$

Let $\Lambda_n^\Diamond$ designate the following set of lattice points:
$$\Lambda_n^\Diamond = \{q : m - p = q, \ |m|^2 = n, \ |p|^2 = n\},$$

and for each $q \in \Lambda_n^\Diamond$, let β_q designate the following number:

$$\beta_q = \sum_{|p|^2=n, |m|^2=n, m-p=q} b_m \bar{b}_p.$$

Then we see from (2.5) that

$$
\begin{aligned}
|B_n(x)|^2 &= \sum_{q\in\Lambda_n^\Diamond} e^{iq\cdot x} \sum_{|p|^2=n, |m|^2=n, m-p=q} b_m \bar{b}_p \\
&= \sum_{q\in\Lambda_n^\Diamond} \beta_q e^{iq\cdot x}.
\end{aligned}
$$

Now, $\theta = (0,0)$ is in $\Lambda_n^\Diamond$. Hence, we observe from this last set of equalities that

(2.7)
$$(4\pi)^{-2} \int_{T_2} |B_n(x)|^4 \, dx = |\beta_\theta|^2 + \sum_{q\in\Lambda_n^\Diamond\setminus\{\theta\}} |\beta_q|^2 .$$

Next, we claim that *given $q \in \Lambda_n^\Diamond\setminus\{\theta\}$, there are at most two pair of lattice points (m^*, p^*) and (m^{**}, p^{**}) such that $m^* - p^* = m^{**} - p^{**} = q$ and $|m^*|^2 = |p^*|^2 = |m^{**}|^2 = |p^{**}|^2 = n$.*

To see that this claim is indeed valid, let $C(\theta, r)$ be the circle centered at θ of radius r and observe

$$\text{if } m^* - p^* = q \text{ with } |m^*|^2 = |p^*|^2 = n,$$

then the line segment $\overrightarrow{p^* m^*}$ is a chord of the circle $C(\theta, \sqrt{n})$ that is both

(i) parallel to $\overrightarrow{\theta q}$ and

(ii) of length $|q|$.

There is at most one other chord of the circle $C(\theta, \sqrt{n})$ with these two properties. Hence, the validity of the above claim is as asserted.

It follows from this claim that

$$|\beta_q|^2 \leq 2 \sum_{|p|^2=n, |m|^2=n, m-p=q} |b_m|^2 |b_p|^2 \quad \text{for } q \in \Lambda_n^{\Diamond} \backslash \{\theta\}.$$

Consequently, we obtain from (2.7) that

$$
\begin{aligned}
(4\pi)^{-2} \int_{T_2} |B_n(x)|^4 \, dx &\leq |\beta_\theta|^2 + 2 \sum_{q \in \Lambda_n^{\Diamond} \backslash \{\theta\}} \sum_{|p|^2=n, |m|^2=n, m-p=q} |b_m|^2 |b_p|^2 \\
&\leq |\beta_\theta|^2 + 2 \sum_{|m|^2=n} \sum_{|p|^2=n} |b_m|^2 |b_p|^2 \\
&\leq 3 \Big(\sum_{|m|^2=n} |b_m|^2 \Big)^2.
\end{aligned}
$$

But then from (2.6), we see that

$$(4\pi)^{-2} \int_{T_2} |B_n(x)|^4 \, dx \leq 3 [(4\pi)^{-2} \int_{T_2} |B_n(x)|^2 \, dx]^2.$$

The conclusion to the lemma follows from this last inequality. ∎

Proof of Theorem 2.2. Suppose the conclusion to the theorem is false. Then there exists $\delta > 0$ and a sequence $\{n_j\}_{j=1}^{\infty}$ such that

$$\sum_{|m|^2=n_j} |b_m|^2 \geq \delta \quad \text{for } j = 1, 2, \dots.$$

Set

$$A_j(x) = \frac{B_{n_j}(x)}{(\sum_{|m|^2=n_j} |b_m|^2)^{\frac{1}{2}}}.$$

Then

$$(2.8) \qquad \lim_{j \to \infty} A_j(x) = 0 \text{ for a.e. } x \in T_2$$

and

$$(2.9) \qquad \int_{T_2} |A_j(x)|^2 \, dx = 4\pi^2 \quad \text{for } j = 1, 2, \dots.$$

However, from (2.8) combined with Egoroff's Theorem, we see that

$$\exists E \subset T_2 \quad \text{s.t. } |E| \leq 10^{-4} \quad \text{and} \quad \lim_{j \to \infty} A_j(x) = 0 \quad \text{uniformly } \forall x \in T_2 \backslash E.$$

Consequently, we obtain from the equality in (2.9) that

$$(2.10) \qquad 4\pi^2 \leq \limsup_{j \to \infty} \int_E |A_j(x)|^2 \, dx \quad \text{where } |E| \leq 10^{-4}.$$

But, by Holder's inequality and Lemma 2.3,

$$\int_E |A_j(x)|^2 \, dx \leq 10^{-2} \left(\int_E |A_j(x)|^4 \, dx \right)^{\frac{1}{2}}$$

$$\leq 10^{-2} \int_{T_2} |A_j(x)|^2 \, dx.$$

Therefore, from (2.9) and (2.10), we see that

$$4\pi^2 \leq 10^{-2} 4\pi^2.$$

This contradiction shows the conclusion of the theorem is valid. ■

Proof of Theorem 2.1.   From a consideration of the series

$$\sum_{m \in \Lambda_2} (b_m + \bar{b}_{-m}) e^{im \cdot x} \quad \text{and} \quad \sum_{m \in \Lambda_2} i(b_m - \bar{b}_{-m}) e^{im \cdot x},$$

we see from the start that it is sufficient to prove the theorem under the additional assumption

$$\bar{b}_m = b_{-m} \quad \forall m \in \Lambda_2.$$

Also, we see that

$$S_R(x) = \sum_{n=0}^{[R^2]} B_n(x),$$

where $B_n(x) = \sum_{|m|^2 = n} b_m e^{im \cdot x}$ and $[R^2]$ is the largest integer $\leq R^2$. Consequently, from (2.1) in the hypothesis of the theorem, we see, in particular, that

$$\lim_{n \to \infty} B_n(x) = 0 \quad \text{for a.e. } x \in T_2.$$

Hence, we obtain from (2.3) in Theorem 2.2 that

$$\lim_{|m| \to \infty} |b_m| = 0.$$

Now, the number of lattice points in the annulus determined by $R - 1$ and R is $O(R)$. Therefore, we see from this last limit that

(2.11) $$\sum_{R-1 \leq |m| < R} |b_m| = o(1)O(R) = o(R).$$

Also, we see from condition (2.1) in the hypothesis of the theorem and from Theorem 1.2 in Appendix B that

$$\lim_{t \to 0} \sum_{m \in \Lambda_2} b_m e^{im \cdot x} e^{-|m|t} = 0 \quad \text{for } x \in T_2 \backslash \{0\}.$$

This last fact coupled with the observation in (2.11) shows that the conditions in the hypothesis of Corollary 1.2 are met. Hence,

$$b_m = 0 \quad \text{for } m \in \Lambda_2,$$

and Theorem 2.1 is established.   ∎

On S_2, the theorem analogous to the one just established for T_2, namely Theorem 2.1, has been open for the last 135 years. There is a possibility that it may be false, for the analogue of Theorem 2.2 is false, as we will demonstrate.

Here is what is known so far about the uniqueness of surface spherical harmonic expansions on S_2, frequently called Laplace series. We use the notation introduced in §3 of Appendix A and let $Y_n(\xi)$ be a surface spherical harmonic of degree n where $\xi \in S_2$. We also set

$$\|Y_n\|_{L^2} = \{\int_{S_2} |Y_n(\xi)|^2 \, dS(\xi)\}^{\frac{1}{2}}.$$

The following result was established in [Sh9, p. 12] using ideas similar to those in the proofs of Theorem 1.1 and Corollary 1.2 but adapted for S_2.

Theorem 2.4.   *Given* $\{Y_n(\xi)\}_{n=0}^{\infty}$, *a sequence of surface spherical harmonics of degree* n *on* S_2, *suppose that*

$$(i)\ \|Y_n\|_{L^2} = o(n^{\frac{1}{2}}) \text{ as } n \to \infty \ \text{ and}$$

$$(ii)\ \lim_{r \to 1} \sum_{n=0}^{\infty} r^n Y_n(\xi) = 0 \quad \text{for } \xi \in S_2 \backslash \{\eta^*\},$$

where $\eta^* = (1, 0, 0)$. *Then*

$$Y_n(\xi) = 0 \quad \forall \xi \in S_2 \text{ and } \forall n.$$

We observe from formula (3.29) in Appendix A that

$$\frac{1 - r^2}{(1 - 2r\eta^* \cdot \xi + r^2)^{3/2}} = \sum_{n=0}^{\infty} (2n + 1) P_n(\eta^* \cdot \xi) r^n \quad \forall \xi \in S_2 \text{ and } 0 \leq r < 1,$$

where $P_n(t)$ is the Legendre polynomial of order n. From this last equality, we obtain that

$$(2.12) \qquad \lim_{r \to 1} \sum_{n=0}^{\infty} r^n (2n + 1) P_n(\eta^* \cdot \xi) = 0 \quad \text{for } \xi \in S_2 \backslash \{\eta\}^*.$$

As we have shown in §3 of Appendix A, $(2n + 1)\, P_n(\eta^* \cdot \xi) = Y_n(\xi)$ is a surface spherical harmonic of degree n. Also, (see [Pi, p. 227]),

$$(2.13) \qquad \int_{S_2} |P_n(\eta^* \cdot \xi)|^2 \, dS(\xi) = 2\pi \int_{-1}^{1} |P_n(t)|^2 \, dt = 4\pi(2n + 1)^{-1}.$$

This sequence $\{(2n+1)P_n(\eta^* \cdot \xi)\}_{n=0}^{\infty}$ demonstrates that Theorem 2.4 is in a certain sense a best possible result because

(a) by (2.13), it meets (i) in the theorem if "o" is replaced with "O";

(b) by (2.12), it meets (ii) in the theorem;

(c) $P_n(1) = 1$ for every n shows that $P_n(\eta^* \cdot \xi)$ is not identically zero.

We will not establish Theorem 2.4 here, but refer the reader to [Sh9, p. 12] for the proof.

The open problem for the last 135 years is the following:

Problem 2.5. *Given $\{Y_n(\xi)\}_{n=0}^{\infty}$ a sequence of surface spherical harmonics of degree n on S_2, suppose that*

$$\lim_{n \to \infty} \sum_{k=0}^{n} Y_k(\xi) = 0 \quad \text{for } \xi \in S_2 \backslash \{\eta^*\},$$

where $\eta^* = (1, 0, 0)$. Then

$$Y_n(\xi) = 0 \quad \forall \xi \in S_2 \text{ and } \forall n.$$

Problem 2.5 is still open because the analogue of the Cantor-Lebesgue type theorem established by Roger Cooke on T_2 (i.e., Theorem 2.2) is not available on S_2, as the following counter-example of Walter Rudin [Ru2, p. 302], demonstrates.

Proposition 2.6. *Given a positive integer J, there exists $\{Y_n(\xi)\}_{n=1}^{\infty}$, a sequence of surface spherical harmonics of degree n on S_2, such that*

$$(i) \; \lim_{n \to \infty} Y_n(\xi) \;\; = \;\; 0 \quad \forall \xi \in S_2, \; and$$

$$(ii) \;\; \frac{\|Y_n\|_{L^2}}{n^J} \;\; \to \;\; \infty \quad as \; n \to \infty.$$

Proof of Proposition 2.6. We observe for every integer $n \geq 1$,

$$u(x_1, x_2, x_3) = x_1 \, \mathrm{Im}(x_2 + ix_3)^{n-1}$$

is harmonic in $\mathbf{R}^3$ and also is a homogeneous polynomial of degree n in the variables x_1, x_2, x_3. Hence, it is a spherical harmonic of degree n. Therefore, with $\xi = (\xi_1, \xi_2, \xi_3) \in S_2$,

$$Y_n(\xi) = n^{J+4} \xi_1 \, \mathrm{Im}(\xi_2 + i\xi_3)^{n-1}$$

is a surface spherical harmonic of degree n.

Using $\xi_1 = \cos\theta$, $\xi_2 = \sin\theta\cos\phi$, $\xi_3 = \sin\theta\sin\phi$, we see that

$$(2.14) \qquad Y_n(\xi) = n^{J+4} \cos\theta (\sin\theta)^{n-1} \sin(n-1)\phi,$$

where $0 \leq \theta \leq \pi$ and $0 \leq \phi < 2\pi$. Since $|\sin \theta| < 1$ for $\theta \neq \pi/2 \mod \pi$ and $\cos \pi/2 = 0$, it follows from (2.14) that

$$\lim_{n \to \infty} Y_n(\xi) = 0 \quad \forall \xi \in S_2.$$

To complete the proof of Proposition 2.6, it remains to show that (ii) above holds. It follows from (2.14) that this will be accomplished once we demonstrate that

$$(2.15) \qquad \int_0^\pi (\cos \theta)^2 (\sin \theta)^{2n-1} d\theta \geq c/n^2 \quad \text{for} \quad n \geq 1,$$

where c is a positive constant independent of n.

Letting I_n designate the integral on the left-hand side of the inequality in (2.15), we see after an integration by parts that

$$I_n = \int_0^\pi (\sin \theta)^{2n+1} d\theta/2n = \int_0^{\pi/2} (\sin \theta)^{2n+1} d\theta/n.$$

Using the fact that $\sin \theta \geq 2\theta/\pi$ for $0 \leq \theta \leq \pi/2$, we obtain from this last equality that

$$I_n \geq \pi \int_0^1 t^{2n+1} dt/2n \quad \text{for} \quad n \geq 1,$$

and the inequality in (2.15) follows. ∎

Exercises.

1. Prove that $x_1 \operatorname{Im}(x_2 + ix_3)^{n-1}$ is a homogeneous polynomial of degree n that is harmonic in $\mathbf{R}^3$.

2. With $\xi_1 = \cos \theta$, $\xi_2 = \sin \theta \cos \phi$, $\xi_3 = \sin \theta \sin \phi$, find the surface spherical harmonic $Y_n(\xi)$, which comes from the spherical harmonic $x_1 \operatorname{Im}(x_2 + ix_3)^{n-1}$.

3. Uniqueness, Number Theory, and Fractals

One problem that Cantor was unable to solve was the following: If a trigonometric series S on T_1 converges to zero in the complement of the usual Cantor set, is S identically zero? In this case, the answer is in the affirmative and is the basis for a very deep result in mathematics connecting number theory and trigonometric series [Zy2, p. 152]. It is the purpose of this section to present a result analogous to this on T_N, $N \geq 2$. Also, instead of using the convergence of trigonometric series to deal with sets of uniqueness, the more sophisticated idea of distributions on the N-torus will be employed. The basis for the material developed here is the paper entitled *Algebraic Integers and Distributions on the N-Torus* (see [Sh5]).

Using the notion of distributions also enables us to obtain uniqueness results for trigonometric series of the form

$$\sum_{m \in \Lambda_N} a_m e^{im \cdot x}$$

without assuming anything about a_m under the spherical norm. In particular, in Theorem 3.2 below, we do not make any assumption like

$$\lim_{|m| \to \infty} |a_m| = 0.$$

Instead, we make assumptions of the following nature: (i) $\{a_m\}_{m \in \Lambda_N}$ is uniformly bounded; and (ii) $\lim_{\min(|m_1|,...,|m_N|) \to \infty} a_m = 0$.

Clearly, the norm used in (ii) is weaker than the spherical norm.

In this section, we present three theorems involving uniqueness results for trigonometric series. The second and third theorems do use the spherical norm. Also, the second theorem is connected with the theory of fractals.

In the sequel, with

$$T_N = \{x : -\pi \le x_j < \pi, \ j = 1, ..., N\},$$

we will need the notion of the torus topology. We observe that $E \subset T_N$ is closed in the torus topology if and only if the set $E^* \subset \mathbf{R}^N$ is a closed set in $\mathbf{R}^N$ where

$$E^* = \cup_{m \in \Lambda_N}\{E + 2\pi m\}.$$

To define the notion of a distribution S on T_N, we first need the class of real functions $\mathcal{D}(T_N)$, called test functions, where

$$\mathcal{D}(T_N) = \{\phi : \phi \in C^\infty(\mathbf{R}^N), \ \phi \text{ is periodic of period } 2\pi \text{ in each variable}\}.$$

As before, we define

$$\|\phi\|_{L^2}^2 = \int_{T_N} |\phi|^2 \, dx,$$

and $\Delta\phi$ to be the Laplacian of ϕ. So, in particular, for $\phi \in \mathcal{D}(T_N)$,

$$(3.1) \qquad \left\|\Delta^k \phi\right\|_{L^2}^2 = (2\pi)^N \sum_{m \in \Lambda_N} |m|^{4k} \left|\widehat{\phi}(m)\right|^2 \quad \text{for } k = 0, 1, 2, ...$$

With $\{\phi_n\}_{n=1}^\infty \subset \mathcal{D}(T_N)$, we say $\phi_n \to 0$ in $\mathcal{D}(T_N)$ provided the following takes place:

$$\lim_{n \to \infty} \left\|\Delta^k \phi_n\right\|_{L^2} = 0 \quad \text{for } k = 0, 1, 2, ...$$

A distribution S on $\mathcal{D}(T_N)$, also called a periodic distribution, is a real linear functional on $\mathcal{D}(T_N)$ with the following property.

Given $\{\phi_n\}_{n=1}^\infty \subset \mathcal{D}(T_N)$. If

$$(3.2) \qquad \phi_n \to 0 \text{ in } \mathcal{D}(T_N), \text{ then } \lim_{n \to \infty} S(\phi_n) = 0.$$

We'll designate this class of distributions by $\mathcal{D}'(T_N)$ and set for $S \in \mathcal{D}'(T_N)$ and $\lambda, \phi \in \mathcal{D}(T_N)$,

$$S(\lambda + i\phi) = S(\lambda) + iS(\phi).$$

$\widehat{S}(m)$ is then defined by the following formula:

(3.3) $(2\pi)^N \widehat{S}(m) = S(e^{-im \cdot x}) = S(\cos m \cdot x) - iS(\sin m \cdot x)$

for $m \in \Lambda_N$.

In the sequel, we will need

(3.4) $\left|\widehat{S}(m)\right|$ is uniformly bounded for $m \in \Lambda_N$,

and

(3.5) $\lim_{\min(|m_1|,...,|m_N|) \to \infty} \left|\widehat{S}(m)\right| = 0.$

In particular, we define the class $\mathcal{A}(T_N) \subset \mathcal{D}'(T_N)$ as follows:

$$\mathcal{A}(T_N) = \left\{S \in \mathcal{D}'(T_N) : \left|\widehat{S}(m)\right| \text{ meets (3.4) and (3.5)}\right\}.$$

If $\phi \in \mathcal{D}(T_N)$, the set

$$Su^{\diamond}(\phi) = \{x \in T_N : \phi(x) \neq 0\}^{\widetilde{}}$$

(where $\widetilde{}$ designates the closure in the torus topology) will be called the support of ϕ. Let $G \subset T_N$ be open in the torus topolgy. $S = 0$ in G means

$$\phi \in \mathcal{D}(T_N) \text{ and } Su^{\diamond}(\phi) \subset G \Longrightarrow S(\phi) = 0.$$

Let $E \subset T_N$ be closed in the torus topology. E is called a set of uniqueness for the class $\mathcal{A}(T_N)$ provided the following holds:

$$\text{if } S \in \mathcal{A}(T_N) \text{ and } S = 0 \text{ in } T_N \backslash E, \text{ then } S \equiv 0.$$

Before proceeding with the statement of the theorem, we establish the following proposition.

Proposition 3.1. *Let $J \geq 2$ be an integer. Suppose $E_j \subset T_N$ is both a set closed in the torus topology and a set of uniqueness for the class $\mathcal{A}(T_N)$ for $j=1,...,J$. Then $\cup_{j=1}^{J} E_j$ is also a set of uniqueness for the class $\mathcal{A}(T_N)$.*

Proof of Proposition 3.1.   It is clear from an induction process that the proof of the proposition will be complete once we show that the proposition is true for the special case $J = 2$.

Suppose, therefore, that the proposition is false for the case $J = 2$. Then there is an $S \in \mathcal{A}(T_N)$ such that $S = 0$ in $T_N \backslash (E_1 \cup E_2)$ but S is not identically zero. Consequently, there is a $\phi_1 \in \mathcal{D}(T_N)$ such that

$$Su^{\diamond}(\phi_1) \subset T_N \backslash E_1 \text{ and } S(\phi_1) \neq 0.$$

Since $Su^\diamond(\phi_1)$ and E_1 are closed sets in the torus topology with an empty intersection, it follows that there is a set G_1 that is open in the torus topology with $E_1 \subset G_1$ and $G_1 \cap Su^\diamond(\phi_1) = \emptyset$.

Now, $\phi_1 S \in \mathcal{D}'(T_N)$ where

$$\phi_1 S(\phi) = S(\phi_1\phi) \quad \forall \phi \in \mathcal{D}(T_N).$$

Also, by Proposition 3.4 in Appendix B, $\phi_1 S \in \mathcal{A}(T_N)$.

Let $\phi \in \mathcal{D}(T_N)$ be such that $Su^\diamond(\phi) \subset T_N \backslash E_2$. Then there is a set G_2 with $E_2 \subset G_2$ that is open in the torus topology and $G_2 \cap Su^\diamond(\phi) = \emptyset$. Therefore,

$$\phi_1(x)\phi(x) = 0 \quad \forall x \in G_1 \cup G_2,$$

and hence,

$$Su^\diamond(\phi_1\phi) \subset T_N \backslash (E_1 \cup E_2).$$

But then by assumption $S(\phi_1\phi) = 0$. So $\phi_1 S(\phi) = 0$ for every $\phi \in \mathcal{D}(T_N)$ which, has its support in $T_N \backslash E_2$. However, as we have observed, $\phi_1 S \in \mathcal{A}(T_N)$ and by assumption E_2 is a set of uniqueness for the class $\mathcal{A}(T_N)$. We conclude that $\phi_1 S \equiv 0$.

In particular, $\phi(x) \equiv 1$ is a function in $\mathcal{D}(T_N)$. So,

$$0 = \phi_1 S(1) = S(\phi_1),$$

and we have arrived at a contradiction. Therefore, the proposition is true for the case $J = 2$, and the proof of the proposition is complete. ∎

Next, $C(\xi)$ with $0 < \xi < 1/2$ will designate the familiar Cantor set in the half open interval $[-\pi, \pi)$, i.e., $t \in C(\xi)$ if and only if

$$t = 2\pi(1 - \xi) \sum_{k=1}^{\infty} \varepsilon_k \xi^{k-1} - \pi,$$

where $\varepsilon_k = 0$ or 1 and not all $\varepsilon_k = 1$. We note that if $0 < \xi_j < 1/2$, $j = 1, ..., N$, then $C(\xi_j) \subset T_1$ is a closed set in the torus topology of T_1 and

$$C(\xi_1) \times \cdots \times C(\xi_N) \subset T_N$$

is a set closed in the torus topology of T_N.

An algebraic number is a complex number γ that satisfies an equation of the form

$$x^n + \alpha_1 x^{n-1} + ... + \alpha_n = 0$$

where the coefficients of the equation are rational numbers. In case all the coefficients of the equation are integers, γ is called an algebraic integer.

Following [Zy2, p.148], we say γ is an $\mathcal{S}$ number if the following three properties hold:

(i) γ is a real algebraic integer;

(ii) $\gamma > 1$;

(iii) if α is a conjugate of γ, then $|\alpha| < 1$.

$\mathcal{S}$ numbers are sometimes referred to as Pisot numbers.

We intend to prove the following theorem [Sh5] regarding $\mathcal{S}$ numbers:

Theorem 3.2. *Let $0 < \xi_j < 1/2$ for $j = 1, ..., N$ where $N \geq 2$. A necessary and sufficient condition that $C(\xi_1) \times \cdots \times C(\xi_N)$ be a set of uniqueness for the class $\mathcal{A}(T_N)$ is that ξ_j^{-1} be an $\mathcal{S}$ number for $j = 1, ..., N$.*

As an immediate corollary to Theorem 3.2 , we have the following result:

Corollary 3.3. *Suppose p_j and q_j are relatively prime positive integers with $p_j/q_j < \frac{1}{2}$ for $j=1,...,N$ where $N \geq 2$. A necessary and sufficient condition that $C(p_1/q_1) \times \cdots \times C(p_N/q_N)$ be a set of uniqueness for the class $\mathcal{A}(T_N)$ is that $p_j = 1$ for $j = 1, ..., N$.*

Proof of the Necessary Condition of Theorem 3.2. With no loss in generality, we can assume from the start that ξ_1^{-1} is not an $\mathcal{S}$ number. We accomplish the proof by showing that, under the assumption that ξ_1^{-1} is not an $\mathcal{S}$ number, there exists a trigonometric series $\sum_{m \in \Lambda_N} a_m e^{im \cdot x}$ with $a_m = \bar{a}_{-m}$, which has the four properties listed below:

(3.6) $\qquad\qquad |a_m|$ is uniformly bounded for $m \in \Lambda_N$.

(3.7) $$\lim_{\min(|m_1|,...,|m_N|) \to \infty} |a_m| = 0.$$

(3.8) $\qquad\qquad \exists m_0 \in \Lambda_N$ such that $a_{m_0} \neq 0$.

If $x_0 \in T_N \backslash C(\xi_1) \times \cdots \times C(\xi_N)$, then $\exists\, r_0 > 0$ such that

(3.9) $$\lim_{t \to 0} \sum_{m \in \Lambda_N} a_m e^{im \cdot x} e^{-|m|^2 t} = 0 \quad \text{uniformly for } x \in B(x_0, r_0).$$

To see that a trigonometric series with the properties stated in (3.6)-(3.9) will establish the necessary condition of the theorem, we first prove that

(3.10) $$S(\phi) = (2\pi)^N \sum_{m \in \Lambda_N} a_m \widehat{\phi}(-m) \quad \text{for } \phi \in \mathcal{D}(T_N)$$

defines a distribution in the class $\mathcal{A}(T_N)$.

To show that $S \in \mathcal{D}'(T_N)$, suppose $\{\phi_n\}_{n=1}^{\infty} \subset \mathcal{D}(T_N)$ and $\phi_n \to 0$ in $\mathcal{D}(T_N)$. Then

(3.11) $$\sum_{1 \leq |m|} \left| a_m \widehat{\phi}_n(-m) \right| \leq \left(\sum_{1 \leq |m|} \frac{|a_m|^2}{|m|^{4N}} \right)^{\frac{1}{2}} \left\| \Delta^N \phi_n \right\|_{L^2}.$$

From (3.6), we see that the first series on the right-hand side of the inequality in (3.11) is finite. Since $\left\|\Delta^N \phi_n\right\|_{L^2} \to 0$, we obtain therefore from (3.11) that

$$\lim_{n\to\infty} \sum_{1\le|m|} \left| a_m \widehat{\phi}_n(-m)\right| = 0.$$

Also, $a_0 \widehat{\phi}_n(0) \to 0$. So, $S \in \mathcal{D}'(T_N)$.

From (3.10), it follows that $\widehat{S}(m) = a_m$, and consequently from (3.7) that $S \in \mathcal{A}(T_N)$.

To show that a series with properties (3.6)-(3.9) will establish the necessary condition of the theorem, it remains to prove that the following holds:

(3.12) $\phi \in \mathcal{D}(T_N)$ and $Su^{\diamond}(\phi) \subset T_N \backslash C(\xi_1) \times \cdots \times C(\xi_N) \implies S(\phi) = 0.$

To accomplish this, we set

$$S(x,t) = \sum_{m\in\Lambda_N} a_m e^{im\cdot x} e^{-|m|^2 t},$$

and apply the Heine-Borel theorem in the torus topology. It then follows from (3.9) and the fact that $Su^{\diamond}(\phi) \subset T_N \backslash C(\xi_1) \times \cdots \times C(\xi_N)$ is compact in the torus topology that

$$\lim_{t\to 0} S(x,t) = 0 \text{ uniformly for } x \in Su^{\diamond}(\phi).$$

But then

$$S(\phi) = \lim_{t\to 0} \int_{T_N} S(x,t)\phi(x)dx = \lim_{t\to 0} \int_{Su^{\diamond}(\phi)} S(x,t)\phi(x)dx = 0,$$

giving the assertion in (3.12).

From (3.8), we have that S is not identically zero, and the necessary condition of the theorem is established.

To complete the proof of the necessary condition of the theorem, it remains to show that ξ_1^{-1} is not an S number, which implies the existence of a series $\sum_{m\in\Lambda_N} a_m e^{im\cdot x}$ with $a_m = \bar{a}_{-m}$ that has properties (3.6)-(3.9).

With $C(\xi) \subset T_1, \ 0 < \xi < 1/2$, we use the familiar Lebesgue-Cantor function (see [Zy1, p. 194], [Sa, p. 101]) associated with $C(\xi)$ to obtain a nonnegative Borel measure ν on T_1 with the property that $\nu[C(\xi)] = 1$ and $\nu[T_1\backslash C(\xi)] = 0$. Also, ν is nonatomic, i.e., $\nu[\{s\}] = 0 \ \forall s \in T_1$. We set

(3.13) $\widehat{\nu}(n) = (2\pi)^{-1} \int_{T_1} e^{-ins} d\nu(s) \quad \text{for } n = 0, \pm 1, \pm 2, \dots.$

It follows from (3.13) that $\overline{\widehat{\nu}}(n) = \widehat{\nu}(-n)$ and also that the sequence $\{\widehat{\nu}(n)\}_{n=-\infty}^{\infty}$ is uniformly bounded. In particular, it is easy to see that this last fact implies that there is a constant c such that

(3.14) $\left| \sum_{n=-\infty}^{\infty} \widehat{\nu}(n) e^{ins} e^{-|n|^2 t} \right| \le ct^{-1/2} \ \forall s \in T_1 \text{ and for } 0 < t < 1.$

From [Zy2, p. 151], we also observe that

$$(3.15) \qquad \xi^{-1} \text{is not an } \mathcal{S} \text{ number} \implies \lim_{|n|\to\infty} \widehat{\nu}(n) = 0.$$

To obtain the coefficients a_m, we let ν_k be the nonnegative Borel measure associated with $C(\xi_k)$ as above for $k = 1, ..., N$, and set

$$(3.16) \qquad a_m = \widehat{\nu}_1(m_1) \cdots \widehat{\nu}_N(m_N) \quad \forall m \in \Lambda_N.$$

It is clear that $a_m = \overline{a}_{-m}$ and that the coefficients a_m meet (3.6). From (3.15), it also follows that $\lim_{|m_1|\to\infty} \widehat{\nu}_1(m_1) = 0$. Hence, from (3.16) we have that the coefficients a_m meet (3.7). Furthermore, it is clear from (3.13) and (3.16) that $a_0 = (2\pi)^{-N}$. Therefore, (3.8) is also met.

So to complete the proof of the necessary condition of the theorem, it only remains to show that $\sum_{m\in\Lambda_N} a_m e^{im\cdot x}$ meets (3.9). We now do this.

We are given $x_0 = (x_{01}, ..., x_{0N}) \in T_N \backslash C(\xi_1) \times \cdots \times C(\xi_N)$. Therefore there exists at least one $x_{0j} \in T_1 \backslash C(\xi_j)$. For simplicity of notation, let us suppose this occurs for $j = 2$, i.e., $x_{02} \in T_1 \backslash C(\xi_2)$. A similar proof will work in case $x_{01} \in T_1 \backslash C(\xi_1)$ or $x_{0j} \in T_1 \backslash C(\xi_j)$ for $j = 3, ..., N$.

Let I_r be the open interval $(-r + x_{02}, x_{02} + r)$. Since $x_{02} \in T_1 \backslash C(\xi_2)$, it follows from the above that there exists $r_0 > 0$ such that $\nu_2(I_{r_0}) = 0$. Hence, we obtain from Theorem 5.3 in Chapter 1 that

$$(3.17) \qquad \lim_{t\to 0} \sum_{n=-\infty}^{\infty} n^{2j} \widehat{\nu}_2(n) e^{ins} e^{-n^2 t} = 0 \quad \text{uniformly for } s \in I_{r_0/2},$$

for $j = 1, ..., N - 1$.

Next, for $s \in T_1$ and for $t > 0$, we set

$$(3.18) \qquad f_j(s,t) = \sum_{n=-\infty}^{\infty} \widehat{\nu}_{0j}(n) e^{ins} e^{-n^2 t} \quad \text{for } j = 1, ..., N,$$

and see that the series in (3.17) is $\partial^j f_2(s,t)/\partial t^j$. Consequently, it follows from the generalized mean-value theorem and (3.17) that

$$(3.19) \qquad \lim_{t\to 0} |f_2(s,t)| / t^{N-1} = 0 \quad \text{uniformly for } s \in I_{r_0/2},$$

Also, we obtain from (3.18) that there exists a constant c such that

$$|f_j(s,t)| \le c t^{-1/2} \quad \text{for } s \in T_1, \text{ for } 0 < t < 1,$$

and for $j = 1, ..., N$.

Observing from (3.16) and (3.18) that

$$\sum_{m\in\Lambda_N} a_m e^{im\cdot x} e^{-|m|^2 t} = f_1(x_1,t) \cdots f_N(x_N,t),$$

we conclude that

$$\left| \sum_{m\in\Lambda_N} a_m e^{im\cdot x} e^{-|m|^2 t} \right| \le c^{N-1} t^{(N-1)/2} f_2(x_2,t)/t^{N-1}$$

for $x \in B(x_0, r_0/2)$ and for $0 < t < 1$.

It follows from this last inequality and (3.19) that

$$\lim_{t \to 0} \sum_{m \in \Lambda_N} a_m e^{im \cdot x} e^{-|m|^2 t} = 0 \text{ uniformly for } x \in B(x_0, r_0/2).$$

So $\sum_{m \in \Lambda_N} a_m e^{im \cdot x}$ meets (3.9), and the proof of the necessary condition is complete. ∎

To prove the sufficiency condition of Theorem 3.2, we will need the following lemma:

Lemma 3.4. *Let $N \geq 2$, and let $E_j \subset T_1$ be closed in the one-dimensional torus topology for $j=1,..,N$. Also, let $\{\lambda_{j,k}\}_{k=1}^\infty$ be a sequence of functions in $\mathcal{D}(T_1)$ having the following four properties:*

(i)$Su^\diamond(\lambda_{j,k}) \subset T_1 \backslash E_j$;

(ii) $\lim_{k \to \infty} \widehat{\lambda}_{j,k}(0) = 1$;

(iii) $\lim_{k \to \infty} \widehat{\lambda}_{j,k}(n) = 0$ for $n \neq 0$;

(iv)$\exists M > 0$ such that $\sum_{n=-\infty}^\infty \left| \widehat{\lambda}_{j,k}(n) \right| \leq M < \infty$,

for $j = 1, ..., N$ and $k = 1, 2, ...$ Furthermore, suppose $S \in \mathcal{A}(T_N)$ has the following property:

(3.20) $S = 0$ in $T_N \backslash E_1 \times \cdots \times E_N$.

Then $S \equiv 0$.

Proof of Lemma 3.4. To prove the lemma, for $i = 1, ...N$, we define

(3.21) $$\eta_{i,k}(x) = \sum_{j_1=1}^N \cdots \sum_{j_i=1}^N \lambda_{j_1,k}(x_{j_1}) \cdots \lambda_{j_i,k}(x_{j_i}),$$

where the sum is over all i-tuples $(j_1,...,j_i)$ from the numbers $(1, ..., N)$ with $j_1 < j_2 < j_3 < \cdots < j_i$, and we set

(3.22) $$\eta_k(x) = \sum_{i=1}^N (-1)^{i+1} \eta_{i,k}(x).$$

In particular, for $N = 3$,

$$\begin{aligned}
\eta_k(x) &= \lambda_{1,k}(x_1) + \lambda_{2,k}(x_2) + \lambda_{3,k}(x_3) - \lambda_{1,k}(x_1)\lambda_{2,k}(x_2) \\
&\quad - \lambda_{1,k}(x_1)\lambda_{3,k}(x_3) - \lambda_{2,k}(x_2)\lambda_{3,k}(x_3) \\
&\quad + \lambda_{1,k}(x_1)\lambda_{2,k}(x_2)\lambda_{3,k}(x_3).
\end{aligned}$$

From (ii) in the hypothesis of the lemma, (3.21), and (3.22), we obtain

$$(3.23) \qquad \lim_{k \to \infty} \widehat{\eta}_k(0) = \sum_{i=1}^{N} (-1)^{i+1} \binom{N}{i} = 1.$$

Likewise, we have for $m_1 \neq 0$ since $\sum_{i=0}^{N-1} (-1)^i \binom{N-1}{i} = 0$, that

$$\widehat{\eta}_k(m_1, 0, ..., 0) = o(1)\widehat{\lambda}_{1,k}(m_1) \text{ as } k \to \infty \text{ uniformly for } |m_1| > 0,$$

and in general that for $m_1 \cdots m_{j_0} \neq 0$, $1 \le j_0 \le N - 1$,

$$\widehat{\eta}_k(m_1, ..., m_{j_0}, 0, ..., 0)$$

$$= o(1)(-1)^{j_0+1}\widehat{\lambda}_{1,k}(m_1) \cdots \widehat{\lambda}_{j_0,k}(m_{j_0}) \text{ as } k \to \infty$$

uniformly for $|m_1 ... m_{j_0}| > 0$ because $\sum_{i=0}^{N-j_0} (-1)^i \binom{N-j_0}{i} = 0$.

On the other hand, $\widehat{\eta}_k(m_1, ..., m_N) = (-1)^{N+1}\widehat{\lambda}_{1,k}(m_1) \cdots \widehat{\lambda}_{N,k}(m_N)$ for $m_1 \cdots m_N \neq 0$.

Summarizing these last two facts, we have that if $m \neq 0$,
(3.24)
$$\widehat{\eta}_k(m) = o(1)(-1)^{r+1}\widehat{\lambda}_{j_1,k}(m_{j_1}) \cdots \widehat{\lambda}_{j_r,k}(m_{j_r}) \quad \text{when} \quad m_1 \cdots m_N = 0$$
$$\text{with } m_{j_1} \cdots m_{j_r} \neq 0 \text{ as } k \to \infty \text{ uniformaly for } |m_{j_1} \cdots m_{j_r}| > 0.$$
$$\widehat{\eta}_k(m) = (-1)^{N+1}\widehat{\lambda}_{1,k}(m_1) \cdots \widehat{\lambda}_{N,k}(m_N) \quad \text{when} \quad m_1 \cdots m_N \neq 0.$$

Next, we see from the hypothesis of the lemma that $E_1 \times \cdots \times E_N \subset T_N$ is a closed set in the torus-topology. Also, we have that $\eta_k \in \mathcal{D}(T_N)$ and from (3.21) and (3.22) that

$$Su^\Diamond(\eta_k) \subset T_N \backslash E_1 \times \cdots \times E_N,$$

for $k = 1, 2, ...$.

It follows from this last fact and (3.20) that

$$S(\eta_k \phi) = 0 \quad \forall \phi \in \mathcal{D}(T_N).$$

Hence the distribution $\eta_k S \equiv 0 \ \forall k$. But then it follows from (3.7) in Appendix B that

$$(3.25) \qquad \sum_{p \in \Lambda_N} \widehat{\eta}_k(p)\widehat{S}(m - p) = 0 \quad \forall m \in \Lambda_N \ \text{ and } \ \forall k.$$

Also, by Proposition 3.3 in Appendix B, the series in (3.25) is absolutely convergent.

Next, we let $\delta_j = -1$ or 1 for $j = 1, ..., N$ and define
$$I_k(\delta_1, ..., \delta_N; m)$$

$$(3.26) \qquad = \sum_{p_1=1}^{\infty} \cdots \sum_{p_N=1}^{\infty} \widehat{\eta}_k(p_1\delta_1, ..., p_N\delta_N)\widehat{S}[m - (p_1\delta_1, ..., p_N\delta_N)].$$

So in particular for $N = 2$, we have $I_k(1, 1; m), I_k(1, -1; m)$, $I_k(-1, 1; m)$, and $I_k(-1, -1; m)$.

It follows from (3.24), (3.25), and (3.26), and (iv) in the hypothesis of the lemma plus the fact that $\widehat{S}(m)$ is uniformly bounded that

$$(3.27) \quad -\widehat{\eta}_k(0)\widehat{S}(m) = \sum_{\delta_1 = \pm 1, \ldots, \delta_N = \pm 1} I_k(\delta_1, \ldots, \delta_N; m) + o(1) \text{ as } k \to \infty.$$

If we show that for each choice of $(\delta_1, \ldots, \delta_N)$,

$$(3.28) \qquad\qquad \lim_{k \to \infty} I_k(\delta_1, \ldots, \delta_N; m) = 0,$$

it will then follow from (3.23) and (3.27) that $\widehat{S}(m) = 0$ for every $m \in \Lambda_N$. But then from Theorem 3.2 in Appendix B, we see that $S \equiv 0$, which is the desired conclusion of the lemma. Therefore the proof of the lemma will be complete once we establish (3.28).

Fix m and let $\varepsilon > 0$ be given. We show (3.28) holds by showing

$$(3.29) \qquad\qquad \limsup_{k \to \infty} |I_k(\delta_1, \ldots, \delta_N; m)| \leq \varepsilon.$$

With M as in condition (iv) in the hypothesis of the lemma, we use (3.5) and choose an integer P sufficiently large so that

$$(3.30) \qquad \left|\widehat{S}(m - p)\right| \leq \varepsilon M^{-N} \quad \text{for} \quad min(|p_1|, \ldots, |p_N|) \geq P.$$

Next, we introduce various sets of lattice points in the positive octant as follows for $j = 1, \ldots, N$:

$$(3.31) \quad \begin{aligned} Q_{j,P} = \{p : 1 \leq p_j \leq P, 1 \leq p_i < \infty, P + 1 \leq p_n < \infty; \\ i = 1, \ldots j - 1 \text{ and } n = j + 1, \ldots N\}. \end{aligned}$$

In case $j = 1$ or N in (3.31), the obvious interpretation is to be given.

Also, we set

$$(3.32) \quad \begin{aligned} Q = \{p : 1 \leq p_i < \infty, i = 1, \ldots, N\}, \\ Q'_P = \{p : P + 1 \leq p_i < \infty, i = 1, \ldots, N\}. \end{aligned}$$

We observe from (3.31) and (3.32) that for $N \geq 2$,

$$(3.33) \qquad\qquad Q = \cup_{j=1}^N Q_{j,P} \cup Q'_P.$$

By hypothesis, $S \in \mathcal{A}(T_N)$. We therefore see from the definition of $\mathcal{A}(T_N)$ and (3.4) that there exists a constant c such that $\left|\widehat{S}(m - p)\right| \leq c \ \forall p$. Consequently, we have from (3.24) that for $p \in Q$,

$$\left|\widehat{\eta}_k(p_1\delta_1, \ldots, p_N\delta_N)\widehat{S}[m - (p_1\delta_1, \ldots, p_N\delta_N)]\right|$$

$$\leq c \left|\widehat{\lambda}_{1,k}(p_1\delta_1) \cdots \widehat{\lambda}_{N,k}(p_N\delta_N)\right|.$$

From condition (iv) in the hypothesis of the lemma, from the definition of $Q_{j,P}$ in (3.31), and from this last inequality, we conclude that

(3.34)
$$\sum_{p\in Q_{j,P}} \left| \widehat{\eta}_k(p_1\delta_1,...,p_N\delta_N)\widehat{S}[m-(p_1\delta_1,...,p_N\delta_N)]\right|$$

$$\leq cM^{N-1}\sum_{p_j=1}^{P}\left|\widehat{\lambda}_{j,k}(p_j\delta_j)\right|,$$

for $j=1,...,N$. But from condition (iii) in the lemma, we have

$$\lim_{k\to\infty}\sum_{p_j=1}^{P}\left|\widehat{\lambda}_{j,k}(p_j\delta_j)\right| = 0.$$

So we infer from the inequality in (3.34) that

$$\lim_{k\to\infty}\sum_{p\in Q_{j,P}}\left|\widehat{\eta}_k(p_1\delta_1,...,p_N\delta_N)\widehat{S}[m-(p_1\delta_1,...,p_N\delta_N)]\right| = 0$$

for $j=1,...,N$. Hence from (3.26) and (3.33), it follows that

(3.35)
$$\limsup_{k\to\infty}|I_k(\delta_1,...,\delta_N;m)|$$
$$\leq \limsup_{k\to\infty}\sum_{p\in Q'_P}\left|\widehat{\eta}_k(p_1\delta_1,...,p_N\delta_N)\widehat{S}[m-(p_1\delta_1,...,p_N\delta_N)]\right|.$$

Next, we see from (3.30) and (3.32) that

$$\left|\widehat{S}[m-(p_1\delta_1,...,p_N\delta_N)]\right|\leq \varepsilon M^{-N} \quad \text{for} \;\; p\in Q'_P$$

and consequently from (3.35) that

(3.36)
$$\limsup_{k\to\infty}|I_k(\delta_1,...,\delta_N;m)|$$
$$\leq \varepsilon M^{-N}\limsup_{k\to\infty}\sum_{p\in Q'_P}|\widehat{\eta}_k(p_1\delta_1,...,p_N\delta_N)|.$$

But from (3.24) and (3.32),

$$\sum_{p\in Q'_P}|\widehat{\eta}_k(p_1\delta_1,...,p_N\delta_N)| = \sum_{p_1=P+1}^{\infty}\cdots\sum_{p_N=P+1}^{\infty}\left|\widehat{\lambda}_{1,k}(p_1\delta_1)\cdots\widehat{\lambda}_{N,k}(p_N\delta_N)\right|.$$

So we conclude from (3.36), condition (iv) in the hypothesis of the lemma, and this last equality that

$$\limsup_{k\to\infty}|I_k(\delta_1,...,\delta_N;m)|\leq \varepsilon.$$

This establishes the inequality in (3.29), and the proof of the lemma is complete. ∎

Proof of the Sufficiency Condition of Theorem 3.2. Let $\{V_k^J\}_{k=1}^\infty$ be a sequence of J-tuples with positive integral entries, i.e., $V_k^J = (v_k^1, ..., v_k^J)$ with v_k^j a positive integer for $j = 1, ..., J$. We say $\{V_k^J\}_{k=1}^\infty$ is a normal sequence provided

$$\lim_{k \to \infty} |b^1 v_k^1 + \cdots + b^J v_k^J| = \infty$$

for every J-tuple $B^J = (b^1, ..., b^J)$ with each entry an integer and at least one entry different from zero.

A set $E \subset T_1$ closed in the one-dimensional torus topology is called an $H^{(J)}$-set provided the following holds:

There is a normal sequence $\{V_k^J\}_{k=1}^\infty$ and a parallelopiped $Q \subset T_J$ where

$$Q = \{x : -\pi < \alpha_j < x_j < \beta_j < \pi, \ j = 1, ..., J\}$$

such that

$$s \in E \implies (sv_k^1, ..., sv_k^J) \in T_J \backslash Q \mod 2\pi \text{ in each entry} \quad \text{for } k = 1, 2, ...,$$

i.e., $(sv_k^1, ..., sv_k^J) \in \mathbf{R}^J \backslash Q^*$ where $Q^* = \cup_{m \in \Lambda_J} \{Q + 2\pi m\}$ and Λ_J is the set of integral lattice points in $\mathbf{R}^J$.

What we want to show is

$$E_j \subset T_1 \text{ is an } H^{(n_j)}\text{-set for } j = 1, ...N,$$
$$\implies E_1 \times \cdots \times E_N \text{ is a set of uniqueness for } \mathcal{A}(T_N).$$

In order to do this, we first prove the following:

Given a set $E \subset T_1$ that is closed in the one-dimensional torus topology and that is also an $H^{(J)}$-set, there is a sequence $\{\lambda_k\}_{k=1}^\infty$ with the following five properties:

(i) $\lambda_k \in \mathcal{D}(T_1)$;

(ii) $Su^\diamond(\lambda_k) \subset T_1 \backslash F$;

(3.37)   (iii) $\lim_{k \to \infty} \widehat{\lambda}_k(0) = 1$;

(iv) $\lim_{k \to \infty} \widehat{\lambda}_k(n) = 0 \ for \ n \neq 0$;

(v) $\exists M > 0 \ such \ that \ \sum_{n=-\infty}^\infty |\widehat{\lambda}_k(n)| \leq M$

where M is a finite number and $k = 1, 2, ...$

To establish (3.37), let $\{V_k^J\}_{k=1}^\infty$ be the normal sequence and $Q \subset T_J$ be the parallelopiped associated with the $H^{(J)}$-set E. Also, let

$$Q_j = (\alpha_j, \beta_j) \subset T_1$$

be the one-dimensional open interval for $j = 1, ...J$ such that

$$Q = Q_1 \times \cdots Q_J.$$

Next, select numbers $\alpha_j', \beta_j', \alpha_j'', \beta_j''$ such that

$$-\pi < \alpha_j < \alpha_j' < \alpha_j'' < \beta_j'' < \beta_j' < \beta_j < \pi, \quad \text{for} \quad j = 1, ..., J,$$

and define the functions of one real variable $\eta_j(s) \in \mathcal{D}(T_1)$ such that

(3.38)
$$\eta_j(s) = 1 \quad \text{in} \quad [\alpha_j'', \beta_j'']$$
$$= 0 \quad \text{in} \quad [-\pi, \pi] \backslash [\alpha_j', \beta_j']$$

and also such that
(3.39)

$$\widehat{\eta}_j(0) = 1 \quad \text{where} \quad \widehat{\eta}_j(n) = (2\pi)^{-1} \int_{-\pi}^{\pi} \eta_j(s) e^{-ins} ds \quad n = 0, \pm 1, \pm 2, ...,$$

for $j = 1, ...J$.

Next, define

(3.40)
$$\lambda_k(s) = \eta_1(v_k^1 s) \cdots \eta_J(v_k^J s)$$

for $k = 1, 2, ...$. Since v_k^j is a positive integer and $\eta_j(s) \in \mathcal{D}(T_1)$, it follows that $\eta_j(v_k^j s) \in \mathcal{D}(T_1)$. Hence, from (3.40), we obtain that $\lambda_k(s) \in \mathcal{D}(T_1)$, and (i) of (3.37) is established.

To establish (ii) of (3.37), set

$$Q_j^* = \cup_{n=-\infty}^{\infty} \{Q_j + 2\pi n\}$$

and observe that

$$Q^* = \cup_{m \in \Lambda} \{Q + 2\pi m\} = Q_1^* \times \cdots \times Q_J^*.$$

Fix k and note that because E is an $H^{(J)}$-set, it follows that for $s \in E$, $(sv_k^1, ..., sv_k^J) \notin Q^*$. Let s_0 be a fixed point in E. Then there exists a j such that $s_0 v_k^j \notin Q_j^*$. For simplicity, say $j = 1$. Then $s_0 v_k^1 \notin Q_1^*$. Also, because Q_1^* is a closed set, $\exists \varepsilon > 0$ such that for $s \in (s_0 - \varepsilon, s_0 + \varepsilon)$, $sv_k^1 \notin Q_1^*$. But then it follows from (3.38) that

$$\eta_1(v_k^1 s) = 0 \quad \text{for} \quad s \in (s_0 - \varepsilon, s_0 + \varepsilon).$$

Hence, we obtain from (3.40) that

$$\lambda_k(s) = 0 \quad \text{for} \quad s \in (s_0 - \varepsilon, s_0 + \varepsilon).$$

We conclude there exists a set G that is open in the torus topology of T_1 such that $E \subset G$ and

$$\lambda_k(s) = 0 \quad \text{for} \quad s \in G.$$

On the other hand, $Su^{\Diamond}(\lambda_k) \subset T_1$ is a set closed in the torus topology of T_1. So we obtain from this last equality that

$$Su^{\Diamond}(\lambda_k) \cap E = \varnothing,$$

and (ii) in (3.37) is established.

To establish the last three items in (3.37), we observe that

$$\eta_j(v_k^j s) = \sum_{n_j=-\infty}^{\infty} \widehat{\eta}_j(n_j) e^{i n_j v_k^j s},$$

for $j = 1, ... J$. Consequently, we see from (3.40) that

$$\lambda_k(s) = \sum_{n=-\infty}^{\infty} \widehat{\lambda}_k(n) e^{ins}$$

where

$$(3.41) \qquad \widehat{\lambda}_k(n) = \sum_{n_1 v_k^1 + \cdots + n_J v_k^J = n} \widehat{\eta}_1(n_1) \cdots \widehat{\eta}_J(n_J)$$

for $k = 1, 2, ...$.

From (3.39) and (3.41), we obtain that

$$\widehat{\lambda}_k(0) = 1 + \widehat{\lambda}'_k(0)$$

where

$$(3.42) \quad \widehat{\lambda}'_k(0) = \sum_{n_1 v_k^1 + \cdots + n_J v_k^J = 0} \widehat{\eta}_1(n_1) \cdots \widehat{\eta}_J(n_J) \quad \text{for} \ \ 0 < |n_1| + \cdots + |n_J|.$$

Let $\varepsilon > 0$ be given. We will establish (iii) in (3.37) by showing

$$(3.43) \qquad \limsup_{k \to \infty} \left| \widehat{\lambda}'_k(0) \right| \leq \varepsilon,$$

and therefore that $\lim_{k \to \infty} \widehat{\lambda}'_k(0) = 0$.

To accomplish (3.43), we first note that there exists a constant c such that

$$(3.44) \qquad \sum_{n_j=-\infty}^{\infty} \left| \widehat{\eta}_j(n_j) \right| \leq c \quad \text{for} \ \ j = 1, ... J.$$

Also, we see there exists an integer $r_0 > 0$ such that

$$(3.45) \qquad \sum_{|n_j|=r_0+1}^{\infty} \left| \widehat{\eta}_j(n_j) \right| \leq \varepsilon c^{-(N-1)} J^{-1} \quad \text{and for} \ \ j = 1, ... J.$$

Next, we set

$$(3.46) \qquad \Lambda_{r_0}^J = \{ (n_1, ..., n_J) \in \Lambda_J \backslash \{0\} : |n_j| \leq r_0 \ \text{for} \ j = 1, ... J \}.$$

Since $\Lambda_{r_0}^J$ is a finite set of nonzero lattice points, we observe from the definition of a normal sequence that there exists a positive integer k_0 such that

$$\left| n_1 v_k^1 + \cdots + n_J v_k^J \right| \geq 1 \ \text{for} \ (n_1, ..., n_J) \in \Lambda_{r_0}^J \ \text{and} \ k \geq k_0.$$

Consequently, if $k \geq k_0$ and $n_1 v_k^1 + \cdots + n_J v_k^J = 0$ then at least one of the n_j is such that $|n_j| \geq r_0 + 1$. Hence, we obtain from (3.42) and (3.44) that

$$
\begin{aligned}
\left| \widehat{\lambda}'_k(0) \right| \leq \{ & \sum_{|n_1|=r_0+1}^{\infty} \sum_{n_2=-\infty}^{\infty} \cdots \sum_{n_J=-\infty}^{\infty} |\widehat{\eta}_1(n_1) \cdots \widehat{\eta}_J(n_J)| \\
& + \sum_{n_1=-\infty}^{\infty} \sum_{|n_2|=r_0+1}^{\infty} \sum_{n_3=-\infty}^{\infty} \cdots \sum_{n_J=-\infty}^{\infty} |\widehat{\eta}_1(n_1) \cdots \widehat{\eta}_J(n_J)| + \cdots \\
& + \sum_{n_1=-\infty}^{\infty} \sum_{n_2=-\infty}^{\infty} \cdots \sum_{n_{J-1}=-\infty}^{\infty} \sum_{|n_J|=r_0+1}^{\infty} |\widehat{\eta}_1(n_1) \cdots \widehat{\eta}_J(n_J)| \} \\
\leq \; & c^{N-1} \sum_{|n_1|=r_0+1}^{\infty} |\widehat{\eta}_1(n_1)| + \cdots + c^{N-1} \sum_{|n_J|=r_0+1}^{\infty} |\widehat{\eta}_J(n_J)|
\end{aligned}
$$

for $k \geq k_0$.

We see from (3.45) and this last inequality that

$$
\left| \widehat{\lambda}'_k(0) \right| \leq \varepsilon \quad \text{for} \quad k \geq k_0.
$$

Therefore (3.43) is indeed true, and (iii) in (3.37) is established.

Next, let $n^* \neq 0$ be an integer, and let $\varepsilon > 0$ be given. We will establish (iv) in (3.37) by showing

$$
(3.47) \qquad\qquad \limsup_{k \to \infty} \left| \widehat{\lambda}_k(n^*) \right| \leq \varepsilon.
$$

where $\widehat{\lambda}_k(n^*)$ is given by (3.41).

To do this, we will again use (3.44), (3.45), and (3.46) with a very similar argument as before and choose $k^* > 0$ so that

$$
\left| n_1 v_k^1 + \cdots + n_J v_k^J \right| \geq |n^*| + 1 \quad \text{for} \quad (n_1, ..., n_J) \in \Lambda_{r_0}^J \quad \text{and} \quad k \geq k^*.
$$

Consequently, if $k \geq k^*$ and $n_1 v_k^1 + \cdots + n_J v_k^J = n^*$ then at least one of the n_j is such that $|n_j| \geq r_0 + 1$. Hence, we obtain from (3.41) and (3.44) that

$$
\begin{aligned}
\left| \widehat{\lambda}_k(n^*) \right| \leq \{ & \sum_{|n_1|=r_0+1}^{\infty} \sum_{n_2=-\infty}^{\infty} \cdots \sum_{n_J=-\infty}^{\infty} |\widehat{\eta}_1(n_1) \cdots \widehat{\eta}_J(n_J)| \\
& + \sum_{n_1=-\infty}^{\infty} \sum_{|n_2|=r_0+1}^{\infty} \sum_{n_3=-\infty}^{\infty} \cdots \sum_{n_J=-\infty}^{\infty} |\widehat{\eta}_1(n_1) \cdots \widehat{\eta}_J(n_J)| + \cdots \\
& + \sum_{n_1=-\infty}^{\infty} \sum_{n_2=-\infty}^{\infty} \cdots \sum_{n_{J-1}=-\infty}^{\infty} \sum_{|n_J|=r_0+1}^{\infty} |\widehat{\eta}_1(n_1) \cdots \widehat{\eta}_J(n_J)| \} \\
\leq \; & c^{N-1} \sum_{|n_1|=r_0+1}^{\infty} |\widehat{\eta}_1(n_1)| + \cdots + c^{N-1} \sum_{|n_J|=r_0+1}^{\infty} |\widehat{\eta}_J(n_J)|
\end{aligned}
$$

for $k \geq k^*$.

We see from (3.45) and this last inequality that

$$\left|\widehat{\lambda}_k(n^*)\right| \leq \varepsilon \quad \text{for} \quad k \geq k^*.$$

This last inequality shows that (3.47) is true and establishes (iv) of (3.37).

It remains to show that (3.37)(v) is true. To do this, we observe from (3.41) that

$$\sum_{n=-\infty}^{\infty} \left|\widehat{\lambda}_k(n)\right| \leq \sum_{n=-\infty}^{\infty} \left| \sum_{n_1 v_k^1 + \cdots + n_J v_k^J = n} \widehat{\eta}_1(n_1) \cdots \widehat{\eta}_J(n_J) \right|$$

$$\leq \sum_{n_1=-\infty}^{\infty} \cdots \sum_{n_J=-\infty}^{\infty} |\widehat{\eta}_1(n_1) \cdots \widehat{\eta}_J(n_J)|$$

$$\leq \sum_{n_1=-\infty}^{\infty} |\widehat{\eta}_1(n_1)| \cdots \sum_{n_J=-\infty}^{\infty} |\widehat{\eta}_J(n_J)|$$

for $k = 1, 2, \ldots$. But then it follows from (3.44) and this last inequality that

$$\sum_{n=-\infty}^{\infty} \left|\widehat{\lambda}_k(n)\right| \leq c^J$$

for $k = 1, 2, \ldots$, and (3.37)(v) is established with $M = c^J$.

From (3.37), we see that if $E_j \subset T_1$ is both closed in the one-dimensional torus topology and an $H^{(n_j)}$-set for $j = 1, \ldots N$, then we have sequences $\{\lambda_{j,k}\}_{k=1}^{\infty}$ that meet the conditions (i)-(iv) in Lemma 3.4. Therefore, it follows from this lemma that the Cartesian product set $E_1 \times \cdots \times E_N$ is a set of uniqueness for the class $\mathcal{A}(T_N)$.

Summarizing, we see that we have established the following result:
(3.48)
$$E_j \subset T_1 \text{ is an } H^{(n_j)} - \text{set for} \quad j = 1, \ldots N,$$
$$\Longrightarrow E_1 \times \cdots \times E_N \text{ is a set of uniqueness for } \mathcal{A}(T_N).$$

Next, let $E_{jl} \subset T_1$ be both closed in the one-dimensional torus topology and an $H^{(n_{jl})}$-set for $l = 1, \ldots, \gamma_j$ and $j = 1, \ldots N$ where n_{jl} is a positive integer. Set

(3.49) $$E_j = \cup_{l=1}^{\gamma_j} E_{jl}$$

for $j = 1, \ldots N$ and observe that

(3.50) $$E_1 \times \cdots \times E_N = \cup_{l_1=1}^{\gamma_1} \cdots \cup_{l_N=1}^{\gamma_N} E_{1l_1} \times \cdots \times E_{Nl_N}.$$

But $E_{1l_1} \times \cdots \times E_{Nl_N}$ is a set of uniqueness for the class $\mathcal{A}(T_N)$ by (3.48). Therefore, from (3.50), $E_1 \times \cdots \times E_N$ is a finite union of sets of uniqueness. Hence, it follows from Proposition 3.1 that

(3.51) $E_1 \times \cdots \times E_N$ is a set of uniqueness for the class $\mathcal{A}(T_N)$,

where E_j is given by (3.49) for $j = 1, ...N$.

Next, we see from [Zy2, pp. 152-3] that because ξ_j^{-1} is an $\mathcal{S}$ number, $C(\xi_j)$ is a finite union of $H^{(n)}$-sets, i.e.,

$$C(\xi_j) = \cup_{l=1}^{\gamma_j} E_{jl}$$

where $E_{jl} \subset T_1$ is both closed in the one-dimensional torus topology and an $H^{(n_{jl})}$-set for $l = 1, ..., \gamma_j$. This last fact is true for $j = 1, ...N$. Therefore, $C(\xi_1) \times \cdots \times C(\xi_N)$ is just like $E_1 \times \cdots \times E_N$ in (3.50). Hence, we conclude from (3.51) that

$$C(\xi_1) \times \cdots \times C(\xi_N) \text{ is a set of uniqueness for the class } \mathcal{A}(T_N),$$

and the sufficiency condition of Theorem 3.2 is established.   ∎

Proof of Corollary 3.3.   To prove the necessary condition of the corollary, we observe from the theorem just established if

$$C(p_1/q_1) \times \cdots \times C(p_N/q_N) \text{ is a set of uniqueness for the class } \mathcal{A}(T_N),$$

then $\frac{q_j}{p_j}$ is an S number for $j = 1, ..., N$ where q_j and p_j are relatively prime positive integers with $q_j > 2p_j$. In particular, $\frac{q_j}{p_j}$ is an algebraic integer. Hence, $p_j = 1$ for $j = 1, ..., N$, and the necessary condition of the corollary is established.

To prove the sufficiency condition of the corollary, we observe that if $q_j > 2$ is a positive integer for $j = 1, ..., N$, then it is an S number. Hence, by Theorem 3.2, $C(1/q_1) \times \cdots \times C(1/q_N)$ is a set of uniqueness for the class $\mathcal{A}(T_N)$, and the sufficiency condition of the corollary is established.   ∎

In the theorem we just established, the sets of uniqueness for the class $\mathcal{A}(T_N)$ were Cartesian product sets. Next, we introduce a strictly smaller class $\mathcal{B}(T_N) \subset \mathcal{A}(T_N)$, which possesses sets of uniqueness that are not Cartesian product sets. Some of these sets arise in the theory of fractals where they are called carpets in two dimensions and fractal foam in three dimensions (see [Man, p. 133]). Examples will be given at the end of this section.

We define the class $\mathcal{B}(T_N) \subset \mathcal{D}'(T_N)$ in the following manner:

$$\mathcal{B}(T_N) = \{S \in \mathcal{D}'(T_N) : \lim_{|m| \to \infty} \widehat{S}(m) = 0\}.$$

It is clear from (3.4) and (3.5) above that $\mathcal{B}(T_N) \subset \mathcal{A}(T_N)$. After the proof of the next theorem, we will present a set $E \subset T_N$ such that E is a set of uniqueness for $\mathcal{B}(T_N)$ but E is not a set of uniqueness for $\mathcal{A}(T_N)$.

We say E is an $H^{\#}$-set provided $E \subset T_N$ is closed in the torus topology and the following holds: There is

(i) a sequence $\{p^k\}_{k=1}^{\infty}$ of integral lattice points $p^k = (p_1^k, ..., p_N^k)$ with

$$p_j^k > 0 \quad \text{and} \quad \lim_{k \to \infty} p_j^k = \infty \text{ for } j = 1, ...N, \quad \text{and}$$

(ii) a parallelopiped $Q \subset T_N$ where

$$Q = \{x : -\pi < \alpha_j < x_j < \beta_j < \pi, j = 1, ..., N\},$$

such that

$$x \in E \Longrightarrow (x_1 p_1^k, ..., x_N p_N^k) \in T_N \backslash Q \mod 2\pi \text{ in each entry}$$

for $k = 1, 2, ...$, i.e., $(x_1 p_1^k, ..., x_N p_N^k) \in \mathbf{R}^N \backslash Q^*$ where

$$Q^* = \cup_{m \in \Lambda_N} \{Q + 2\pi m\}$$

and Λ_N is the set of integral lattice points in $\mathbf{R}^N$.

We intend to prove the following theorem for $H^{\#}$-sets [Sh3]:

Theorem 3.5. *Let $E \subset T_N$ be closed in the torus topology and an $H^{\#}$-set. Then E is a set of uniqueness for the class $\mathcal{B}(T_N)$.*

To prove Theorem 3.5, we will need the following lemma that is similar to Lemma 3.4 but has a much easier proof:

Lemma 3.6. *Let $E \subsetneq T_N$ be closed in the torus topology. Also, let $\{\lambda_k\}_{k=1}^{\infty}$ be a sequence of functions in $\mathcal{D}(T_N)$ having the following four properties:*

(i) $Su^{\diamond}(\lambda_k) \subset T_N \backslash E \quad \forall k;$

(ii) $\lim_{k \to \infty} \widehat{\lambda}_k(0) = \gamma_0 > 0;$

(iii) $\lim_{k \to \infty} \widehat{\lambda}_k(m) = 0 \quad for \quad m \in \Lambda_N, \ m \neq 0;$

(iv) $\exists M > 0 \ \ such \ that \ \ \sum_{m \in \Lambda_N} \left| \widehat{\lambda}_k(m) \right| \leq M < \infty \quad \forall k.$

Suppose, furthermore, that $S \in \mathcal{B}(T_N)$ and that $S = 0$ in $T_N \backslash E$. Then $S \equiv 0$.

Proof of Lemma 3.6. To prove the lemma, we observe from (i) that for each k, the support of λ_k is in the open set $T_N \backslash E$. In addition, $S = 0$ in $T_N \backslash E$. Consequently,

$$S(\lambda_k \phi) = 0 \quad \forall \phi \in \mathcal{D}(T_N) \text{ and } \forall k.$$

Therefore, the distribution $\lambda_k S \equiv 0 \ \ \forall k$. Hence, from Proposition 3.3 in Appendix B, we obtain that

(3.52) $0 = \sum_{p \in \Lambda_N} \widehat{\lambda}_k(p) \widehat{S}(m - p) \quad \text{for } m \in \Lambda_N \text{ and } \forall k.$

Let m_0 be an arbitrary but fixed integral lattice point. If we can show

$$(3.53) \qquad\qquad \widehat{S}(m_0) = 0.$$

it will follow that $\widehat{S}(m) = 0$ for $m \in \Lambda_N$. Theorem 3.2 in Appendix B then tells us that

$$S(\phi) = 0 \qquad \forall \phi \in \mathcal{D}(T_N)$$

and completes the proof of the theorem.

To establish (3.53), we invoke the equality in (3.52) and see that

$$(3.54) \qquad -\widehat{\lambda}_k(0)\widehat{S}(m_0) = \sum_{p \neq 0} \widehat{\lambda}_k(p)\widehat{S}(m_0 - p) \qquad \forall k.$$

Let $\varepsilon > 0$ be given. Then because m_0 is a fixed integral lattice point and $S \in \mathcal{B}(T_N)$, we have the existence of an $s_0 > 1$ such that

$$\left|\widehat{S}(m_0 - p)\right| \leq \varepsilon \gamma_0 / M \quad \text{for} \quad |p| > s_0.$$

From (iv) and (3.54), we then infer that

$$\left|\widehat{\lambda}_k(0)\widehat{S}(m_0)\right| \leq \sum_{1 \leq |p| \leq s_0} \left|\widehat{\lambda}_k(p)\widehat{S}(m_0 - p)\right| + \varepsilon \gamma_0 \quad \forall k.$$

But there are only a finite number of lattice points p involved in the summand of this last inequality. Consequently, on passing to the limit as $k \to \infty$, we obtain from (ii) and (iii) and this last inequality that

$$\gamma_0 \left|\widehat{S}(m_0)\right| \leq \varepsilon \gamma_0.$$

Hence, $\left|\widehat{S}(m_0)\right| \leq \varepsilon$. Since $\varepsilon > 0$ is arbitrary, we have that the equality in (3.53) is indeed true. ∎

Proof of Theorem 3.5. Since E is closed in the torus topology and an $H^\#$-set, it is easy to see that E is a proper subset of T_N, i.e., $E \subsetneq T_N$. Therefore, to establish the theorem, it is sufficient to show the existence of a sequence of functions $\{\lambda_k\}_{k=1}^\infty$ in $\mathcal{D}(T_N)$ having properties (i)-(iv) in Lemma 3.6.

In order to do this, we choose numbers $\alpha_j', \beta_j', \alpha_j'', \beta_j''$ such that

$$-\pi < \alpha_j < \alpha_j' < \alpha_j'' < \beta_j'' < \beta_j' < \beta_j < \pi, \qquad \text{for} \quad j = 1, \dots N,$$

where α_j and β_j are the numbers used in the definition of an $H^\#$-set and set

$$Q' = (\alpha_1', \beta_1') \times \cdots \times (\alpha_N', \beta_N') \text{ and } Q'' = (\alpha_1'', \beta_1'') \times \cdots \times (\alpha_N'', \beta_N'').$$

Then $Q'' \subsetneq Q' \subsetneq Q$ where $Q = (\alpha_1, \beta_1) \times \cdots \times (\alpha_N, \beta_N)$. Also, for $x \in E$

$$(3.55) \qquad (p_1^k x_1, \dots, p_N^k x_N) \notin Q^* \quad \forall k,$$

where $Q^* = \cup_{m \in \Lambda_N}\{Q + 2\pi m\}$ and p_j^k are positive integers for $j = 1, \dots, N$ with $p_j^k \to \infty$.

Proceeding with the proof of the theorem, we select a function $\lambda(x)$, which is in class $C^\infty(\mathbf{R}^N)$ and periodic of period 2π in each variable with the following properties:

$$(i)\ \lambda(x) = 1\ \text{ for } x \in Q'';$$

(3.56)
$$(ii)\ \lambda(x) = 0\ \text{ for } x \in T_N \backslash Q';$$

$$(iii)\ \lambda(x) \geq 0\ \text{ for } x \in T_N.$$

To obtain the sequence $\{\lambda_k\}_{k=1}^\infty$ in $\mathcal{D}(T_N)$, which is alluded to in the paragraph above, we let p_j^k be the positive integers in (3.55) and define

$$(3.57) \qquad \lambda_k(x) = \lambda(p_1^k x_1, ..., p_N^k x_N)\ \ \forall x \in \mathbf{R}^N\ \text{ and }\ \forall k.$$

Then, it is clear from the properties of $\lambda(x)$ that for each k, $\lambda_k(x) \in \mathcal{D}(T_N)$.

To verify that (i) of Lemma (3.6) holds for the sequence, we fix k and observe from (3.55) that given $x_0 \in E$, $\exists s_0 > 0$ then

$$x \in B(x_0, s_0) \Longrightarrow (p_1^k x_1, ..., p_N^k x_N) \notin Q'^*$$

where $Q'^* = \cup_{m \in \Lambda_N} \{Q' + 2\pi m\}$. Hence, we obtain from (3.56)(ii) and (3.57) that there is a set $G \subset T_N$ open in the torus topology with $E \subset G$ such that

$$\lambda_k(x) = 0\ \text{ for }\ x \in G.$$

Since $Su^\diamond(\lambda_k) \subset T_N$ is a closed set in the torus topology, we conclude from this last equality that $Su^\diamond(\lambda_k) \cap E = \varnothing$ and (i) of Lemma (3.6) is established for the sequence $\{\lambda_k\}_{k=1}^\infty$.

To show that the other parts of the lemma prevail for the sequence, we see that $\lambda \in \mathcal{D}(T_N)$ implies that

$$\lambda(x) = \sum_{m \in \Lambda_N} \widehat{\lambda}(m) e^{i(m_1 x_1 + \cdots + m_N x_N)}$$

with the series converging absolutely and uniformly for $x \in \mathbf{R}^N$. Consequently, it follows from (3.57) that

$$(3.58) \qquad \lambda_k(x) = \widehat{\lambda}(0) + \sum_{m \neq 0} \widehat{\lambda}(m) e^{i(p_1^k m_1 x_1 + \cdots + p_N^k m_N x_N)}\ \ \ \forall k,$$

where the series converges absolutely and uniformly for $x \in T_N$.

We recall that $p_j^k > 0$. Consequently, we obtain from (3.58) that

$$\widehat{\lambda}_k(0) = \widehat{\lambda}(0)\ \ \forall k.$$

Also, we see from (3.56) that $\widehat{\lambda}(0) > 0$. So (ii) of Lemma 3.6 is established for the sequence $\{\lambda_k\}_{k=1}^\infty$.

To show that (iii) of the lemma is valid, we let m_0 be an arbitrary but fixed integral lattice point with $m_0 \neq 0$. Since $p_j^k \to \infty$ for $j = 1, ..., N$, it follows that $\exists k_0 > 0$ such that for $k \geq k_0$,

$$\left(\sum_{j=1}^{N} \left| p_j^k m_j \right|^2 \right)^{1/2} \geq \min \left(p_1^k, ..., p_N^k \right) \geq |m_0| + 1$$

for all $m \neq 0$. But then we obtain from (3.58) that

$$\widehat{\lambda}_k(m_0) = 0 \quad \text{for} \quad k \geq k_0.$$

This establishes (iii) of Lemma 3.6 for the sequence $\{\lambda_k\}_{k=1}^{\infty}$.

It remains to show that (iv) of Lemma 3.6 holds for the sequence $\{\lambda_k\}_{k=1}^{\infty}$. To see that this is indeed the case, we observe from (3.58) that

$$\sum_{m \in \Lambda_N} \left| \widehat{\lambda}_k(m) \right| \leq \sum_{m \in \Lambda_N} \left| \widehat{\lambda}(m) \right| \quad \forall k,$$

which gives (iv) of Lemma 3.6 with $M = \sum_{m \in \Lambda_N} \left| \widehat{\lambda}(m) \right|$. So all the conditions in the hypothesis of Lemma 3.6 hold for the sequence $\{\lambda_k\}_{k=1}^{\infty}$ and the proof of the theorem is complete. ∎

Next, with $N \geq 2$, we present a set $E \subset T_N$, which is closed in the torus topology and is a set of uniqueness for the class $\mathcal{B}(T_N)$, but is not a set of uniqueness for the class $\mathcal{A}(T_N)$.

With $0 < \xi < 1/2$, let $C(\xi)$ designate the familiar Cantor set on the half-open interval $[-\pi, \pi) = T_1$ used in the statement of Theorem 3.2. The set E that will qualify for our example is

$$E = C(1/3) \times C(2/5) \times \cdots \times C(2/5).$$

It is clear that E is not a set of uniqueness for $\mathcal{A}(T_N)$ because $5/2$ is not an algebraic integer, and therefore not an S number.

To demonstrate that E is a set of uniqueness for $\mathcal{B}(T_N)$, we observe that $E \subset F$ where

$$F = C(1/3) \times T_1 \times \cdots \times T_1.$$

We will show F is a set of uniqueness for the class $\mathcal{B}(T_N)$, which then implies that E is also a set of uniqueness for the class $\mathcal{B}(T_N)$.

What remains to show by Theorem 3.5 is that F is an $H^{\#}$-set. For the set Q in the definition of an $H^{\#}$-set, we use

$$Q = (-\pi/3, \pi/3) \times \cdots \times (-\pi/3, \pi/3).$$

For our sequence of integral lattice points, we take $p^k = (3^k, ..., 3^k)$. Then given $x = (x_1, ..., x_N) \in F$, we have to demonstrate that for each k, there exists an integer j_k with $1 \leq j_k \leq N$ such that

$$3^k x_{j_k} \notin (-\pi/3, \pi/3) \quad \text{mod } 2\pi \quad \forall k.$$

We will take $j_k = 1$ for every k, and show that for $t \in C(1/3)$,

(3.59) $$3^k t \notin (-\pi/3, \pi/3) \quad \text{mod } 2\pi \quad \forall k.$$

Recall that $C(1/3)$ is the classical Cantor set on the interval $[-\pi, \pi)$. So

$$(3.60) \qquad t = 2\pi \sum_{j=1}^{\infty} 2\varepsilon_j 3^{-j} - \pi \text{ where } \varepsilon_j = 0 \text{ or } 1,$$

and not all $\varepsilon_j = 1$. We set $s = t + \pi$, and see that (3.59) will be satisfied provided that

$$3^k s - \pi \notin (-\pi/3, \pi/3) \mod 2\pi \quad \forall k,$$

which is the same as

$$3^k s \notin (2\pi/3, 4\pi/3) \mod 2\pi \quad \forall k.$$

This last fact is the same as

$$3^k s/2\pi \notin (1/3, 2/3) \mod 1 \quad \forall k.$$

But this last statement is obvious from the representation given in (3.60). Hence F is indeed an $H^{\#}$-set, and our example is complete.

Before giving examples of $H^{\#}$-sets that are not Cartesian product sets, we will establish the following corollary (which for dimension $N = 1$, is established in [Zy1, p. 318]).

Corollary 3.7. *Let $E \subset T_N$ be a set closed in the torus topology and also an $H^{\#}$-set. Then, E is a set of N-dimensional Lebesgue measure zero.*

Proof of Corollary 3.7. Suppose, to the contrary, that E has positive N-dimensional Lebesgue measure. Let χ_E designate the characteristic function (also called the indicator function) of E. Then

$$(3.61) \qquad \widehat{\chi}_E(0) > 0,$$

and from the Riemann-Lebesgue Lemma (Corollary 2.4 in Chapter 1), it follows that

$$\lim_{|m| \to \infty} \widehat{\chi}_E(m) = 0.$$

For $\phi \in \mathcal{D}(T_N)$, we define

$$(3.62) \qquad S(\phi) = \int_{T_N} \chi_E(x)\phi(x)dx.$$

Clearly, $S \in \mathcal{D}'(T_N)$, and from this last limit, we see that it is also in class $\mathcal{B}(T_N)$.

Next, suppose $\phi \in \mathcal{D}(T_N)$ with

$$Su^{\Diamond}(\phi) \subset T_N \backslash E.$$

Then, from (3.62), it follows that $S(\phi) = 0$. But E is a set of uniqueness for the class $\mathcal{B}(T_N)$. Consequently, $S \equiv 0$. In particular,

$$\widehat{\chi}_E(0) = \widehat{S}(0) = 0.$$

This contradicts the inequality in (3.61) and completes the proof to the corollary. ∎

The final theorem we establish in this section, which is partially motivated by [Sh6] (see also [AW2] and [AW3, p. 10]), is the following:

Theorem 3.8.   *Let $0 < \xi_j < 1/2$ for $j = 1, ..., N$ where $N \geq 2$. A necessary and sufficient condition that $C(\xi_1) \times \cdots \times C(\xi_N)$ be a set of uniqueness for the class $\mathcal{B}(T_N)$ is that at least one of $\xi_1^{-1}, ..., \xi_N^{-1}$ must be an $\mathcal{S}$ number.*

Proof of Theorem 3.8. We establish the necessary condition first. We are given $\xi_1^{-1}, ..., \xi_N^{-1}$, which are N positive numbers greater than 2 and none of which are $\mathcal{S}$ numbers. Will show that this implies the existence of a a trigonometric series $\sum_{m \in \Lambda_N} a_m e^{im \cdot x}$ with $a_m = \bar{a}_{-m}$, which has the four properties listed below.

(3.63) $|a_m|$ is uniformly bounded for $m \in \Lambda_N$.

(3.64) $$\lim_{|m| \to \infty} |a_m| = 0.$$

(3.65) $\exists m_0 \in \Lambda_N$ such that $a_{m_0} \neq 0$.

If $x_0 \in T_N \backslash C(\xi_1) \times \cdots \times C(\xi_N)$, then $\exists r_0 > 0$ such that

(3.66) $$\lim_{t \to 0} \sum_{m \in \Lambda_N} a_m e^{im \cdot x} e^{-|m|^2 t} = 0 \quad \text{uniformly for } x \in B(x_0, r_0).$$

Just as in the proof of the necessary condition of Theorem 3.2 (see (3.10)-(3.12)), the existence of a series with properties (3.63)-(3.66) will imply the necessary condition for Theorem 3.8. In particular, (3.64) shows that we are now dealing with the class $\mathcal{B}(T_N)$.

To demonstrate the existence of a series with properties (3.63)-(3.66), we proceed as follows:

With $C(\xi_j) \subset T_1$, $0 < \xi_j < 1/2$, we use the familiar Lebesgue-Cantor function (see [Zy1, p. 194], [Sa, p. 101]) associated with $C(\xi_j)$ to obtain a nonnegative Borel measure ν_j on T_1 with the property that $\nu_j[C(\xi_j)] = 1$ and $\nu_j[T_1 \backslash C(\xi_j)] = 0$. Also, ν_j is nonatomic, i.e., $\nu_j[\{s\}] = 0 \ \forall s \in T_1$. We set

(3.67) $$\widehat{\nu}_j(n) = (2\pi)^{-1} \int_{T_1} e^{-ins} d\nu_j(s) \quad \text{for } n = 0, \pm 1, \pm 2, ...$$

It follows from (3.67) that $\overline{\widehat{\nu}_j(n)} = \widehat{\nu}_j(-n)$ and also that the sequence $\{\widehat{\nu}_j(n)\}_{n=-\infty}^{\infty}$ is uniformly bounded. In particular, it is easy to see that this last fact implies that there is a constant c such that

$$(3.68) \qquad \left| \sum_{n=-\infty}^{\infty} \widehat{\nu}_j(n) e^{ins} e^{-|n|^2 t} \right| \leq ct^{-1/2} \quad \forall s \in T_1 \text{ and for } 0 < t < 1.$$

From [Zy2, p. 151], we also observe that

$$(3.69) \qquad \xi_j^{-1} \text{is not an } \mathcal{S} \text{ number} \implies \lim_{|n| \to \infty} \widehat{\nu}_j(n) = 0.$$

All this was for $j = 1, ..., N$. To obtain the coefficients a_m, we set

$$(3.70) \qquad a_m = \widehat{\nu}_1(m_1) \cdots \widehat{\nu}_N(m_N) \quad \forall m \in \Lambda_N.$$

It is clear from (3.67) and (3.70) that $a_m = \overline{a}_{-m}$ and that the coefficients a_m meet (3.63). This last fact joined with (3.69) and (3.70) shows that the coefficients a_m also meet (3.64). Also, we see from (3.67) joined with (3.70) that $a_0 = (2\pi)^{-N}$. So the coefficients a_m also meet (3.65). It remains to show that the coefficients a_m also meet (3.66). The proof for this fact proceeds exactly as before using (3.17)-(3.19). Hence the necessary condition for Theorem 3.8 is established.

To show that the sufficiency condition holds, for ease of notation, we will prove the theorem for the case $N = 3$. A similar proof prevails for $N = 2$ and for $N \geq 4$.

We are given $C(\xi_1) \times C(\xi_2) \times C(\xi_3)$ where at least one of $\xi_1^{-1}, \xi_2^{-1}, \xi_3^{-1}$ is an $\mathcal{S}$ number. Without loss of generality, we shall suppose ξ_1^{-1} is an $\mathcal{S}$ number. Hence, it follows from [Zy2, pp. 152-3.] that $C(\xi_1)$ is a finite union of $H^{(J)}$-sets. Since the analogue of Proposition 3.1 holds for the class $\mathcal{B}(T_N)$, to show that $C(\xi_1) \times C(\xi_2) \times C(\xi_3)$ is a set of uniqueness for the class $\mathcal{B}(T_3)$, it is sufficient to show that

$$E \times C(\xi_2) \times C(\xi_3) \text{ is a set of uniqueness for the class } \mathcal{B}(T_3)$$

where $E \subset T_1$ is an $H^{(J)}$-set that is closed in the torus topology. This last fact, in turn, will follow if we demonstrate that

$$(3.71) \qquad E \times T_1 \times T_1 \text{ is a set of uniqueness for the class } \mathcal{B}(T_3)$$

where $E \subset T_1$ is an $H^{(J)}$-set closed in the torus topology and J is a positive integer.

So once we show that (3.71) holds the proof of the sufficiency condition, the theorem will be complete. We now do this.

Since $E \times T_1 \times T_1$ is a set closed in the torus topology of T_3, it follows from Lemma 3.6 that (3.71) will be established once we show the existence of a sequence of functions $\{\lambda_k\}_{k=1}^{\infty}$ in $\mathcal{D}(T_3)$ with the following four properties:

(3.72)
$$(i) Su^\lozenge(\lambda_k) \subset T_3 \setminus E \times T_1 \times T_1 \quad \forall k;$$

$$(ii) \lim_{k \to \infty} \widehat{\lambda}_k(0) = \gamma_0 > 0;$$

$$(iii) \lim_{k \to \infty} \widehat{\lambda}_k(m) = 0 \quad for \quad m \in \Lambda_3, \; m \neq 0;$$

$$(iv) \exists M > 0 \; such \; that \; \sum_{m \in \Lambda_3} \left| \widehat{\lambda}_k(m) \right| \leq M < \infty \quad \forall k$$

where Λ_3 represents the set of integral lattice points in $\mathbf{R}^3$.

In order to obtain the sequence $\{\lambda_k\}_{k=1}^\infty$, we use the fact that $E \subset T_1$ is an $H^{(J)}$-set (see the definition above (3.37)) and let $\{V_k^J\}_{k=1}^\infty$ be the normal sequence and $Q \subset T_J$ be the parallelopiped associated with the $H^{(J)}$-set E. Also, let

$$Q_j = (\alpha_j, \beta_j) \subset T_1$$

be the one-dimensional open interval for $j = 1, ...J$ such that

$$Q = Q_1 \times \cdots Q_J.$$

Next, select numbers $\alpha_j', \beta_j', \alpha_j'', \beta_j''$ such that

$$-\pi < \alpha_j < \alpha_j' < \alpha_j'' < \beta_j'' < \beta_j' < \beta_j < \pi, \quad for \quad j = 1, ..., J,$$

and define functions of one real variable $\eta_j(s) \in \mathcal{D}(T_1)$ such that

$$\eta_j(s) = 1 \quad in \quad [\alpha_j'', \beta_j'']$$

(3.73)
$$= 0 \quad in \quad [-\pi, \pi] \setminus [\alpha_j', \beta_j']$$

and also such that

(3.74) $\qquad \widehat{\eta}_j(0) = 1 \quad where \quad \widehat{\eta}_j(n) = (2\pi)^{-1} \int_{-\pi}^{\pi} \eta_j(s) e^{-ins} ds$

for $n = 0, \pm 1, \pm 2, ...,$ and for $j = 1, ...J$. Also, set

(3.75) $\qquad \eta_{J+1}(s) = \eta_{J+2}(s) = \eta_1(s).$

Next, with $x = (x_1, x_2, x_3)$, define

(3.76) $\qquad \lambda_k(x) = \eta_1(v_k^1 x_1) \cdots \eta_J(v_k^J x_1) \eta_{J+1}(3^k x_2) \eta_{J+2}(3^k x_3).$

for $k = 1, 2, ...$ Since v_k^j is a positive integer, it follows that $\lambda_k \in \mathcal{D}(T_3)$. It remains to show that $\{\lambda_k\}_{k=1}^\infty$ meets the conditions (3.72)(i)-(iv).

To establish (i) of (372), set

$$Q_j^* = \cup_{n=-\infty}^\infty \{Q_j + 2\pi n\}$$

and observe that

$$Q^* = \cup_{m \in \Lambda_J} \{Q + 2\pi m\} = Q_1^* \times \cdots \times Q_J^*$$

where Λ_J represents the set of integral lattice points in $\mathbf{R}^J$.

Fix k and note that because E is an $H^{(J)}$-set, it follows that for $x_1 \in E$, $(v_k^1 x_1, ..., v_k^J x_1) \notin Q^*$. Let x_1^0 be a fixed point in E. Then there exists a j such that $v_k^j x_1^0 \notin Q_j^*$. For simplicity, say $j = 1$. Then $v_k^j x_1^0 \notin Q_1^*$. Also, because Q_1^* is a closed set, $\exists \varepsilon > 0$ such that for $x_1 \in (x_1^0 - \varepsilon, x_1^0 + \varepsilon)$, $v_k^1 x_1 \notin Q_1^*$. But then it follows from (3.73) that

$$\eta_1(v_k^1 x_1) = 0 \quad \text{for} \quad x_1 \in (x_1^0 - \varepsilon, x_1^0 + \varepsilon).$$

Hence, we obtain from (3.75) and (3.76) that

$$\lambda_k(x) = 0 \quad \text{for} \quad x_1 \in (x_1^0 - \varepsilon, x_1^0 + \varepsilon) \text{ and } x_2, x_3 \in T_1.$$

We conclude there exists a set G, which is open in the torus topology of T_3 such that $E \times T_1 \times T_1 \subset G$ and

$$\lambda_k(x) = 0 \quad \text{for} \quad x \in G.$$

On the other hand, $Su^\diamond(\lambda_k) \subset T_3$ is a set closed in the torus topology of T_3. So we obtain from this last equality that

$$Su^\diamond(\lambda_k) \cap E \times T_1 \times T_1 = \varnothing,$$

and (i) in (3.72) is established.

To establish the last three items in (3.72), we observe from (3.74) that

$$(3.77) \qquad \eta_j(s) = \sum_{n=-\infty}^{\infty} \widehat{\eta}_j(n) e^{ins} \quad \text{for} \quad s \in T_1,$$

and that there is a constant c such that

$$(3.78) \qquad \sum_{n=-\infty}^{\infty} \left| \widehat{\eta}_j(n) \right| \leq c \quad \text{for} \quad j = 1, ..., J+2.$$

It then follows from (3.76) that

$$(3.79) \qquad \lambda_k(x) = \sum_{m \in \Lambda_J} \widehat{\lambda}_k(m) e^{im \cdot x}$$

where $\widehat{\lambda}_k(m) = 0$ unless there are integers p_{J+1}, p_{J+2} such that

$$m_2 = 3^k p_{J+1}, \ m_3 = 3^k p_{J+2},$$

and there is a $p = (p_1, .., p_J) \in \Lambda_J$ such that

$$m_1 = v_k^1 p_1 + \cdots + v_k^J p_J.$$

If this is the case, then it follows from (3.76) and (3.77) that

$$(3.80) \quad \widehat{\lambda}_k(m) = \widehat{\eta}_{J+1}(p_{J+1}) \widehat{\eta}_{J+2}(p_{J+2}) \Big[\sum_{v_k^1 p_1 + \cdots + v_k^J p_J = m_1} \widehat{\eta}_1(p_1) \cdots \widehat{\eta}_1(p_J) \Big].$$

It is clear from (3.77) and (3.80) that

$$\sum_{m \in \Lambda^3} \left| \widehat{\lambda}_k(m) \right| \leq \sum_{p_1=-\infty}^{\infty} |\widehat{\eta}_1(p_1)| \cdots \sum_{p_{J+2}=-\infty}^{\infty} \left| \widehat{\eta}_{J+2}(p_{J+2}) \right| \leq c^{J+2},$$

which establishes (3.72)(iv) with $M = c^{J+2}$.

Next, we see from (3.74) and (3.80) that

$$(3.81) \qquad \widehat{\lambda}_k(0) = 1 + \widehat{\lambda}'_k(0)$$

where

$$(3.82) \quad \widehat{\lambda}'_k(0) = \sum_{p_1 v_k^1 + \cdots + p_J v_k^J = 0} \widehat{\eta}_1(p_1) \cdots \widehat{\eta}_J(p_J) \quad \text{for} \ \ 0 < |p_1| + \cdots + |p_J| \, .$$

Let $\varepsilon > 0$ be given. We will establish (ii) in (3.72) by showing

$$(3.83) \qquad \limsup_{k \to \infty} \left| \widehat{\lambda}'_k(0) \right| \le \varepsilon,$$

and therefore that $\lim_{k \to \infty} \widehat{\lambda}'_k(0) = 0$.

To accomplish (3.83), we see there exists an integer $r_0 > 0$ such that

$$(3.84) \qquad \sum_{|p_j| = r_0 + 1}^{\infty} \left| \widehat{\eta}_j(p_j) \right| \le \varepsilon c^{-(N-1)} J^{-1} \quad \text{and for} \ \ j = 1, \ldots J$$

where c is the constant in (3.78). Next, we set

$$(3.85) \qquad \Lambda_{r_0}^J = \{ (p_1, \ldots, p_J) \in \Lambda_J \backslash \{0\} : |p_j| \le r_0 \ \text{for} \ j = 1, \ldots J \},$$

and use exactly the same reasoning and same computation used in the paragraph below (3.45) to obtain the inequality in (3.83) from (3.84) and (3.85). Since ε is an arbitrary positive number, it follows then from (3.83) that

$$\lim_{k \to \infty} \widehat{\lambda}'_k(0) = 0.$$

We consequently conclude from (3.81) that

$$\lim_{k \to \infty} \widehat{\lambda}_k(0) = 1,$$

and (ii) of (3.72) is established.

It only remains to show that (3.72) (iii) is valid. To do this, let

$$m^* \in \Lambda_3 \backslash \{0\}$$

be a fixed lattice point, and let $\varepsilon > 0$ be given. We will show that (3.72) (iii) holds by demonstrating that

$$(3.86) \qquad \limsup_{k \to \infty} \left| \widehat{\lambda}_k(m^*) \right| \le \varepsilon.$$

If $m_3^* \ne 0$, then according to the discussion in between (3.79) and (3.80), if no integer p_{J+2} exists such that $m_3^* = 3^k p_{J+2}$, then

$$\widehat{\lambda}_k(m^*) = 0.$$

Obviously, since $m_3^* \ne 0$ is fixed, if k is sufficiently large, there is no such integer p_{J+2}. Consequently,

$$m_3^* \ne 0 \Longrightarrow \lim_{k \to \infty} \widehat{\lambda}_k(m^*) = 0.$$

Similar reasoning applies when $m_2^* \neq 0$. So, in particular, (3.86) is established in these two cases, and therefore we need only consider the situation when

$$m^* = (m_1^*, 0, 0) \text{ where } m_1^* \neq 0.$$

In this case, it follows from (3.74), (3.75), and (3.80) that

$$(3.87) \qquad \widehat{\lambda}_k(m^*) = \sum_{v_k^1 p_1 + \cdots + v_k^J p_J = m_1^*} \widehat{\eta}_1(p_1) \cdots \widehat{\eta}_1(p_J)].$$

With $\Lambda_{r_0}^J$ defined in (3.85), we choose k_0 so that

$$k \geq k_0 \implies \left| v_k^1 p_1 + \cdots + v_k^J p_J \right| \geq |m_1^*| + 1.$$

Consequently, if $k \geq k_0$ and $v_k^1 p_1 + \cdots + v_k^J p_J = m_1^*$, then for at least one p_j,

$$|p_j| \geq r_0 + 1.$$

Hence, we obtain from (3.78) and (3.87) that

$$\begin{aligned}
\left| \widehat{\lambda}_k(m^*) \right| &\leq \Big\{ \sum_{|p_1|=r_0+1}^{\infty} \sum_{p_2=-\infty}^{\infty} \cdots \sum_{p_J=-\infty}^{\infty} |\widehat{\eta}_1(p_1) \cdots \widehat{\eta}_J(p_J)| \\
&\quad + \sum_{p_1=-\infty}^{\infty} \sum_{|p_2|=r_0+1}^{\infty} \sum_{p_3=-\infty}^{\infty} \cdots \sum_{p_J=-\infty}^{\infty} |\widehat{\eta}_1(p_1) \cdots \widehat{\eta}_J(p_J)| + \cdots \\
&\quad + \sum_{p_1=-\infty}^{\infty} \sum_{p_2=-\infty}^{\infty} \cdots \sum_{p_{J-1}=-\infty}^{\infty} \sum_{|p_J|=r_0+1}^{\infty} |\widehat{\eta}_1(p_1) \cdots \widehat{\eta}_J(p_J)| \Big\} \\
&\leq c^{N-1} \sum_{|p_1|=r_0+1}^{\infty} |\widehat{\eta}_1(p_1)| + \cdots + c^{N-1} \sum_{|p_J|=r_0+1}^{\infty} |\widehat{\eta}_J(n_{p_J})|
\end{aligned}$$

for $k \geq k_0$.

We infer from (3.84) and this last set of inequalities that

$$\left| \widehat{\lambda}_k(m^*) \right| \leq \varepsilon \quad \text{for } k \geq k_0.$$

This establishes (3.86), which implies that (3.72)(iii) is valid, and completes the proof of the sufficiency condition to the theorem. ∎

The sets of uniqueness for the class $\mathcal{B}(T_N)$ that we dealt with in Theorem 3.8 were all cartesian product sets. From Theorem 3.5, we also have that $H^\#$-sets are sets of uniqueness for the class $\mathcal{B}(T_N)$, and what is interesting is that there are $H^\#$-sets that are not Cartesian product sets. It turns out that these sets, which we will discuss, also arise in the mathematical theory of fractals. The examples presented here are from the article *Fractals and Distributions on the N-torus* [Sh3].

For ease of notation, we will work on the unit N-torus, T_N^1, which we define as

$$T_N^1 = \{x : 0 \le x_j < 1, \ j = 1, ..., N\}.$$

If $E \subset T_N^1$ is closed in the torus topology of T_N^1, we say it is an $H^\#$-set on T_N^1 provided

(i) there is a sequence $\{p^k\}_{k=1}^\infty$ of integral lattice points $p^k = (p_1^k, ..., p_N^k)$ with

$$p_j^k > 0 \ \text{ and } \ lim_{k \to \infty} p_j^k = \infty \ \text{ for } j = 1, ...N, \ \text{ and}$$

(ii) a parallelopiped $Q \subset T_N^1$ where

$$Q = \{x : 0 < \alpha_j < x_j < \beta_j < 1, \ j = 1, ..., N\},$$

such that

$$x \in E \implies (x_1 p_1^k, ..., x_N p_N^k) \in T_N^1 \backslash Q \mod 1 \text{ in each entry,}$$

for $k = 1, 2, ...$

The first example of an $H^\#$-set that is not a Cartesian product set that we look at arises in dimension $N = 3$, and is referred to by Mandelbrot as triadic fractal foam [Man, p. 133]. We define it on $\bar{T}_3^1$, the closed unit cube in $\mathbf{R}^3$, and refer to it as TFF. The $H^\#$-set of our example will then be

(3.88) $$E = TFF \cap T_3^1.$$

To define TFF, subdivide $\bar{T}_3^1$ into 27 closed congruent cubes by cutting $\bar{T}_3^1$ with planes parallel to the three axes, i.e., $x_j = 1/3, 2/3$ for $j = 1, 2, 3$. Each cube has a distinguished point within it, namely $x^{j_1,1}$, which is the point with the smallest Euclidean norm in each cube. Each $x^{j_1,1}$ corresponds to a unique triple

(3.89) $$x^{j_1,1} \longleftrightarrow (\varepsilon_1, \delta_1, \zeta_1)$$

with $x^{j_1,1} = (\varepsilon_1/3, \delta_1/3, \zeta_1/3)$ where $\varepsilon_1, \delta_1, \zeta_1$ runs through the numbers $0, 1, 2$ with one caveat: we do not allow the triple with $\varepsilon_1 = \delta_1 = \zeta_1 = 1$ since we are going to remove the open cube corresponding to this point. We shall define an ordering on different triples of the nature $(\varepsilon_1, \delta_1, \zeta_1) \ne (\varepsilon_1', \delta_1', \zeta_1')$ as follows:

(3.90) $$(\varepsilon_1, \delta_1, \zeta_1) \prec (\varepsilon_1', \delta_1', \zeta_1') \text{ means}$$

(i) $\varepsilon_1 < \varepsilon_1'$ or (ii) $\varepsilon_1 = \varepsilon_1'$ and $\delta_1 < \delta_1'$ or (iii) $\varepsilon_1 = \varepsilon_1'$ and $\delta_1 = \delta_1'$ and $\zeta_1 < \zeta_1'$. This also imposes an ordering on $\{x^{j_1,1}\}$ via (3.89).

Now we have 26 triples, and we count them out according to this $\prec$-ordering, giving us $\{x^{j_1,1}\}_{j_1=1}^{26}$. Thus, $x^{1,1} = (0,0,0)$, $x^{2,1} = (0,0,1/3)$, $x^{3,1} = (0,0,2/3)$, $x^{4,1} = (0,1/3,0), ..., x^{26,1} = (2/3,2/3,2/3)$. The closed cube, which has $x^{j_1,1}$ as its distinguished point, will have the label $I^{j_1,1}$. We then define $I^1 \subset \bar{T}_3$ to be the closed set

$$I^1 = \cup_{j_1=1}^{26} I^{j_1,1}.$$

In each of the 26 cubes, which have sides of length $1/3$, we now perform the same operation as above, obtaining $(26)^2$ cubes, which now have sides of length $(1/3)^2$. Each of these last mentioned cubes has a distinguished point

$$x^{j_2,2} = x^{j_1,1} + (\varepsilon_2/3^2, \delta_2/3^2, \zeta_2/3^2)$$

where $\varepsilon_2, \delta_2, \zeta_2$ runs through the numbers $0, 1, 2$, and we do not allow the triple with $\varepsilon_2 = \delta_2 = \zeta_2 = 1$. These triples have an ordering imposed on them by (3.89), which, in turn, gives an ordering on $\{x^{j_2,2}\}$ defined as follows:

$x^{j_2,2} \prec x^{j_2',2}$ means

$$(3.91) \qquad (i) x^{j_1,1} \prec x^{j_1',1} \text{ or } (ii) x^{j_1,1} = x^{j_1',1} \text{ and } (\varepsilon_2, \delta_2, \zeta_2) \prec (\varepsilon_2', \delta_2', \zeta_2').$$

We then count out the $(26)^2$ points according to this ordering and obtain $\{x^{j_2,2}\}_{j_2=1}^{(26)^2}$. The closed cube contains $x^{j_2,2}$ as its distinguished point, and we will call it $I^{j_2,2}$. We then define $I^2 \subset I^1 \subset \bar{T}_3$ to be the closed set

$$(3.92) \qquad I^2 = \cup_{j_2=1}^{(26)^2} I^{j_2,2}.$$

In each of the $(26)^2$ cubes that have sides of length $(1/3)^2$, we now perform the same operation as before obtaining $(26)^3$ cubes with each having sides of length $(1/3)^3$. We get distinguished points in each of these cubes and put an ordering on them similar to the procedure in (3.91) to obtain $\{x^{j_3,3}\}_{j_3=1}^{(26)^3}$. Next, in a procedure similar to (3.92), we get the closed set I^3 with $I^3 \subset I^2 \subset I^1 \subset \bar{T}_3$.

Continuing in this manner, we get the decreasing sequence of closed sets $\{I^n\}_{n=1}^\infty$ with $I^{n+1} \subset I^n \subset \bar{T}_3$ where each I^n consists of $(26)^n$ cubes each with sides of length $(1/3)^n$. The closed set TFF is then defined to be

$$(3.93) \qquad TFF = \cap_{n=1}^\infty I^n.$$

With E defined by (3.88) where TFF is defined by (3.93), we see that E is closed in the torus sense because every point in the boundary of $\bar{T}_3^1$ is contained in TFF. We will demonstrate that E is an $H^\#$-set by showing that

$$(3.94) \qquad x \in E \Rightarrow (3^k x_1, 3^k x_2, 3^k x_3) \in E \mod 1 \text{ in each variable}$$

for k, a positive integer, where $x = (x_1, x_2, x_3)$. For recall, the first open cube removed above had sides of length $1/3$ and a distinguished point $(1/3, 1/3, 1/3)$. So if we take the Q in the definition of an $H^\#$-set to be

$$Q = (4/9, 5/9) \times (4/9, 5/9) \times (4/9, 5/9),$$

it will follow from (3.94) that

$$x \in E \Rightarrow (3^k x_1, 3^k x_2, 3^k x_3) \notin Q \mod 1 \text{ in each variable}.$$

Therefore, once (3.94) is established, it will follow that E is indeed an $H^\#$-set

To show that (3.94) holds, it is clearly sufficient to show that it holds in the special case when k=1, i.e.,

(3.95) $x \in E \Rightarrow (3x_1, 3x_2, 3x_3) \in E \mod 1$ in each entry.

It follows from the definition of TFF in (3.93) that given $x_o \in TFF$, $\exists \{x_o^{j_n,n}\}_{n=1}^{\infty}$, where each $x_o^{j_n,n}$ is a distinguished point of one of the $(26)^n$ cubes in I^n of sides $(1/3)^n$ such that

$$\left| x_o^{j_n,n} - x_o \right| \to 0 \text{ as } n \to \infty.$$

Consequently, since E is closed in the torus topology, to show that (3.95) holds, it is sufficient to show that it holds in the special case when x is a distinguished point $x^{j_n,n}$.

If $x = x^{j_1,1}$, then it follows from the enumeration of the 26 such points given below (3.90) that the conclusion in (3.95) holds. Hence from the above, (3.95) will hold if we show the following:

Given $x^{j_n,n} = (x_1^{j_n,n}, x_2^{j_n,n}, x_3^{j_n,n})$, a distinguished point in an $I^{j_n,n}$, then

(3.96) $(3x_1^{j_n,n}, 3x_2^{j_n,n}, 3x_3^{j_n,n}) = x^{j_{n-1},n-1} \mod 1$ in each entry

for $n \geq 2$ where $x^{j_{n-1},n-1}$ is a distinguished point in an $I^{j_{n-1},n-1}$.

It is clear from the representation of $x^{j_2,2}$ given above (3.91) that

$$x^{j_2,2} = \left(\frac{\varepsilon_1}{3} + \frac{\varepsilon_2}{3^2}, \frac{\delta_1}{3} + \frac{\delta_2}{3^2}, \frac{\zeta_1}{3} + \frac{\zeta_2}{3^2} \right)$$

where $\varepsilon_i, \delta_i, \zeta_i$ runs through the numbers $0, 1, 2$, and we do not allow $\varepsilon_i = \delta_i = \zeta_i = 1$ for i=1, 2. Exactly similar reasoning shows that

$$x^{j_n,n} = \left(\sum_{i=1}^{n} \frac{\varepsilon_i}{3^i}, \sum_{i=1}^{n} \frac{\delta_i}{3^i}, \sum_{i=1}^{n} \frac{\zeta_i}{3^i} \right)$$

where now $\varepsilon_i = \delta_i = \zeta_i = 1$ is not allowed for $i = 1, ..., n$. From this last equality, we see that

$$(3x_1^{j_n,n}, 3x_2^{j_n,n}, 3x_3^{j_n,n}) = \left(\varepsilon_1 + \sum_{i=1}^{n-1} \frac{\varepsilon_{i+1}}{3^i}, \delta_1 + \sum_{i=1}^{n-1} \frac{\delta_{i+1}}{3^i}, \zeta_1 + \sum_{i=1}^{n-1} \frac{\zeta_{i+1}}{3^i} \right).$$

But ε_1, δ_1, and ζ_1 are each nonnegative integers, and we conclude from this last computation that (3.96) does indeed hold. Hence E defined by (3.88) is an $H^{\#}$-set, and our example is complete.

Our next example of an $H^{\#}$-set that is not a Cartesian product set will take place in dimension $N = 2$. We will call it a generalized carpet and refer to it as GC_{pq} where $p \geq 3$ and $q \geq 3$ and both are integers. The set GC_{pq} will be a subset of the closed unit square

$$\bar{T}_2^1 = \{x = (x_1, x_2) : 0 \leq x_j \leq 1, \ j = 1, 2\}.$$

In particular, when $p = q = 3$, GC_{pq} is the set referred to in the literature as the Sierpinski carpet [Man, p. 142]. Fractal sets related to GC_{pq} are also discussed in Falconer's book [Fal, p. 129].

To define GC_{pq}, subdivide $\bar{T}_2^1$ into pq closed congruent rectangles by cutting $\bar{T}_2$ with lines parallel to the two axes, as follows:

$$x_1 = 1/p, 2/p, ..., (p-1)/p; \qquad x_2 = 1/q, 2/q, ..., (q-1)/q.$$

Each rectangle has a distinguished point within it, namely $x^{j_1,1}$, which is the point with the smallest Euclidean norm in each rectangle. Each $x^{j_1,1}$ corresponds to a unique double

(3.97) $$x^{j_1,1} \longleftrightarrow (\varepsilon_1, \delta_1)$$

with $x^{j_1,1} = (\varepsilon_1/p, \delta_1/q)$ where ε_1 and δ_1 run through the numbers $0, 1, ...,$ $p-1$ and $0, 1, ..., q-1$, respectively. There is a caveat, however: the doubles with $\varepsilon_1 = 1, ..., p-2$, and simultaneously $\delta_1 = 1, ..., q-2$ are not allowed, for the rectangles corresponding to these points will be removed, i.e., the middle $(p-2)(q-2)$ rectangles will be deleted.

For example, when $p = 5$ and $q = 4$, each of the 6 rectangles with a $\cdot$ in it in the following diagram will be removed:

An ordering on different doubles of the nature $(\varepsilon_1, \delta_1) \neq (\varepsilon_1', \delta_1')$ is then defined as follows:

(3.98) $$(\varepsilon_1, \delta_1) \prec (\varepsilon_1', \delta_1') \text{ means}$$

(i) $\varepsilon_1 < \varepsilon_1'$ or (ii) $\varepsilon_1 = \varepsilon_1'$ and $\delta_1 < \delta_1'$.

This also imposes an ordering on $\{ x^{j_1,1} \}_{j_1=1}^{\gamma}$ via (3.97) where γ is the integer

$$\gamma = pq - (p-2)(q-2).$$

In particular, we see that $x^{1,1} = (0,0)$, $x^{2,1} = (0, 1/q)$, $x^{3,1} = (0, 2/q), ...,$ $x^{\gamma,1} = ((p-1)/p, (q-1)/q)$. We also observe that $x^{q,1} = (0, (q-1)/q)$, $x^{q+1,1} = (1/p, 0)$, and $x^{q+2,1} = (1/p, (q-1)/q)$. The closed rectangle that has $x^{j_1,1}$ as its distinguished point, will have the label $I^{j_1,1}$. We then define $I^1 \subset \bar{T}_2^1$ to be the closed set

$$I^1 = \cup_{j_1=1}^{\gamma} I^{j_1,1}.$$

In each of the γ closed rectangles, which have sides of length $1/p$ and $1/q$, we now perform the same operation as above, obtaining γ^2 closed rectangles, which now have sides of length $(1/p)^2$ and $(1/q)^2$. Each of these last mentioned rectangles has a distinguished point within it, namely $x^{j_2,2}$, where

$$x^{j_2,2} = x^{j_1,1} + (\varepsilon_2/p^2, \delta_2/q^2),$$

and where ε_2 and δ_2 run through the numbers $0, 1, ..., p - 1$ and $0, 1, ...,$ $q - 1$, respectively. Also, we do not allow the doubles with $\varepsilon_2 = 1, ..., p - 2$, and simultaneously $\delta_2 = 1, ..., q - 2$ because the corresponding rectangles have been removed.

The doubles $(\varepsilon_2, \delta_2) \neq (\varepsilon_2', \delta_2')$ have an ordering imposed upon them by (3.98), which, in turn, imposes an ordering on the distinguished points given by $x^{j_2,2} \prec x^{j_2',2}$ akin to the ordering given in (3.91). We count out the γ^2 points according to this ordering and obtain $\{x^{j_2,2}\}_{j_2=1}^{\gamma^2}$. The closed rectangle of sides $(1/p)^2$ and $(1/q)^2$ containing $x^{j_2,2}$ as its distinguished point, will be called $I^{j_2,2}$. We then define $I^2 \subset I^1 \subset \bar{T}_2^1$ to be the closed set

$$I^2 = \cup_{j_2=1}^{\gamma^2} I^{j_2,2}.$$

Continuing in this manner, we get the decreasing sequence of closed sets $\{I^n\}_{n=1}^{\infty}$ with $I^{n+1} \subset I^n \subset \bar{T}_2^1$ where each I^n consists of γ^n rectangles each with sides of length $(1/p)^n$ and $(1/q)^n$. The set GC_{pq} is then defined to be

$$(3.99) \qquad GC_{pq} = \cap_{n=1}^{\infty} I^n.$$

Next, we define E to be the set

$$(3.100) \qquad E = GC_{pq} \cap T_2^1,$$

and observe that E is closed in the torus sense because every point in the boundary of $\bar{T}_2$ is contained in GC_{pq}.

We want to show that E is an $H^{\#}$-set. So we take Q to be the open rectangle

$$Q = (\frac{1}{p} + \frac{1}{p^2}, \frac{2}{p} - \frac{1}{p^2}) \times (\frac{1}{q} + \frac{1}{q^2}, \frac{2}{q} - \frac{1}{q^2}).$$

We will demonstrate that

$$(3.101) \qquad x \in E \Rightarrow (p^k x_1, q^k x_2) \in E \quad \text{mod } 1 \text{ in each variable}$$

for k, a positive integer and with $x = (x_1, x_2)$. Once (3.77) is established, then it follows from the way that E was constructed that

$$x \in E \Rightarrow (p^k x_1, q^k x_2) \notin Q \quad \text{mod } 1 \text{ in each variable}$$

$\forall k$, and consequently, that E is indeed an $H^{\#}$-set.

To show that (3.77) holds, it is clearly sufficient to show that it holds in the special case when $k = 1$, i.e.,

$$(3.102) \qquad x \in E \Rightarrow (p x_1, q x_2) \in E \quad \text{mod } 1 \text{ in each variable}.$$

Using the same argument that we used after (3.95), we see that to show that E is an $H^{\#}$-set, it is sufficient to show that (3.102) holds for the special case when $x = x^{j_n,n}$, a distinguished point in one of the closed rectangles $I^{j_n,n}$ with sides $(1/p)^n$ and $(1/q)^n$.

If $x = x^{j_1,1}$, then it follows from the enumeration of such points below (3.98) that (3.102) does indeed hold. Hence, to show that E is an $H^{\#}$-set, it only remains to establish the fact that (3.102) holds when $x = x^{j_n,n}$ for $n \geq 2$. This will be accomplished if we show the following fact prevails:

Given $x^{j_n,n} = (x_1^{j_n,n}, x_2^{j_n,n})$ a distinguished point in an $I^{j_n,n}$, then

(3.103) $\qquad (px_1^{j_n,n}, qx_2^{j_n,n}) = x^{j_{n-1},n-1} \mod 1$ in each variable

for $n \geq 2$ where $x^{j_{n-1},n-1}$ is a distinguished point in an $I^{j_{n-1},n-1}$.

It is clear from the representation of $x^{j_2,2}$ given above that

$$x^{j_2,2} = \left(\frac{\varepsilon_1}{p} + \frac{\varepsilon_2}{p^2}, \frac{\delta_1}{q} + \frac{\delta_2}{q^2} \right)$$

where ε_i and δ_i run through the numbers $0, ..., p-1$, and $0, ..., q-1$, respectively, and we do not allow $\varepsilon_i = 1, ..., p-2$ and simultaneously $\delta_i = 1, ..., q-2$ for i=1, 2. Exactly similar reasoning shows that

(3.104) $\qquad x^{j_n,n} = \left(\sum_{i=1}^{n} \frac{\varepsilon_i}{p^i}, \sum_{i=1}^{n} \frac{\delta_i}{q^i} \right)$

where ε_i and δ_i are exactly as before with now $i = 1, ..., n$.

From (3.104), we see that

$$(px_1^{j_n,n}, qx_2^{j_n,n}) = \left(\varepsilon_1 + \sum_{i=1}^{n-1} \frac{\varepsilon_{i+1}}{p^i}, \delta_1 + \sum_{i=1}^{n-1} \frac{\delta_{i+1}}{q^i} \right).$$

But ε_1 and δ_1 are each nonnegative integers, and we conclude from this last equality and (3.104) that (3.103) does indeed hold. Hence, E defined by (3.100) is an $H^{\#}$-set, and our example is complete.

Exercises.

1. In dimension $N = 2$, use the method of sequences to define the notion $E \subset T_2$ is a closed set in the torus topology where

$$T_2 = \{x : -\pi \leq x_j < \pi, \; j = 1, 2\}.$$

Prove $E \subset T_2$ is a closed set in the torus topology if and only if E^* is a closed set in $\mathbf{R}^2$ where

$$E^* = \cup_{m \in \Lambda_2} \{E + 2\pi m\}.$$

2. With $m \in \Lambda_2$, find a sequence $[a_m]_{m \in \Lambda_2}$ such that

$$\lim_{\min(|m_1||m_2|) \to \infty} a_m = 0$$

but $\lim_{|m| \to \infty} a_m = 0$ is false.

3. In dimension $N = 4$, find $\eta_k(x)$ where $\eta_k(x)$ is defined in (3.22) and prove that

$$\widehat{\eta}_k(m) = (-1)^{N+1} \widehat{\lambda}_{1,k}(m_1) \cdots \widehat{\lambda}_{N,k}(m_4) \quad \text{when} \quad m_1 \cdots m_4 \neq 0.$$

4. In dimension $N = 3$, show that with $P = 8$,

$$Q = \cup_{j=1}^{3} Q_{j,P} \cup Q'_P$$

where $Q_{j,P}$ is defined in (3.31) and Q and Q'_P are defined in (3.32).

5. Let $C(\xi)$ designate the familiar Cantor set on the half-open interval $[-\pi, \pi) = T_1$. Prove that in dimension $N = 2$, $C(1/5) \times C(2/5)$ is an $H^{\#}$-set.

4. Further Results and Comments

1. Zygmund proved an extension of Theorem 2.2, which is Cooke's elegant result. Zygmund in [Zy3] established the following theorem, which for two dimensions is a good generalization of the classical Cantor-Lebesgue lemma.

Theorem. *Given the series $\sum_{m \in \Lambda_2} b_m e^{im \cdot x}$ where $\bar{b}_m = b_{-m}$ and dimension $N = 2$, set $B_n(x) = \sum_{|m|^2 = n} b_m e^{im \cdot x}$, and suppose*

$$\lim_{n \to \infty} B_n(x) = 0 \quad \text{for } x \in E$$

where $E \subset T_2$ and $|E| > 0$. Then

$$\lim_{n \to \infty} \sum_{|m|^2 = n} |b_m|^2 = 0.$$

2. A uniqueness theorem for harmonic functions making use of some of the ideas in Theorem 1.1 appeared in 2002 in the American Mathematical Monthly [Sh17]. It was the following result:

Theorem. *Let $u(x)$ be harmonic in $B(0,1)$ where $B(0,1) \subset R^2$ is the unit 2-ball. Set $U(r, \theta) = u(r \cos \theta, r \sin \theta)$ and suppose*

$$(i) \lim_{r \to 1} U(r, \theta) = 0 \quad for \quad -\pi \le \theta < \pi.$$

$$(ii) \max_{-\pi \le \theta < \pi} |U(r, \theta)| = o\left(\frac{1}{(1-r)^2}\right) \quad as \ r \to 1.$$

Then $u(x)$ is identically zero in $B(0,1)$.

This result is false if (i) is replaced by (i') where

$$(i') \lim_{r \to 1} U(r, \theta) = 0 \quad for \quad -\pi \le \theta < 0 \text{ and } 0 < \theta < \pi,$$

as the familiar function $P(r, \theta) = \frac{1 - r^2}{1 - 2r \cos \theta + r^2}$ illustrates.

Likewise, the theorem is false if (ii) is replaced by (ii') where

$$(ii') \max_{-\pi \le \theta < \pi} |U(r, \theta)| = O\left(\frac{1}{(1-r)^2}\right) \quad as \ r \to 1,$$

as the function

$$\frac{\partial P}{\partial \theta}(r, \theta) = -\frac{2r(1 - r^2) \sin \theta}{(1 - 2r \cos \theta + r^2)^2}$$

illustrates. It is clear that $lim_{r \to 1} \frac{\partial P}{\partial \theta} (r, \theta) = 0 \ \ \forall \theta$ and that

$$\lim_{r \to 1} \frac{\partial P}{\partial \theta} (r, 1 - r) = 1.$$

So, indeed (ii) cannot be replaced by (ii').

For the details of the proof of the above theorem, we refer the reader to the monthly article [Sh17].

The main thrust of the monthly article is that both Riemann and Gauss were capable of conjecturing the above theorem but probably would not have been capable of proving it. The main tool needed in its proof is the Baire Category Theorem, which was not discovered until 1899, which was 33 years after the death of Riemann.

CHAPTER 4

Positive Definite Functions

1. Positive Definite Functions on $\mathbf{S}_{N-1}$

Spherical harmonic functions can be used to solve problems in discrete geometry. In particular, Oleg Musin gave a new proof for the kissing number $k(3)$ for spheres using positive definite functions and spherical harmonics in his paper [Mu], which appeared in 2006 in the journal Discrete and Computational Geometry. In three dimensions, the kissing number problem is to show that no more than twelve white billiard balls can simultaneously kiss (touch) a black billiard ball. This problem goes back to Isaac Newton in 1694 and was not completely solved until 1953. Musin has given a new proof that $k(3) = 12$, which makes strong use of part of Schoenberg's theorem involving surface spherical harmonics on S_2.

Since Schoenberg's theorem [Sch, p. 101] does not appear in the books [AAR], [ABR], or [EMOT], we will present it here on S_{N-1}, $N \geq 3$, making use of the Gegenbauer polynomials. These polynomials and the theorem on S_{N-1} evidently are also useful in dealing with the higher dimensional kissing numbers and possibly other problems in discrete geometry. For all this, we refer the reader to the 2004 article in the Notices of the AMS by Pfender and Ziegler, [PZ].

Let $f(t)$ be a real-valued continuous function defined on the interval $[-1, 1]$. We say f is positive definite on S_{N-1} if the following prevails for every positive integer n:

$$(1.1) \qquad \sum_{j=1}^{n} \sum_{k=1}^{n} f(\xi^j \cdot \xi^k) b_j b_k \geq 0,$$

for $\xi^1, ..., \xi^n \in S_{N-1}$ and for numbers $b_1, ..., b_n \in \mathbf{R}$.

We intend to establish a theorem connecting a positive definite f on S_{N-1} with its Gegenbauer-Fourier series. So for $f \in C([-1, 1])$, using Appendix A, (3.41)-(3.44), we set

$$(1.2) \qquad a_n = \tau_n^{-1} \int_{-1}^{1} C_n^\nu(t) f(t) (1 - t^2)^{\nu - \frac{1}{2}} dt,$$

where $\nu = (N - 2)/2$ and

$$(1.3) \qquad \tau_n = C_n^\nu(1) \frac{\Gamma(\frac{1}{2}) \Gamma(\nu + \frac{1}{2})}{(\nu + n) \Gamma(\nu)}.$$

In Appendix A, Theorem 3.5, we also show that the system of Gegenbauer polynomials, $\{C_n^\nu(t)\}_{n=0}^\infty$, is a complete orthogonal system for

$$L_\mu^2([-1,1])$$

where $\mu = (1-t^2)^{\nu-\frac{1}{2}}$. Here, we establish the following theorem for positive definite functions on S_{N-1} (which is due to Schoenberg [Sch] for $N = 3$).

Theorem 1.1. *Let* $f \in C([-1,1])$, *and suppose* $\nu = (N-2)/2$ *where* $N \geq 3$. *Define* a_n *by the formula in* (1.2). *Then a necessary and sufficient condition that f be positive definite on* S_{N-1} *is that*

(i) $a_n \geq 0 \ \forall n$,

(ii) $f(t) = \displaystyle\lim_{n\to\infty} \sum_{j=0}^n a_j C_j^\nu(t)$ *uniformly for* $t \in [-1,1]$.

Proof of Theorem 1.1. We first show that for each nonnegative integer n, $C_n^\nu(t)$ is a positive definite function on S_{N-1}. To accomplish this, we let $\{Y_{j,n}(\xi)\}_{j=1}^{\mu_{n,N}}$ be an orthonormal set of surface spherical harmonics of degree n, as in (3.8) of Appendix A. It then follows from the addition formula for surface spherical harmonics given in (3.9) of Appendix A that

$$(1.4) \qquad\qquad C_n^\nu(\xi \cdot \eta) = \gamma_N \sum_{l=1}^{\mu_{n,N}} Y_{l,n}(\xi) Y_{l,n}(\eta),$$

where γ_N is a positive constant. Consequently,

$$\sum_{j=1}^M \sum_{k=1}^M C_n^\nu(\xi^j \cdot \xi^k) b_j b_k = \gamma_N \sum_{l=1}^{\mu_{n,N}} \left(\sum_{j=1}^M Y_{l,n}(\xi^j) b_j\right)^2 \geq 0,$$

and we conclude from (1.1) that indeed

$$(1.5) \qquad\qquad C_n^\nu(t) \text{ is a positive definite function on } S_{N-1},$$

for every n.

To establish the sufficiency condition of the theorem, we observe from the definition given for positive definite functions in (1.1) that a finite linear combination of positive definite functions is clearly positive definite. Since by hypothesis (i), $a_n \geq 0$, it follows from (1.5) that

$$\sum_{j=0}^n a_j C_j^\nu(t) \text{ is a positive definite function.}$$

Since the uniform limit of a sequence of positive definite functions is positive definite, we see from hypothesis (ii) in the theorem that $f(t)$ is also positive definite, and the sufficiency condition of the theorem is established.

To prove the necessary condition, we first observe that if $g \in C([-1,1])$ and is also positive definite on S_{N-1}, then

$$(1.6) \qquad \int_{S_{N-1}} \int_{S_{N-1}} g(\eta \cdot \xi) dS(\eta) dS(\xi) \geq 0.$$

To see that this is the case, we use the definition of the Riemann integral and take a sequence of partitions $\{\mathcal{P}_n\}_{n=1}^{\infty}$ of S_{N-1} where

$$\mathcal{P}_n = \{P_j^n\}_{j=1}^{\alpha_n} \quad \text{and} \quad S_{N-1} = \cup_{n=1}^{\alpha_n} P_j^n$$

and the diameter of P_j^n goes to zero as $n \to \infty$. Also, $\alpha_n \to \infty$. In each P_j^n, we choose a $\xi^{j,n} \in P_j^n$. Then

$$\int_{S_{N-1}} g(\eta \cdot \xi) dS(\xi) = \lim_{n \to \infty} \sum_{j=1}^{\alpha_n} g(\eta \cdot \xi^{j,n}) \left| P_j^n \right|$$

where $\left| P_j^n \right|$ designates the $N-1$-volume of P_j^n. Likewise,

$$(1.7) \qquad \begin{aligned} &\int_{S_{N-1}} \int_{S_{N-1}} g(\eta \cdot \xi) dS(\eta) dS(\xi) = \\ &\lim_{n \to \infty} \sum_{j=1}^{\alpha_n} \sum_{k=1}^{\alpha_n} g(\xi^{j,n} \cdot \xi^{k,n}) \left| P_j^n \right| |P_k^n|. \end{aligned}$$

Since g is a positive definite function, it follows that the double sum on the right-hand side of the equality in (1.7) is nonnegative. Consequently, the limit in (1.7) is nonnegative, and (1.6) is established.

Next, with f as our given positive definite function and $\eta \in S_{N-1}$, we see, after introducing a spherical coordinate system as in §3 of Chapter 1 with η as the pole, that

$$\begin{aligned} \int_{S_{N-1}} f(\eta \cdot \xi) C_n^{\nu}(\eta \cdot \xi) \ dS(\xi) &= |S_{N-2}| \int_0^{\pi} f(\cos \theta) C_n^{\nu}(\cos \theta)(\sin \theta)^{2\nu} d\theta \\ &= |S_{N-2}| \int_{-1}^{1} f(t) C_n^{\nu}(t)(1 - t^2)^{\nu - \frac{1}{2}} dt. \end{aligned}$$

Consequently, we obtain from (1.2) that

$$(1.8) \qquad a_n = \gamma_n^* \int_{S_{N-1}} \int_{S_{N-1}} f(\eta \cdot \xi) C_n^{\nu}(\eta \cdot \xi) dS(\eta) dS(\xi),$$

where γ_n^* is a positive constant.

However,

$$(1.9) \qquad f(t) C_n^{\nu}(t) \text{ is a positive definite function on } S_{N-1}.$$

So it follows from (1.6) and (1.8) that

$$(1.10) \qquad a_n \geq 0 \ \forall n,$$

and condition (i) in the necessary part of the theorem is established.

To see that the statement in (1.9) is true, we observe from (1.4) that

$$\sum_{j=1}^{M} \sum_{k=1}^{M} f(\xi^j \cdot \xi^k) C_n^\nu(\xi^j \cdot \xi^k) b_j b_k$$

$$= \gamma_n \sum_{l=1}^{\mu_{n,N}} \{ \sum_{j=1}^{M} \sum_{k=1}^{M} f(\xi^j \cdot \xi^k) b_j b_k Y_{l,n}(\xi^j) Y_{l,n}(\xi^k) \}.$$

By positive definiteness, the double sum inside the braces on the right-hand side in this last equality is nonnegative for every l. Hence, the left-hand side is nonnegative, and the statement in (1.9) is indeed true.

Next, we see from (3.30) in Appendix A and from (1.2) and (1.3) above, that the series

$$(1.11) \qquad \sum_{n=0}^{\infty} a_n C_n^\nu(1) r^n = h(r)$$

converges uniformly for $0 \le r \le r_0$ where $r_0 < 1$.

Recalling that $\nu = \frac{N-2}{2}$, from (1.2) and (1.3), we also see that

$$(1.12) \qquad a_n C_n^\nu(1) = \beta_\nu \frac{(N-2+2n)}{N-2} \int_{-1}^{1} f(t) C_n^\nu(t)(1-t^2)^{\nu-\frac{1}{2}} dt,$$

where β_ν is a positive constant. So from (3.29) in Appendix A, we obtain that

$$h(r) \le \beta_\nu \|f\|_{L^\infty} \int_{-1}^{1} \frac{1-r^2}{(1-2rt+r^2)^{N/2}} (1-t^2)^{\nu-\frac{1}{2}} dt$$

for $0 < r < 1$.

Since $C_0^\nu(t) \equiv 1$, we see from Proposition 3.3 and (3.41) in Appendix A that the integral in this last inequality is less than or equal to 2 for $0 < r < 1$. Hence,

$$(1.13) \qquad h(r) \le 2\beta_\nu \|f\|_{L^\infty} \quad \text{for } 0 < r < 1.$$

But each term in the series defining $h(r)$ in (1.11) is nonnegative. So we conclude from (1.13) that

$$\sum_{n=0}^{\infty} a_n C_n^\nu(1) < \infty.$$

Next, from (3.30) in Appendix A, we have that $|C_n^\nu(t)| \le C_n^\nu(1)$ for $t \in [-1,1]$. Consequently, it follows from this last inequality that the series

$$\sum_{n=0}^{\infty} a_n C_n^\nu(t)$$

converges uniformly for $t \in [-1,1]$.

From Corollary 3.6 in Appendix A, we see furthermore that the series converges uniformly to $f(t)$. So,

$$f(t) = \lim_{n\to\infty} \sum_{j=0}^{n} a_j C_j^\nu(t) \quad \text{uniformly for } t \in [-1,1],$$

and condition (ii) in the theorem is established. ∎

Exercises.

1. Given $\xi \in S_2$ is of the form $\xi = (\cos\theta, \sin\theta\cos\phi, \sin\theta\sin\phi)$, prove directly that $P_1(t) = t$ is positive definite on S_2.

2. Prove that

$$Y_1(\theta, \phi) = \cos\theta\sin\theta\cos\phi,$$
$$Y_2(\theta, \phi) = \cos\theta\sin\theta\sin\phi,$$
$$Y_3(\theta, \phi) = \cos^2\theta - (\sin\theta\cos\phi)^2,$$
$$Y_4(\theta, \phi) = \cos^2\theta - (\sin\theta\sin\phi)^2$$

constitute a set of four surface spherical harmonics of degree 2 that are orthogonal on S_2.

3. Find a nonzero surface spherical harmonics of degree 2, $Y_5(\theta, \phi)$, which is orthogonal to $Y_j(\theta, \phi)$ on S_2. given in Exercise 2, $j = 1, 2, 3, 4$. Using this new set of 5 in conjunction with Theorem 3.4 of Appendix A, prove that $P_2(t) = \frac{1}{2}(3t^2 - 1)$ is positive definite on S_2.

4. Prove that $\int_{-1}^{1} \frac{1-r^2}{(1-2rt+r^2)^{N/2}}(1-t^2)^{\nu-\frac{1}{2}}dt \leq 2$ for $0 < r < 1$ where $\nu = \frac{N-2}{2}$ and $N \geq 3$.

5. Given $f \in C([-1, 1])$ and that $f(t)$ is positive definite on S_3, let $\eta^* = (1, 0, 0, 0)$ and $g(\xi) = f(\eta^* \cdot \xi)$ for $\xi \in S_3$. Find a function

$$u(x) \in C^\infty(B(0,1)) \cap C(\overline{B}(0,1))$$

such that

$$\Delta u(x) = 0 \qquad \forall x \in B(0,1)$$
$$u(\xi) = g(\xi) \qquad \forall \xi \in S_3,$$

where

$$\Delta u = \frac{\partial^2 u}{\partial x_1^2} + \frac{\partial^2 u}{\partial x_2^2} + \frac{\partial^2 u}{\partial x_3^2} + \frac{\partial^2 u}{\partial x_4^2},$$

$B(0,1)$ is the open unit 4-ball, and $\overline{B}(0,1)$ is its closure.

2. Positive Definite Functions on $\mathbf{T}_N$

As before, we say $f \in C(T_N)$ provided that $f(x)$ is a real-valued function in $C(\mathbf{R}^N)$, and f is periodic of period 2π in each variable. All the positive definite functions that we deal with in this section will be in $C(T_N)$. Motivated by Schoenberg's theorem on S_{N-1}, we shall establish an analogous result for positive definite functions on T_N.

In particular, $f(x) \in C(T_N)$ will be called a positive definite function on T_N if it meets the following two conditions:

(i) $f(x) = f(-x) \; \forall x \in T_N$;

(ii) $\forall n \geq 1$, given $x^1, ..., x^n \in T_N$ and $b_1, ..., b_n \in \mathbf{R}$,

(2.1)
$$\sum_{j=1}^{n} \sum_{k=1}^{n} f(x^j - x^k) b_j b_k \geq 0.$$

If f meets condition (i) above, we will call f an even function. We recall that

(2.2)
$$\widehat{f}(m) = (2\pi)^{-N} \int_{T_N} e^{-im \cdot x} f(x) dx$$

for $m \in \Lambda_N$ where Λ_N is the set of integral lattice points in $\mathbf{R}^N$.

Theorem 2.1. *Let f be a real-valued function and suppose $f \in C(T_N)$, $N \geq 1$. Define $\widehat{f}(m)$ by the formula in (2.2). Then a necessary and sufficient condition that f be positive definite on T_N is that*

(i) $\widehat{f}(m) \geq 0$ *for $m \in \Lambda_N$,*

(ii) $f(x) = \lim_{R \to \infty} \sum_{|m| \leq R} \widehat{f}(m) e^{im \cdot x}$ *uniformly for $x \in T_N$.*

This theorem is essentially due to S. Bochner (see [Ru3, p. 19]). The proof that we give here is based on the proof of Schoenberg's theorem given in §1.

Proof of Theorem 2.1.   We first show that for each fixed $m \in \Lambda_N$, $\cos(m \cdot x)$ is a positive definite function on T_N. To see this, let $x^1, ..., x^n \in T_N$ and $b_1, ..., b_n \in \mathbf{R}$ and observe that

$$\cos[m \cdot (x^j - x^k)] = \cos(m \cdot x^j) \cos(m \cdot x^k) + \sin(m \cdot x^j) \sin(m \cdot x^k).$$

Consequently,

$$\sum_{j=1}^{n} \sum_{k=1}^{n} \cos[m \cdot (x^j - x^k)] b_j b_k = [\sum_{j=1}^{n} b_j \cos(m \cdot x^j)]^2 + [\sum_{j=1}^{n} b_j \sin(m \cdot x^j)]^2$$
$$\geq 0.$$

So indeed $\cos(m \cdot x)$ is positive definite on T_N.

Next, we observe that if $g \in C(T_N)$ is positive definite on T_N, then

(2.3)
$$\int_{T_N} \int_{T_N} g(x - y) dx dy \geq 0.$$

To see that this is the case, we use the definition of the Riemann integral and take a sequence of partitions $\{\mathcal{P}_n\}_{n=1}^{\infty}$ of T_N where

$$\mathcal{P}_n = \{P_j^n\}_{j=1}^{\alpha_n} \quad \text{and} \quad T_N = \cup_{n=1}^{\alpha_n} P_j^n$$

and the diameter of P_j^n goes to zero as $n \to \infty$. Also, $\alpha_n \to \infty$. In each P_j^n, we choose a $x^{j,n} \in P_j^n$. Then

$$\int_{T_N} g(x-y)dx = \lim_{n \to \infty} \sum_{j=1}^{\alpha_n} g(x^{j,n}-y)\left|P_j^n\right|$$

where $\left|P_j^n\right|$ designates the N-volume of P_j^n. Likewise,

(2.4)

$$\int_{T_N} \int_{T_N} g(x-y)dxdy =$$

$$\lim_{n \to \infty} \sum_{j=1}^{\alpha_n} \sum_{k=1}^{\alpha_n} g(x^{j,n}-x^{k,n})\left|P_j^n\right||P_k^n|.$$

Since g is a positive definite function, it follows that the double sum on the right-hand side of the equality in (2.4) is nonnegative. Consequently, the limit in (2.4) is nonnegative, and we see that (2.3) is true.

To establish the sufficiency condition of the theorem, we observe from the fact that f is a real-valued function,

$$\widehat{f}(-m) = \overline{\widehat{f}(m)}.$$

Consequently, it follows from (i) in the hypothesis of the theorem that

(2.5) $\widehat{f}(m) = \widehat{f}(-m) \quad \forall m \in \Lambda_N.$

Next, we set

(2.6) $F_R(x) = \sum_{|m| \le R} \widehat{f}(m)e^{im \cdot x},$

and observe from (2.5) that also

$$F_R(x) = \sum_{|m| \le R} \widehat{f}(m)e^{-im \cdot x}.$$

Adding these last two equalities and dividing by two, we obtain that

(2.7) $F_R(x) = \sum_{|m| \le R} \widehat{f}(m)\cos(m \cdot x) \quad \forall R > 0.$

From (ii) in the hypothesis of the theorem and (2.6), it follows that

(2.8) $f(x) = \lim_{R \to \infty} F_R(x) \quad \text{uniformly for } x \in T_N.$

We have already established that $\cos(m \cdot x)$ is positive definite. Since $\widehat{f}(m) \ge 0$, we have that

$$\widehat{f}(m)\cos(m \cdot x) \text{ is positive definite on } T_N.$$

But a finite linear combination of positive definite functions is positive definite. So it follows from (2.7) that

$$F_R(x) \text{ is positive definite on } T_N.$$

This last fact joined with the statement in (2.8) establishes the sufficiency condition of the theorem.

To establish the necessary condition of the theorem, we observe from the formula in (2.2) and the fact that $\sin(m \cdot x)$ is an odd function and $f(x)$ is an even function that

$$(2.9) \qquad \widehat{f}(m) = (2\pi)^{-N} \int_{T_N} f(x) \cos(m \cdot x) dx \quad \forall m \in \Lambda_N.$$

Next, we observe that

$$(2.10) \qquad f(x) \cos(m \cdot x) \text{ is positive definite on } T_N.$$

To establish this fact, let $x^1, ..., x^n \in T_N$ and $b_1, ..., b_n \in \mathbf{R}$. Then

$$\sum_{j=1}^{n} \sum_{k=1}^{n} f(x^j - x^k) \cos[m \cdot (x^j - x^k)] b_j b_k$$

$$= \sum_{j=1}^{n} \sum_{k=1}^{n} f(x^j - x^k) \cos(m \cdot x^j) b_j \cos(m \cdot x^k) b_k$$

$$+ \sum_{j=1}^{n} \sum_{k=1}^{n} f(x^j - x^k) \sin(m \cdot x^j) b_j \sin(m \cdot x^k) b_k.$$

Since $f(x)$ is positive definite on T_N, it follows that both of the sums on the right-hand side of this last equality are nonnegative. Consequently, the left-hand side is also nonnegative and (2.10) is established.

With $g(x) = f(x) \cos(m \cdot x)$, we observe from (2.10) and (2.3) that

$$(2.11) \qquad \int_{T_N} \int_{T_N} f(x - y) \cos[m \cdot (x - y)] dx dy \geq 0 \quad \forall m \in \Lambda_N.$$

But because of periodicity and (2.9)

$$\int_{T_N} f(x - y) \cos[m \cdot (x - y)] dx = (2\pi)^N \widehat{f}(m) \quad \forall y \in T_N.$$

We therefore obtain from (2.11) that

$$(2.12) \qquad (2\pi)^{2N} \widehat{f}(m) \geq 0 \qquad \forall m \in \Lambda_N,$$

and condition (i) in the conclusion of the theorem is established.

To show that condition (ii) holds, let $\sigma_n^{\Diamond}(f, x)$ represent the iterated Fejer partial sum of $f(x)$ as in (2.6) of Chapter 1, i.e.,

$$(2.13) \quad \sigma_n^{\Diamond}(f, x) = \sum_{m_1=-n}^{n} \cdots \sum_{m_N=-n}^{n} \widehat{f}(m) e^{im \cdot x} (1 - \frac{|m_1|}{n+1}) \cdots (1 - \frac{|m_N|}{n+1}).$$

Then

$$\sigma_n^{\Diamond}(f,0) = (2\pi)^{-N} \int_{T_N} f(-y) K_n^{\Diamond}(y) dy$$

where $K_n^{\Diamond}(y)$ is the iterated Fejer kernel given in (2.5) of Chapter 1.

Since $(2\pi)^{-N} \int_{T_N} K_n^{\Diamond}(y) dy = 1$ and $K_n^{\Diamond}(y) \geq 0$, we obtain from (2.12), (2.13), and this last equality that

(2.14) $$0 \leq \sigma_n^{\Diamond}(f,0) \leq \|f\|_{L^\infty(T_N)} \quad \forall n.$$

Let n^* be any fixed positive integer. It then follows from (2.13) and (2.14) that

$$\sum_{m_1=-n^*}^{n^*} \cdots \sum_{m_N=-n^*}^{n^*} \widehat{f}(m)(1 - \frac{|m_1|}{n+1}) \cdots (1 - \frac{|m_N|}{n+1}) \leq \|f\|_{L^\infty(T_N)} \quad \forall n > n^*.$$

Taking the limit as $n \to \infty$ of the left-hand side of this last inequality gives

$$\sum_{m_1=-n^*}^{n^*} \cdots \sum_{m_N=-n^*}^{n^*} \widehat{f}(m) \leq \|f\|_{L^\infty(T_N)}$$

Since n^* is an arbitrary positive integer and $\widehat{f}(m) \geq 0$, we conclude that

$$\lim_{R \to \infty} \sum_{|m| \leq R} \widehat{f}(m) \quad \text{exists and is finite.}$$

Using $\widehat{f}(m) \geq 0$ once again, we see that this last statement joined with Corollary 2.3 in Chapter 1 gives condition (ii) in the statement of the theorem. ∎

Exercises.

1. With $f \in C(T_N)$ and $\widehat{f}(m)$ given by (2.2), prove that if

$$\lim_{R \to \infty} \sum_{|m| \leq R} \widehat{f}(m) e^{im \cdot x} = g(x) \quad \text{uniformly for } x \in T_N,$$

then $f(x) = g(x) \ \forall x \in T_N$.

2. Prove that the function $t \sum_{m \in \Lambda_N} [t^2 + |x + 2\pi m|^2]^{-(N+1)/2}$ for $t > 0$ and $x \in T_N$ is positive definite on T_N.

3. Prove that the function $t^{-N/2} \sum_{m \in \Lambda_N} e^{-|x+2\pi m|^2/4t}$ for $t > 0$ and $x \in T_N$ is positive definite on T_N.

3. Positive Definite Functions on $S_{N_1-1} \times T_N$

In this section, we shall work on the space $S_{N_1-1} \times T_N$. Here, $N_1 \geq 3$ will be a positive integer. Also,

$$\nu_1 = \frac{N_1 - 2}{2}.$$

Using the methods employed in the last two sections, we intend to get an analogous result for functions positive definite on a sphere cross a torus.

To present our result, we let I designate the closed interval on the real line

$$I = [-1, 1].$$

We will say $f(t, x) \in C(I \times T_N)$ provided the following holds:

$(i) f(t, x)$ is a real-valued function;

(3.1) $(ii) f(t, x) \in C(I \times \mathbf{R}^N)$;

$(iii)\ \forall t \in I,\ f(t, x)$ is periodic of period 2π
in each component of the x-variable.

We will say $f \in C(I \times T_N)$ is positive definite on $S_{N_1-1} \times T_N$ provided the following two facts hold:

(3.2) $\forall t \in I,\ f(t, x) = f(t, -x)$ for every $x \in T_N$;

(3.3) $\forall n_1 \geq 1,\ \sum_{k=1}^{n_1} \sum_{j=1}^{n_1} b_j b_k f(\xi^j \cdot \xi^k, x^j - x^k) \geq 0$

for $(\xi^j, x^j) \in S_{N_1-1} \times T_N$ and $b_j \in \mathbf{R},\ j = 1, ..., n_1$.

Next, for $f \in C(I \times T_N)$, we set

(3.4) $\widehat{f}(n, m) = (2\pi)^{-N} \tau_n^{-1} \int_I \int_{T_N} f(t, x)(1 - t^2)^{\nu_1 - \frac{1}{2}} C_n^{\nu_1}(t) e^{-im \cdot x} dx dt,$

for $n \geq 0$, $m \in \Lambda_N$, and τ_n is defined in (1.3) with ν replaced by ν_1.

We shall establish the following theorem:

Theorem 3.1.   *Let $f(t, x)$ be a real-valued function, and suppose $f \in C(I \times T_N), N \geq 1$. Define $\widehat{f}(n, m)$ by the formula in (3.4). Then a necessary and sufficient condition that f be positive definite on $S_{N_1-1} \times T_N$, $N_1 \geq 3$, is that the following two conditions hold:*

$(i) \widehat{f}(n, m) \geq 0$ *for* $n \geq 0$ *and* $m \in \Lambda_N$;

$(ii)\ f(t, x) = \lim_{n \to \infty} \sum_{k=0}^{n} \sum_{|m| \leq n} \widehat{f}(k, m) C_k^{\nu_1}(t) e^{im \cdot x}$

uniformly for $(t, x) \in I \times T_N$.

To prove the above theorem, we will need the following lemma:

Lemma 3.2. *Let $f(t, x)$ be a real-valued function, and suppose $f \in C(I \times T_N)$, $N \geq 1$. Define $\widehat{f}(n, m)$ by the formula in (3.4). Suppose also that*

(3.5) $$\widehat{f}(n, m) = 0 \ \text{ for } \ n \geq 0 \ \text{ and } \ m \in \Lambda_N.$$

Then $f(t, x) = 0$ for $(t, x) \in I \times T_N$.

Proof of Lemma 3.2. To prove the lemma for n, a nonnegative integer, set

(3.6) $$h_n(t) = (1 - t^2)^{\nu_1 - \frac{1}{2}} C_n^{\nu_1}(t).$$

Then it follows from (3.4) and (3.5) that

(3.7) $$\int_{T_N} [\int_I f(t, x) h_n(t) dt] e^{-im \cdot x} dx = 0 \quad \text{for } n \geq 0 \text{ and } m \in \Lambda_N.$$

For each fixed n, it is easy to see that

$$\int_I f(t, x) h_n(t) dt \in C(T_N).$$

But then it follows from (3.7) and Corollary 2.3 in Chapter 1 that

(3.8) $$\int_I f(t, x) \, h_n(t) dt = 0 \quad \text{for } x \in T_N \text{ and } n \geq 0.$$

Next, because

$$f(t, x) \in C(I) \ \text{ for fixed } x \in T_N,$$

it follows from Theorem 3.5 in Appendix A in conjunction with (3.6) and (3.8) that $f(t, x) = 0$ *for* $t \in I$ and $x \in T_N$. ∎

Proof of Theorem 3.1. We first show for $n \geq 0$ and $m \in \Lambda_N$ that

(3.9) $$C_n^{\nu_1}(t) \cos(m \cdot x) \text{ is positive definite on } S_{N_1-1} \times T_N.$$

To see that this is the case, let $(\xi^j, x^j) \in S_{N_1-1} \times T_N$ for $j = 1, ..., n_1$. Then using (1.4) above, we see that

$$C_n^{\nu_1}(\xi^j \cdot \xi^k) \cos[m \cdot (x^j - x^k)] =$$

$$\gamma_n \sum_{l=1}^{\mu_{n,N}} Y_{l,n}(\xi^j) Y_{l,n}(\xi^k) \cos(m \cdot x^j) \cos(m \cdot x^k)$$

$$+ \gamma_n \sum_{l=1}^{\mu_{n,N}} Y_{l,n}(\xi^j) Y_{l,n}(\xi^k) \sin(m \cdot x^j) \sin(m \cdot x^k).$$

where $\gamma_n > 0$. Consequently, for $b_j \in \mathbf{R}$, $j = 1, ..., n_1$,

$$\sum_{j=1}^{n_1} \sum_{k=1}^{n_1} b_j b_k C_n^{\nu_1}(\xi^j \cdot \xi^k) \cos[m \cdot (x^j - x^k)] =$$

$$\gamma_N \sum_{l=1}^{\mu_{n,N}} [\sum_{j=1}^{n_1} Y_{l,n}(\xi^j) b_j \cos(m \cdot x^j)]^2 + \gamma_N \sum_{l=1}^{\mu_{n,N}} [\sum_{j=1}^{n_1} Y_{l,n}(\xi^j) b_j \sin(m \cdot x^j)]^2,$$

and the statement in (3.9) is established.

To establish the sufficiency condition of the theorem, we observe from (3.4) and the fact that

$$f(t,x) \in C(I \times T_N) \quad \text{is real-valued}$$

that

$$\widehat{f}(n,-m) = \overline{\widehat{f}(n,m)} \quad \forall n \geq 0 \quad \text{and} \quad \forall m \in \Lambda_N.$$

But by assumption (i) of the theorem, $\widehat{f}(n,m)$ is also real-valued for $m \in \Lambda_N$. So we conclude

$$(3.10) \qquad \widehat{f}(n,-m) = \widehat{f}(n,m) \quad \forall n \geq 0 \quad \text{and} \quad \forall m \in \Lambda_N.$$

Next, we set

$$(3.11) \qquad F_n(t,x) = \sum_{k=0}^{n} \sum_{|m| \leq n} \widehat{f}(k,m) C_k^{\nu_1}(t) e^{im \cdot x}$$

for $(t,x) \in I \times T_N$ and obtain from (3.10) that also

$$F_n(t,x) = \sum_{k=0}^{n} \sum_{|m| \leq n} \widehat{f}(k,m) C_k^{\nu_1}(t) e^{-im \cdot x}.$$

As a consequence of these two different representations of $F_n(t,x)$, it follows that

$$(3.12) \qquad F_n(t,x) = \sum_{k=0}^{n} \sum_{|m| \leq n} \widehat{f}(k,m) C_k^{\nu_1}(t) \cos(m \cdot x)$$

for $(t,x) \in I \times T_N$. Since $\widehat{f}(k,m) \geq 0$, we obtain from (3.9) that

$$\widehat{f}(k,m) C_k^{\nu_1}(t) \cos(m \cdot x) \text{ is positive definite on } S_{N_1-1} \times T_N.$$

But then from (3.12), it follows that

$$(3.13) \qquad\qquad F_n(t,x) \text{ is positive definite on } S_{N_1-1} \times T_N$$

for every positive integer n. By assumption (ii) of the theorem,

$$(3.14) \qquad\qquad \lim_{n \to \infty} F_n(t,x) = f(t,x)$$

uniformly for $(t,x) \in I \times T_N$.

So from (3.13), we have that

$$(3.15) \qquad\qquad f(t,x) \text{ is positive definite on } S_{N_1-1} \times T_N$$

and the sufficiency condition of the theorem is established.

To prove the necessary condition of the theorem, we first show that if $g(t, x) \in C(I \times T_N)$ is a real-valued function that is positive definite on $S_{N_1-1} \times T_N$, then

$$(3.16) \qquad \int_{S_{N_1-1} \times T_N} [\int_{S_{N_1-1} \times T_N} g(\xi \cdot \eta, x - y) dS(\xi) dx] dS(\eta) dy \geq 0.$$

To see that this is the case, we use the definition of the Riemann integral and take a sequence of partitions $\{\mathcal{P}_n\}_{n=1}^\infty$ of $S_{N_1-1} \times T_N$ where

$$\mathcal{P}_n = \{P_j^n\}_{j=1}^{\alpha_n} \quad \text{and} \quad S_{N_1-1} \times T_N = \cup_{n=1}^{\alpha_n} P_j^n$$

and the diameter of P_j^n goes to zero as $n \to \infty$. Also, $\alpha_n \to \infty$. In each P_j^n, we choose a $\xi^{j,n} \times x^{j,n} \in P_j^n$. Then

$$\int_{S_{N_1-1} \times T_N} g(\xi \cdot \eta, x - y) dS(\xi) dx = \lim_{n \to \infty} \sum_{j=1}^{\alpha_n} g(\xi^{j,n} \cdot \eta, x^{j,n} - y) |P_j^n|$$

where $|P_j^n|$ designates the volume of P_j^n. Likewise,

$$(3.17) \quad \begin{aligned} &\int_{S_{N_1-1} \times T_N} [\int_{S_{N_1-1} \times T_N} g(\xi \cdot \eta, x - y) dS(\xi) dx] dS(\eta) dy = \\ &\lim_{n \to \infty} \sum_{j=1}^{\alpha_n} \sum_{k=1}^{\alpha_n} g(\xi^{j,n} \cdot \xi^{k,n}, x^{j,n} - x^{k,n}) |P_j^n| |P_k^n|. \end{aligned}$$

Since g is a positive definite function on $S_{N_1-1} \times T_N$, it follows that the double sum on the right-hand side of the equality in (3.17) is nonnegative. So, the limit in (3.17) is nonnegative, and (3.16) is established.

Next, from (3.2) and (3.4), we see that

$$(3.18) \qquad \hat{f}(n, m) = \delta_n \int_I \int_{T_N} f(t, x)(1 - t^2)^{\nu_1 - \frac{1}{2}} C_n^{\nu_1}(t) \cos(m \cdot x) dx dt$$

where δ_n is a positive constant.

Introducing a spherical coordinate system as in §3 of Chapter 1 with η as the pole, we also have that

$$\begin{aligned} &\int_{S_{N_1-1}} f(\xi \cdot \eta, x) C_n^{\nu_1}(\xi \cdot \eta) \ dS(\xi) \\ &= |S_{N_1-2}| \int_0^\pi f(\cos \theta, x) C_n^{\nu_1}(\cos \theta)(\sin \theta)^{2\nu_1} d\theta \\ &= \zeta_{N_1} \int_I f(t, x) C_n^{\nu_1}(t)(1 - t^2)^{\nu_1 - \frac{1}{2}} dt, \end{aligned}$$

where ζ_{N_1} is a positive constant.

So from the periodicity of f in the x-variable and (3.18),

$$\int_{T_N}[\int_{S_{N_1-1}} f(\xi \cdot \eta, x-y)C_n^{\nu_1}(\xi \cdot \eta) \ dS(\xi)]\cos[m \cdot (x-y)]dx$$

$$= \zeta_{N_1}\int_{T_N}[\int_I f(t, x-y)C_n^{\nu_1}(t)(1-t^2)^{\nu_1-\frac{1}{2}}dt]\cos[m \cdot (x-y)]dx$$

$$= \zeta_{N_1}\int_{T_N}[\int_I f(t, x)C_n^{\nu_1}(t)(1-t^2)^{\nu_1-\frac{1}{2}}dt]\cos(m \cdot x)dx$$

$$= \zeta_{N_1}\delta_n^{-1}\widehat{f}(n, m).$$

Therefore,

$$(3.19)\ \widehat{f}(n, m) = \gamma_n^* \int_{S_{N_1-1}\times T_N} f(\xi \cdot \eta, x-y)C_n^{\nu_1}(\xi \cdot \eta)\cos[m \cdot (x-y)]dS(\xi)dx$$

for $\eta \in S_{N_1-1}$ and $y \in T_N$ where γ_n^* is a positive constant. But then it follows from (3.19) that

$\widehat{f}(n, m)$ is a positive constant multiple of

$$(3.20)$$
$$\int_{S_{N_1-1}\times T_N}\{\int_{S_{N_1-1}\times T_N} f(\xi \cdot \eta, x-y)C_n^{\nu_1}(\xi \cdot \eta)\cos[m \cdot (x-y)]dS(\xi)dx\}dS(\eta)dy$$

for $n \geq 0$ and for $m \in \Lambda_N$.

We next claim that

$$(3.21)\qquad f(t, x)C_n^{\nu_1}(t)\cos(m \cdot x) \text{ is positive definite on } S_{N_1-1} \times T_N.$$

For this claim, we see that the condition in (3.2) is clearly met. So it only remains to show that the condition in (3.3) holds.

To show that this is the case, we invoke (1.4) once again and obtain

$$C_n^{\nu_1}(\xi \cdot \eta)\cos[m \cdot (x-y)]$$

$$= \gamma_n \sum_{l=1}^{\mu_{n,N_1}} Y_{l,n}(\xi)Y_{l,n}(\eta)\cos(m \cdot x)\cos(m \cdot y)$$

$$+\gamma_n \sum_{l=1}^{\mu_{n,N_1}} Y_{l,n}(\xi)Y_{l,n}(\eta)\sin(m \cdot x)\sin(m \cdot y)$$

where $\gamma_n > 0$. So, it follows from this last equality that

$$(3.22)\qquad \sum_{k=1}^{n_1}\sum_{j=1}^{n_1} b_j b_k f(\xi^j \cdot \xi^k, x^j - x^k)C_n^{\nu_1}(\xi^j \cdot \xi^k)\cos[m \cdot (x^j - x^k)]$$

$$= \gamma_n \sum_{l=1}^{\mu_{n,N_1}}[A(n_1, l) + B(n_1, l)],$$

where

$$A(n_1, l)$$
$$= \sum_{k=1}^{n_1} \sum_{j=1}^{n_1} b_j b_k f(\xi^j \cdot \xi^k, x^j - x^k) Y_{l,n}(\xi^j) Y_{l,n}(\xi^k) \cos(m \cdot x^j) \cos(m \cdot x^k)$$
$$= \sum_{k=1}^{n_1} \sum_{j=1}^{n_1} b_j Y_{l,n}(\xi^j) \cos(m \cdot x^j) b_k Y_{l,n}(\xi^k) \cos(m \cdot x^k) f(\xi^j \cdot \xi^k, x^j - x^k),$$

and

$$B(n_1, l)$$
$$= \sum_{k=1}^{n_1} \sum_{j=1}^{n_1} b_j b_k f(\xi^j \cdot \xi^k, x^j - x^k) Y_{l,n}(\xi^j) Y_{l,n}(\xi^k) \sin(m \cdot x^j) \sin(m \cdot x^k).$$
$$= \sum_{k=1}^{n_1} \sum_{j=1}^{n_1} b_j Y_{l,n}(\xi^j) \sin(m \cdot x^j) b_k Y_{l,n}(\xi^k) \sin(m \cdot x^k) f(\xi^j \cdot \xi^k, x^j - x^k).$$

It is clear from the fact that $f(t, x)$ is positive definite on $S_{N_1-1} \times T_N$ that both

$$A(n_1, l) \geq 0 \quad \text{and} \quad B(n_1, l) \geq 0$$

for $l = 1, ..., \mu_{n,N_1}$.

Therefore, the sum on the right-hand side in (3.22) is nonnegative. But this, in turn, implies that

$$\sum_{k=1}^{n_1} \sum_{j=1}^{n_1} b_j b_k f(\xi^j \cdot \xi^k, x^j - x^k) C_n^{\nu_1}(\xi^j \cdot \xi^k) \cos[m \cdot (x^j - x^k)] \geq 0.$$

Consequently, condition (3.3) holds, and indeed $f(t, x) C_n^{\nu_1}(t) \cos(m \cdot x)$ is positive definite on $S_{N_1-1} \times T_N$. The claim in (3.21) is substantiated.

Next, we set

$$g(t, x) = f(t, x) C_n^{\nu_1}(t) \cos(m \cdot x),$$

and observe from (3.16) and (3.21) that the double integral in (3.20) is nonnegative. Since $\widehat{f}(n, m)$ is a positive constant multiple of this double integral, it follows that

(3.23) $$\widehat{f}(n, m) \geq 0 \quad \text{for } n \geq 0 \text{ and } m \in \Lambda_N,$$

and condition (i) in Theorem 3.1 is established.

To complete the proof of the necessary condition, it remains to show that condition (ii) holds. In order to accomplish this, we set

(3.24) $$h(r, s) = \sum_{n=0}^{\infty} \sum_{m \in \Lambda_N} \widehat{f}(n, m) C_n^{\nu_1}(1) r^n e^{-|m|s},$$

where $0 < r < 1$ and $0 < s < \infty$.

Also, we see from (1.3) and (3.4) that for fixed m,

$$(2\pi)^N \widehat{f}(n,m) C_n^{\nu_1}(1)$$

$$= \beta_{\nu_1} \frac{(N_1 - 2 + 2n)}{N_1 - 2} \int_{T_N} e^{-im \cdot x} dx \int_{-1}^{1} f(t) C_n^{\nu_1}(t)(1 - t^2)^{\nu_1 - \frac{1}{2}} dt$$

where β_{ν_1} is a positive constant. So from (3.29) in Appendix A, we obtain that

$$(2\pi)^N \sum_{n=0}^{\infty} \widehat{f}(n,m) C_n^{\nu_1}(1) r^n$$

$$= \beta_{\nu_1} \int_{T_N} e^{-im \cdot x} dx \int_I f(t,x) \frac{1 - r^2}{(1 - 2rt + r^2)^{N_1/2}} (1 - t^2)^{\nu_1 - \frac{1}{2}} dt.$$

For fixed r, we treat

$$\int_I f(t,x) \frac{1 - r^2}{(1 - 2rt + r^2)^{N_1/2}} (1 - t^2)^{\nu_1 - \frac{1}{2}} dt$$

as a function of x in $C(T_N)$ and obtain from this last equality and (4.5) in Chapter 1 that

$h(r,s)$ is a constant (independent of r and s) multiple of

$$\int_I \frac{1 - r^2}{(1 - 2rt + r^2)^{N_1/2}} (1 - t^2)^{\nu_1 - \frac{1}{2}} dt \int_{\mathbf{R}^N} s[s^2 + |x|^2]^{-(N+1)/2} f(t,x) dx,$$

where $h(r,s)$ is defined in (3.24). Consequently,

$$(3.25) \qquad |h(r,s)| \le \beta^* \|f\|_{L^\infty} \int_I (1 - t^2)^{\nu_1 - \frac{1}{2}} dt,$$

for $0 < r < 1$ and $0 < s < \infty$ where β^* is a constant.

In obtaining the inequality in (3.25), we have made use of the fact (easily checked) that

$$\int_{\mathbf{R}^N} s[s^2 + |x|^2]^{-(N+1)/2} dx = \beta_N \quad \text{for } 0 < s < \infty,$$

where β_N is a constant, and from (3.29) and (3.41) in Appendix A with $C_0^{\nu_1}(t) \equiv 1$, that

$$\int_I \frac{1 - r^2}{(1 - 2rt + r^2)^{N_1/2}} (1 - t^2)^{\nu_1 - \frac{1}{2}} dt = \int_I (1 - t^2)^{\nu_1 - \frac{1}{2}} dt$$

for $0 < r < 1$.

Now each of the terms in the series defining $h(r,s)$ in (3.24) is nonnegative. Hence, from (3.25), we obtain that

$$\sum_{n=0}^{\infty} \sum_{m \in \Lambda_N} \widehat{f}(n,m) C_n^{\nu_1}(1) r^n e^{-|m|s} \le \beta^{\#} < \infty,$$

for $0 < r < 1$ and $0 < s < \infty$ where $\beta^{\#}$ is a constant.

Let n^* be any fixed positive integer. Then, it follows from this last inequality that

$$(3.26) \qquad \sum_{n=0}^{n^*} \sum_{|m| \leq n^*} \widehat{f}(n,m) C_n^{\nu_1}(1) r^n e^{-|m|s} \leq \beta^{\#}$$

for $0 < r < 1$ and $0 < s < \infty$.

Passing to the limit as $r \to 1$ and $s \to 0$, we see from (3.26) that

$$\sum_{n=0}^{n^*} \sum_{|m| \leq n^*} \widehat{f}(n,m) C_n^{\nu_1}(1) \leq \beta^{\#} < \infty$$

for every positive integer n^*.

Using once again the fact that each term of this last series is nonnegative, we conclude that

$$(3.27) \qquad \lim_{n \to \infty} \sum_{k=0}^{n} \sum_{|m| \leq n} \widehat{f}(k,m) C_n^{\nu_1}(1) \text{ exists and is finite.}$$

Next, we observe from (3.30) in Appendix A, that

$$\left| \widehat{f}(k,m) C_n^{\nu_1}(t) e^{im \cdot x} \right| \leq \widehat{f}(k,m) C_n^{\nu_1}(1)$$

$\forall k \geq 0$ and $\forall m \in \Lambda_N$ and $\forall (t,x) \in I \times T_N$.

But then it follows from (3.27), Lemma 3.2, and this last inequality that

$$\lim_{n \to \infty} \sum_{k=0}^{n} \sum_{|m| \leq n} \widehat{f}(k,m) C_n^{\nu_1}(t) e^{im \cdot x} = f(t,x)$$

uniformly for $(t,x) \in I \times T_N$. So condition (ii) in Theorem 3.1 is established, and the proof of the necessary condition of the theorem is complete. ∎

Exercises.

1. Given $\xi \in S_2$ is of the form $\xi = (\cos\theta, \sin\theta\cos\phi, \sin\theta\sin\phi)$ and that $P_1(t) = t$, prove directly that $P_1(t)\cos m \cdot x$ is positive definite on $S_2 \times T_N$ where $m \in \Lambda_N$.

2. Given $f \in C(I \times T_N)$ and $\widehat{f}(k,m)$ defined by (3.4), prove that if

$$\lim_{n \to \infty} \sum_{k=0}^{n} \sum_{|m| \leq n} \widehat{f}(k,m) C_n^{\nu_1}(t) e^{im \cdot x} = g(t,x)$$

uniformly for $(t,x) \in I \times T_N$, then $f(t,x) = g(t,x)$ for $(t,x) \in I \times T_N$.

3. Given $f \in C(I \times T_2)$ and $f(t,x)$ are positive definite on $S_2 \times T_2$, let $\eta^* = (1,0,0)$ and $g(\xi,x) = f(\eta^* \cdot \xi, x)$ for $\xi \in S_2$. Find a function

$u(y, x, s) \in C^{\infty}(B(0,1) \times T_2 \times \mathbf{R}_+)$ with $u(y, x, s)$ continuous in the closure of $B(0,1) \times T_2 \times \mathbf{R}_+$ such that

$$Lu(y, x, s) = 0 \qquad \forall (y, x, s) \in B(0,1) \times T_2 \times \mathbf{R}_+$$
$$u(\xi, x, 0) = g(\xi, x) \qquad \forall (\xi, x) \in S_2 \times T_2,$$

where

$$Lu = \frac{\partial^2 u}{\partial y_1^2} + \frac{\partial^2 u}{\partial y_2^2} + \frac{\partial^2 u}{\partial y_3^2} + \frac{\partial^2 u}{\partial x_1^2} + \frac{\partial^2 u}{\partial x_2^2} + \frac{\partial^2 u}{\partial s^2},$$

$B(0,1)$ is the open unit 3-ball, and $\mathbf{R}_+ = \{s : s > 0\}$.

4. Further Results and Comments

1. Bochner introduced a notion of generalized analyticity for complex-valued functions in $L^1(T_N), N \geq 2$. In particular, $f \in A_v^+$, where $v \in \mathbf{R}^N$ and $|v| = 1$, provided $f \in L^1(T_N)$ and

$$\widehat{f}(m) = 0 \quad \text{if} \ m \cdot v \leq 0.$$

Set $A_v = A_v^+ \cup A_{-v}^+$. It is clear that the class A_v generalizes the notion of analyticity to higher dimensions.

Also, Bochner introduced the notion of a strict generalized analytic class B_v as follows: $f \in B_v^+$ provided (i) $f \in A_v^+$ and

$$(ii) \ \exists \gamma \text{ with } 0 < \gamma < 1 \text{ such that } \widehat{f}(m) = 0 \quad \text{if} \ m \cdot v < \gamma |m| \ .$$

Set $B_v = B_v^+ \cup B_{-v}^+$.

In a similar manner, we define the class of generalized analytic measures on T_N obtaining the classes $\mathcal{A}_v$ and $\mathcal{B}_v$. So, in particular, $\mu \in \mathcal{A}_v^+$ provided μ is a finite-valued complex Borel measure on T_N and

$$\widehat{\mu}(m) = 0 \quad \text{if} \ m \cdot v \leq 0.$$

Also, $\mu \in \mathcal{B}_v^+$ provided $\mu \in \mathcal{A}_v^+$ and

$$\exists \gamma \text{ with } 0 < \gamma < 1 \text{ such that } \widehat{\mu}(m) = 0 \quad \text{if} \ m \cdot v < |m|.$$

Bochner (see [Boc2] or [Ru3, p.201]) obtained the following generalization of a well-known theorem of F. and M. Riesz ([Ho, p. 47] or [Ru3, p.198]) on T_1 to T_N for $N \geq 2$.

Theorem. *Let $\mu \in \mathcal{B}_v$. Then μ is absolutely continuous on T_N, i.e., there exists $f \in L^1(T_N)$ such that if $E \subset T_N$ is a Borel set,*

$$\mu(E) = \int_E f \, dx.$$

This result is false in general if $\mu \in \mathcal{A}_v$ (see [Ho, p. 60]). Another proof of Bochner's theorem was found by Helson and Lowdenschlager [HeLo].

2. There is also another theorem of F. and M. Riesz [RR] dealing with analytic functions in the unit disk, namely the following:

Let $f(z)$ be a bounded function that is analytic for $|z| < 1$. Suppose $\lim_{r \to 1} f\left(re^{i\theta}\right) = 0$ *for $\theta \in E$ where $E \subset T_1$ and $|E| > 0$. Then $f(z)$ is identically zero for $|z| < 1$.*

There is an interesting generalization of this result to T_N for $N \geq 2$ involving the class B_v introduced above.

For the purpose of the statement of this generalization, we will say $f \in L^\infty(T_N)$ vanishes at x_0 provided

$$\lim_{r \to 0} f_{[r]}(x_0) = 0,$$

where $f_{[r]}(x_0) = |B(0,r)|^{-1} \int_{B(0,r)} f(x_0 + x)\, dx$.

In order to present this generalization, we have to introduce the following one parameter subgroup G_v of T_N. With $v \in \mathbf{R}^N$ and $|v| = 1$,

$$G_v = \{x : -\pi \leq x_j < \pi, x_j \equiv tv_j \bmod 2\pi,\ j = 1, ..., N,\ -\infty < t < \infty\}.$$

A set $E \subset G_v$ is said to a set of positive linear measure if the following holds: Define $E^* \subset \mathbf{R}$ as follows:

$$E^* = \{t : \exists x \in E \text{ such that } x_j \equiv tv_j \bmod 2\pi \text{ for } j = 1, ..., N\}.$$

Then E is said to be a set of positive linear measure provided E^* is a set of positive one-dimensional Lebesgue measure.

f vanishes on a subset $E \subset G_v$ provided f vanishes at every point $x \in E$.

The following generalization of the F. and M. Riesz theorem holds [Sh18].

Theorem. *A necessary and sufficient condition that every $f \in B_v$, which is in $L^\infty(T_N)$ and vanishes on a subset of G_v of positive linear measure be equal to zero almost everywhere on T_N, is that v be linear independent with respect to rational coefficients.*

Also, a counter-example is given that shows that the sufficiency condition of the theorem is false in general for $f \in A_v \cap L^\infty(T_N)$.

CHAPTER 5

Nonlinear Partial Differential Equations

1. Reaction-Diffusion Equations on the N-Torus

In this chapter, we show the power of Fourier series in several variables in solving problems in nonlinear partial differential equations.

The techniques presented in this section come from the paper by the author [Sh11] that appeared in the Indiana University Mathematics journal in the year 2009.

Operating in N-dimensional Euclidean space $\mathbf{R}^N$, $N \geq 1$, and as before, letting T_N be the N-dimensional torus

$$T_N = \{x : -\pi \leq x_n < \pi, \ n = 1, ..., N\},$$

we shall deal with the following reaction-diffusion system with periodic boundary conditions and zero initial conditions:

$$(1.1) \quad \begin{cases} \frac{\partial u_j}{\partial t} - \Delta u_j = f_j(x, t, u_1, ..., u_J) & \text{in } T_N \times (0, T) \\ \\ u_j(x, 0) = 0 \end{cases}$$

for $j = 1, ..., J$.

A system of equations of the form (1.1) is generally referred to in the literature as a reaction-diffusion system and occurs in mathematical biology and many other places in applied mathematics (see [Sm, pp. 208-210], [Mur, pp. 375-379], and [EK, p. 426]).

With $s = (s_1, ..., s_J)$ and $T_N \times (0, T) = \tilde{\Omega}$, we assume in the above that

$$(1.2) \qquad f_j(x, t, s) \text{ is a Caratheodory function, i.e.,}$$

$f_j(x, t, s)$ is measurable in (x, t) for $s \in \mathbf{R}^J$ and continuous in s for a.e. $(x, t) \in \tilde{\Omega}$.

Also, we assume the following two conditions:

$$(1.3) \qquad \begin{aligned} &\forall R > 0, \ \exists \alpha_R(x, t) \in L^1(\tilde{\Omega}) \text{ such that} \\ \\ &\sup_{|s_j| < R} |f_j(x, t, s_1, ..., s_J)| \leq \alpha_R(x, t) \end{aligned}$$

for $s_k \in R$, $k \neq j$, $k = 1, ..., J$ for $j = 1, ..., J$.

$$(1.4) \qquad \begin{aligned} &s_j f_j(x, t, s) \\ \\ &\leq C_1 |s_j|^2 + C_2(x, t) |s_j| + C_3(x, t) \quad \forall s \in \mathbf{R} \end{aligned}$$

and for a.e. $(x, t) \in \tilde{\Omega}$ where $j = 1, ..., J$.

Also, $C_1 > 0$ with C_2 and C_3 nonnegative functions in $L^2(\tilde{\Omega})$ and $L^1(\tilde{\Omega})$, respectively.

$g_1(x) \in C^\infty(T_N)$ will mean that $g_1(x) \in C^\infty(\mathbf{R}^N)$ and is periodic of period 2π in each of the variables x_n for $n = 1, ..., N$.

Likewise, $g_1(x) \in C^2(T_N)$ will mean that $g_1(x) \in C^2(\mathbf{R}^N)$ and is periodic of period 2π in each of the variables x_n for $n = 1, ..., N$.

We introduce the Hilbert space $H^1(T_N)$ as follows: $H^1(T_N)$ is the closure of the set of functions in $C^\infty(T_N)$ under the norm generated by the following real inner product:

$$< g_1, h_1 >_{H^1} = \int_{T_N} [g_1 h_1 + \nabla g_1 \cdot \nabla h_1] dx \quad \text{for } g_1, h_1 \in C^\infty(T_N).$$

So if $u_1, v_1 \in H^1(T_N)$, then both u_1 and v_1 are in the familiar Sobolev space $W^{1,2}(T_N)$ (see [Ev, pp. 241-257] for the theory of Sobolev spaces) and

$$(1.5) \quad < u_1, v_1 >_{H^1} = \int_{T_N} [u_1 v_1 + \nabla u_1 \cdot \nabla v_1] dx \quad \text{for } u_1, v_1 \in H^1(T_N).$$

In this section, we establish two theorems for reaction-diffusion systems. The first theorem will deal with a one-sided condition placed on $f_j(x, t, s)$ as in (1.4) above. The second theorem will deal with the two-sided condition placed on $f_j(x, t, s)$ as in (1.98) below.

We establish the following theorem for our reaction-diffusion system (1.1).

Theorem 1.1. *Let $T_N \times (0, T) = \tilde{\Omega}$. With $s = (s_1, ..., s_J)$, assume that $f_j(x, t, s)$ satisfies the conditions stated in (1.2), (1.3), and (1.4). Then there exists*

$$u \in [L^2(0, T; H^1(T_N))]^J \cap [L^\infty(0, T; L^2(T_N))]^J$$

with both

$$u_j f_j(x, t, u) \text{ and } f_j(x, t, u) \in L^1(\tilde{\Omega}) \qquad \text{for } j = 1, ..., J,$$

such that $u = (u_1, ..., u_J)$ is a generalized periodic solution of the reaction-diffusion system (1.1).

For periodic solutions, this theorem is a two-way improvement over [BN, Th. V.1, p. 302]. In the first place, this result deals with systems. Secondly, for $J = 1$, the condition in (1.4) entails a far weaker assumption than the corresponding one made in this last named reference. In particular, this last named reference assumes

$$f_1(x, t, s_1) \leq C_2(x, t) + C_3(x, t)/s_1 \qquad \text{for } s_1 > 0.$$

On the other hand, it is assumed in (1.4) above that

$$f_1(x, t, s_1) \leq C_1 s_1 + C_2(x, t) + C_3(x, t)/s_1 \qquad \text{for } s_1 > 0,$$

i.e., linear growth is allowed in this last inequality but not in the previous one. A similar situation prevails for $s_1 < 0$. Everything else about the assumptions remains the same.

Also, the techniques employed here are completely different from those used in [BN].

Theorem 1.1 is motivated by and similar to the theorem in the author's manuscript [Sh11], which appeared in the Indiana University Mathematics Journal.

$u_1 \in L^2(0, T; H^1(T_N))$ will mean that

$$u_1(x, t) \text{ is measureable in } \tilde{\Omega} = T_N \times (0, T),$$

and that

$$u_1(x, t) \in H^1(T_N) \text{ for a.e. } t \in (0, T).$$

Also, it will mean that $\nabla u_1(x, t)$ is measureable in $\tilde{\Omega}$ and that

$$\int_0^T \|u_1(\cdot, t)\|_{H^1}^2 \, dt < \infty.$$

What is meant by the statement in the theorem that $u = (u_1, ..., u_J)$ is a generalized periodic solution of (1.1) under the assumption that u possesses the properties enumerated in the theorem and $f_j(x, t, u) \in L^1(\tilde{\Omega})$ is the following (where E$\subset (0, T)$ is a set of measure zero):

$$(1.6) \qquad (i) \int_0^T \int_{T_N} \left(-u_j \frac{\partial \theta}{\partial t} - u_j \Delta \theta\right) = \int_{\tilde{\Omega}} f_j(x, t, u)\theta \quad \forall \theta \in \widetilde{C}_c^\infty(\tilde{\Omega}),$$

$$(ii) \lim_{t \to 0} \|u_j(\cdot, t)\|_{L^2} = 0 \quad t \in (0, T) \backslash E,$$

for $j = 1, ..., J$.

The assertion $\theta \in \widetilde{C}_c^\infty(\tilde{\Omega})$ means the following:

(i) $\theta \in C^\infty[\mathbf{R}^N \times (0, T)]$;

(ii) for each fixed $t \in (0, T)$, $\theta(x, t) \in C^\infty(T_N)$;

(iii) $\exists t_1, t_2$ with $0 < t_1 < t_2 < T$ such that
$\theta(x, t) = 0$ for $0 < t < t_1$ and $t_2 < t < T$.

We say $f_1 \in L^p(T_N), 1 \le p < \infty$, provided f_1 is a real-valued (unless explicitly stated otherwise) Lebesgue measurable function defined on $\mathbf{R}^N$ of period 2π in each variable such that

$$\int_{T_N} |f_1|^p \, dx < \infty.$$

As before, we denote the set of integral lattice points in $\mathbf{R}^N$ by Λ_N, and for $m \in \Lambda_N$, $x \in \mathbf{R}^N$, $m \cdot x$ will designate the usual dot product

$$m \cdot x = m_1 x_1 + \cdots + m_N x_N.$$

For $f_1 \in L^1(T_N)$, we set

$$\widehat{f_1}(m) = (2\pi)^{-N} \int_{T_N} e^{-im \cdot x} f_1(x) dx,$$

and $< \cdot, \cdot >_{L^2}$ will designate the usual real inner product in $L^2(T_N)$. Also,

$$< \nabla u_1, \nabla v_1 >_{L^2} = \int_{T_N} [\nabla u_1 \cdot \nabla v_1] dx \quad \text{for } u_1, v_1 \in H^1(T_N).$$

The first lemma that we deal with is the following:

Lemma 1.2.   *Let* $g_1 \in L^2(T_N)$. *Then*

$$g_1 \in H^1(T_N) \iff \sum_{m \in \Lambda_N} |m|^2 |\widehat{g_1}(m)|^2 < \infty.$$

Furthermore, if $g_1 \in H^1(T_N)$, *then*

$$(1.7) \qquad < \nabla g_1, \nabla g_1 >_{L^2} = (2\pi)^N \sum_{m \in \Lambda_N} |m|^2 |\widehat{g_1}(m)|^2 .$$

Proof of Lemma 1.2.   Given $g_1 \in H^1(T_N)$, let $\{g_1^n\}_{n=1}^{\infty}$ be the sequence of elements in $C^{\infty}(T_N)$ that tends to g_1 in the H^1-norm. Since the Fourier coefficients of g_1^n are $O(|m|^{-(N+5)})$ as $|m| \to \infty$, the analog of the equality in (1.7) clearly holds for g_1^n, and we have

$$< \nabla g_1^n, \nabla g_1^n >_{L^2} = (2\pi)^N \sum_{m \in \Lambda_N} |m|^2 |\widehat{g_1^n}(m)|^2 .$$

Passing to the limit as $n \to \infty$ on both sides of this last equality, we obtain the finiteness of the right-hand side of (1.7). The equality in (1.7) comes from Parsevaal's theorem (Corollary 2.5 in Chapter 1).

To establish the lemma in the other direction, we are given $g_1 \in L^2(T_N)$ and that the right-hand side of the equality in (1.7) is finite. We set

$$(1.8) \qquad g_1^n = \sum_{|m| \le n} \widehat{g_1}(m) e^{im \cdot x}$$

and obtain from (1.5) that the sequence is Cauchy in the H^1-norm. Hence, there exists $h_1 \in H^1(T_N)$ such that

$$\lim_{n \to \infty} \|g_1^n - h_1\|_{H^1} = 0.$$

On the other hand, we obtain from (1.8) and Corollary 2.5 in Chapter 1 that

$$\lim_{n \to \infty} \|g_1^n - g_1\|_{L^2} = 0.$$

But then $g_1 = h_1$ almost everywhere. Therefore, $g_1 \in H^1(T_N)$.   ∎

As a matter of convenience, from now on, we shall assume that the T that occurs in (1.1) and in the statement of the theorem has the value 2π. Thus, from now on, we assume

(1.9) $\qquad\qquad (0,T) = (0, 2\pi)$ and $\tilde{\Omega} = T_N \times (0, 2\pi).$

We next recall the well-known fact that (see Corollary 2.3 in Chapter 1)

(1.10) $\quad \{(2\pi)^{-(N+1)/2} e^{ikt} e^{im\cdot x}\}_{k=-\infty}^{\infty}$ with $m \in \Lambda_N$ is a CONS for $L^2(\tilde{\Omega})$,

i.e., a complete orthonormal system for $L^2(\tilde{\Omega})$.

We define the Fourier coefficients for $f_1(x,t) \in L^2(\tilde{\Omega})$ with respect to the CONS in (1.10) as follows:

(1.11) $\qquad\qquad \widehat{f_1}(m,k) = (2\pi)^{-N-1} \int_{\tilde{\Omega}} f_1(x,t) e^{-ikt} e^{-im\cdot x} dx dt.$

Also, we set

(1.12) $\qquad\qquad S_M f_1(x,t) = \sum_{|m|\leq M} \sum_{k=-M}^{M} \widehat{f_1}(m,k) e^{ikt} e^{im\cdot x}$

and observe from (1.9) that if $f_1 \in L^2(\tilde{\Omega})$, then from Parsevaal's theorem (Corollary 2.5 in Chapter 1),

(1.13) $\qquad\qquad \lim_{M\to\infty} \|S_M f_1 - f_1\|_{L^2(\tilde{\Omega})} = 0.$

In the next lemma, we shall study a linear periodic initial value problem that is a variation of the nonlinear one set forth in (1.1) above when $J = 1$. In particular, with $f_1(x,t) \in L^2(\tilde{\Omega})$, we shall study the following problem:

(1.14) $\qquad (i)\ \dfrac{\partial u_1(x,t)}{\partial t} - \Delta u_1 + u_1 = f_1(x,t) \qquad$ in $T_N \times (0, 2\pi) = \tilde{\Omega},$

$\qquad (ii)\ u_1(x,t) \in L^2(0, 2\pi; H^1(T_N))$

$\qquad (iii)\ u_1(x,0) = 0 \qquad x \in \Omega.$

Condition (ii) is the boundary value part of the problem and asserts that we are dealing with periodic boundary values. Condition (iii) is the initial value part.

In order to study this periodic IVP, we shall need another Hilbert space. To do this, we introduce the pre-Hilbert space

(1.15)
$\tilde{C}^2(\tilde{\Omega}) = \{u_1 \in C^2[\mathbf{R}^N \times (0, 2\pi)]$ and
$\qquad\qquad$ for each fixed $t \in (0, 2\pi)$, $u_1(x,t) \in C^2(T_N)\}.$

In particular, we see if

$$u_1(x,t) = \sum_{j=1}^{n} \xi_j(t)\psi_j(x),$$

where $\xi_j \in C^2[0, 2\pi]$ and $\psi_j \in C^\infty(T_N)$, then $u_1 \in \tilde{C}^2(\tilde{\Omega})$.

Assuming (1.9), we endow $\tilde{C}^2(\tilde{\Omega})$ with an inner product $< \cdot, \cdot >_{\tilde{H}^1}$ defined as follows:
(1.16)

$$< u_1, v_1 >_{\tilde{H}^1} = \int_0^{2\pi} [< u_{1t}, v_{1t} >_{L^2} + < u_1, v_1 >_{L^2} + < \nabla u_1, \nabla v_1 >_{L^2}] dt$$

for $u_1, v_1 \in \tilde{C}^2(\tilde{\Omega})$ where $u_{1t} = \frac{\partial u_1}{\partial t}$. We complete the space $\tilde{C}^2(\tilde{\Omega})$ using the method of Cauchy sequences, and call the resulting Hilbert space $\tilde{H}^1(\tilde{\Omega})$.

Also, if u_1 and w_1 are in $L^2(\tilde{\Omega})$ and

$$\int_{\tilde{\Omega}} u_1 \frac{\partial \phi}{\partial t} dx dt = - \int_{\tilde{\Omega}} w_1 \phi dx dt \quad \forall \phi \in \tilde{C}_c^\infty(\tilde{\Omega}),$$

we shall refer to w_1 as u_{1t} and call it the weak derivative of u_1 with respect to t in $\tilde{\Omega}$. A similar definition prevails for the weak derivative of u_1 with respect to x_j in $\tilde{\Omega}$.

In particular, we see that if $u_1(x, t) \in \tilde{H}^1(\tilde{\Omega})$, then $u_1 \in L^2(\tilde{\Omega})$ and also that the following three conditions hold (see [Ev, pp.242, 244]):

(1.17)   (i) $u_1(x, t) \in W^{1,2}(\tilde{\Omega})$,

(ii) $u_1(x, t) \in L^2(0, 2\pi; H^1(T_N))$,

(iii) $u_{1t}(x, t) \in L^2(\tilde{\Omega})$.

We say $u_1(x, t)$ is a weak solution of the initial value problem (1.14) if $u_1(x, t) \in \tilde{H}^1(\tilde{\Omega})$ and the following two additional conditions are met (where $E \subset (0, 2\pi)$ is a set of measure zero):
(1.18)
$$(i) \lim_{t \to 0} \|u_1(\cdot, t)\|_{L^2} = 0 \quad t \in (0, 2\pi) \backslash E;$$

$$(ii) < \frac{\partial u_1}{\partial t}(\cdot, t), \psi(\cdot) >_{L^2} + < u_1(\cdot, t), \psi(\cdot) >_{H^1} = < f_1(\cdot, t), \psi(\cdot) >_{L^2}$$

$\forall \psi \in H^1(T_N)$ and for a.e. $t \in (0, 2\pi)$.

In the next lemma, we deal with weak solutions of the linear periodic IVP (1.14).

Lemma 1.3. *Suppose $f_1(x, t) \in L^2(\tilde{\Omega})$. Then there exists $u_1(x,t) \in \tilde{H}^1(\tilde{\Omega})$ which is a weak solution of the periodic initial value problem (1.14), i.e., both (1.18)(i) and (ii) hold.*

The above lemma is, in a way, the analogue of [Ev, Th.5, p.356] for the linear periodic IVP (1.14) but with a different proof.

Proof of Lemma 1.3. We first establish Lemma 1.3 for $S_M f_1(x,t)$ defined in (1.12). To do this, we set

$$(1.19) \qquad u_{1M}^{(1)}(x,t) = \sum_{|m| \leq M} \sum_{k=-M}^{M} \frac{\widehat{f_1}(m,k)}{ik + |m|^2 + 1} e^{ikt} e^{im \cdot x},$$

$$(1.20) \qquad u_{1M}^{(2)}(x,t) = \sum_{|m| \leq M} \sum_{k=-M}^{M} \frac{\widehat{f_1}(m,k)}{ik + |m|^2 + 1} e^{-(|m|^2+1)t} e^{im \cdot x}.$$

It follows from the definition of $\tilde{C}^2(\tilde{\Omega})$ in (1.15) that

$$(1.21) \qquad u_{1M}^{(j)}(x,t) \in \tilde{C}^2(\tilde{\Omega}) \text{ for } j = 1,2,$$

and we observe from (1.19), (1.20), and Parsevaal's theorem that

$$\left\| u_{1M}^{(1)}(\cdot,t) - u_{1M}^{(2)}(\cdot,t) \right\|_{L^2}^2$$

$$= (2\pi)^N \sum_{|m| \leq M} \left| \sum_{k=-M}^{M} \frac{\widehat{f}(m,k)}{ik+|m|^2+1} \left(e^{ikt} - e^{-(|m|^2+1)t} \right) \right|^2$$

for $t \in (0, 2\pi)$. Consequently, we obtain that

$$(1.22) \qquad \lim_{t \to 0} \left\| u_{1M}^{(1)}(\cdot,t) - u_{1M}^{(2)}(\cdot,t) \right\|_{L^2}^2 = 0.$$

Defining

$$(1.23) \qquad u_{1M}(x,t) = u_{1M}^{(1)}(x,t) - u_{1M}^{(2)}(x,t),$$

we conclude from (1.21) and (1.22) that

$$(1.24) \qquad u_{1M}(x,t) \in \tilde{C}^2(\tilde{\Omega}) \text{ and } \lim_{t \to 0} \| u_{1M}(\cdot,t) \|_{L^2}^2 = 0.$$

Next, as an easy calculation shows, we see from (1.19) and (1.20) and from (1.24) that

$$\frac{\partial u_{1M}}{\partial t} - \Delta u_{1M} + u_{1M} = S_M f_1(x,t) \qquad \text{for } (x,t) \in \tilde{\Omega}.$$

Consequently, (1.18) (ii) holds with respect to u_{1M} and $S_M f_1(x,t)$, and we have that

$$(1.25) \qquad \text{Lemma 1.3 holds for } S_M f_1(x,t).$$

It remains to show that Lemma 1.3 holds for $f_1(x,t)$. In order to do this, we observe from a calculation using the integral test for series that

$$(1.26) \qquad \sum_{k=1}^{\infty} \frac{1}{k^2 + a^2} \leq \frac{\pi}{2} a^{-1} \ \forall a > 0,$$

and also from (1.10) and Parsevaal's theorem that

$$(1.27) \qquad \sum_{m \in \Lambda_N} \sum_{k=-\infty}^{\infty} \left| \widehat{f}_1(m,k) \right|^2 < \infty.$$

Next, we set

$$(1.28) \qquad u_1^{(1)}(x,t) = \sum_{m \in \Lambda_N} \sum_{k=-\infty}^{\infty} \frac{\widehat{f}_1(m,k)}{ik + |m|^2 + 1} e^{ikt} e^{im \cdot x}$$

and observe from (1.27) that both

$$\sum_{m \in \Lambda_N} \sum_{k=-\infty}^{\infty} \left| \frac{\widehat{f}_1(m,k)}{ik + |m|^2 + 1} \right|^2 < \infty \text{ and } \sum_{m \in \Lambda_N} \sum_{k=-\infty}^{\infty} \left| \frac{ik \widehat{f}_1(m,k)}{ik + |m|^2 + 1} \right|^2 < \infty .$$

Therefore, it follows from (1.10), (1.28), and the Riesz-Fischer theorem [Zy1, p. 127] that both

$$(1.29) \qquad u_1^{(1)}(x,t) \text{ and } u_{1t}^{(1)}(x,t) \text{ are in } L^2(\tilde{\Omega})$$

where $u_{1t}^{(1)}$ is the weak derivative of $u_1^{(1)}$ in $L^2(\tilde{\Omega})$.

In a similar manner, we see from (1.26) and Schwarz's inequality that

$$(1.30)$$
$$\sum_{m \in \Lambda_N} |m|^2 \left| \sum_{k=-\infty}^{\infty} \frac{\widehat{f}_1(m,k)}{ik + |m|^2 + 1} e^{ikt} \right|^2 \le (\pi + 1) \sum_{m \in \Lambda_N} \sum_{k=-\infty}^{\infty} \left| \widehat{f}_1(m,k) \right|^2 < \infty$$

for $0 < t < 2\pi$.

We conclude from Lemma 1.2 and this last inequality that

$$u_1^{(1)}(x,t) \in L^2(0, 2\pi; H^1(T_N)).$$

Also, it follows from (1.26) coupled with this last fact and (1.29) that

$$(1.31) \qquad u_1^{(1)}(x,t) \in \tilde{H}^1(\tilde{\Omega}) \text{ and } \lim_{M \to \infty} \left\| u_{1M}^{(1)} - u_1^{(1)} \right\|_{\tilde{H}^1} = 0$$

where $u_M^{(1)}(x,t)$ is defined in (1.19).

Next, we set

$$(1.32) \qquad u_1^{(2)}(x,t) = \sum_{m \in \Lambda_N} \left(\sum_{k=-\infty}^{\infty} \frac{\widehat{f}(m,k)}{ik + |m|^2 + 1} \right) e^{-(|m|^2 + 1)t} e^{im \cdot x}$$

where the coefficient of $e^{im \cdot x}$ involving the k-sum is well-defined by (126), (127), and Schwarz's inequality. Also, since the multiple series involved converge absolutely for $0 < t < 2\pi$, we see that $u_1^{(2)}(x,t) \in C^\infty[\mathbf{R}^N \times (0, 2\pi)]$. Furthermore, it is clear from Parsevaal's theorem that

$$\left\| u_1^{(2)}(\cdot, t) \right\|_{L^2}^2 = (2\pi)^N \sum_{m \in \Lambda_N} \left| \sum_{k=-\infty}^{\infty} \frac{\widehat{f}(m,k)}{ik + |m|^2 + 1} \right|^2 e^{-2(|m|^2 + 1)t} \text{ for } t \in (0, 2\pi),$$

and we conclude from a calculation like that used in (1.30) along with Lemma 1.2 that

$$(1.33) \qquad u_1^{(2)}(x,t) \in L^2(\widetilde{\Omega}) \cap L^2(0, 2\pi; H^1(T_N)).$$

Furthermore,

$$\left\| u_{1t}^{(2)}(\cdot, t) \right\|_{L^2}^2 = (2\pi)^N \sum_{m \in \Lambda_N} (|m|^2 + 1)^2 \left| \sum_{k=-\infty}^{\infty} \frac{\widehat{f}(m,k)}{ik + |m|^2 + 1} \right|^2 e^{-2(|m|^2+1)t}$$

for $t \in (0, 2\pi)$, and we conclude from a similar calculation to that used in (1.30) that $u_{1t}^{(2)} \in L^2(\widetilde{\Omega})$. Hence, we obtain from (1.33) that

$$(1.34) \qquad u_1^{(2)}(x,t) \in \widetilde{H}^1(\widetilde{\Omega}) \text{ and also that } \lim_{M \to \infty} \left\| u_{1M}^{(2)} - u_1^{(2)} \right\|_{\widetilde{H}^1} = 0$$

where $u_{1M}^{(2)}(x,t)$ is defined in (1.20).

Next, we set

$$(1.35) \qquad u_1(x,t) = u_1^{(1)}(x,t) - u_1^{(2)}(x,t),$$

and observe from (1.31) and (1.34) that

$$u_1(x,t) \in \widetilde{H}^1(\widetilde{\Omega}).$$

To complete the proof to Lemma 1.3, it remains to show that (1.18)(i) and (ii) hold for $u_1(x,t)$.

To show the former, we observe from (1.28) and (1.32) that
(1.36)

$$\left\| u_1^{(1)}(\cdot, t) - u_1^{(2)}(\cdot, t) \right\|_{L^2}^2 = (2\pi)^N \sum_{m \in \Lambda_N} \left| \sum_{k=-\infty}^{\infty} \frac{\widehat{f}_1(m,k)}{ik + |m|^2 + 1} (e^{ikt} - e^{-(|m|^2+1)t}) \right|^2$$

for $t \in (0, 2\pi) \backslash E$ where E is of measure zero.

Using (1.26), we obtain that

$$\left| \sum_{k=-\infty}^{\infty} \frac{\widehat{f}_1(m,k)}{ik + |m|^2 + 1} (e^{ikt} - e^{-(|m|^2+1)t}) \right|^2 \leq 4(\pi+1)(|m|^2+1)^{-1} \sum_{k=-\infty}^{\infty} \left| \widehat{f}_1(m,k) \right|^2,$$

and we conclude from (1.36) and the Lebesgue dominated convergence theorem applied to series that

$$\lim_{t \to 0} \left\| u_1^{(1)}(\cdot, t) - u_1^{(2)}(\cdot, t) \right\|_{L^2}^2 = 0 \text{ for } t \in (0, 2\pi) \backslash E.$$

This fact coupled with (1.35) tells us that indeed (1.18)(i) holds for $u_1(x,t)$.

To show that (1.18)(ii) holds for $u_1(x,t)$, we first observe from (1.31) and (1.34) that for a subsequence

$$\lim_{M \to \infty} \left\| \frac{\partial u_{1M}^{(j)}}{\partial t}(\cdot, t) - \frac{\partial u_1^{(j)}}{\partial t}(\cdot, t) \right\|_{L^2} = 0 \text{ for a.e. } t \in (0, 2\pi) \text{ for } j = 1, 2.$$

Hence, we see from (1.35) that for a subsequence

$$(1.37) \qquad \lim_{M \to \infty} \left\| \frac{\partial u_{1M}}{\partial t}(\cdot, t) - \frac{\partial u_1}{\partial t}(\cdot, t) \right\|_{L^2} = 0 \text{ for a.e. } t \in (0, 2\pi).$$

Likewise, we see from (1.31) and (1.34) that for a subsequence

$$(1.38) \qquad \lim_{M \to \infty} \| u_{1M}(\cdot, t) - u_1(\cdot, t) \|_{H^1} = 0 \text{ for a.e. } t \in (0, 2\pi),$$

and also that for a subsequence

$$(1.39) \qquad \lim_{M \to \infty} \| S_M f_1(\cdot, t) - f_1(\cdot, t) \|_{L^2} = 0 \text{ for a.e. } t \in (0, 2\pi).$$

Also, we obtain from (1.25) and (1.18)(ii)

$$< \frac{\partial u_{1M}}{\partial t}(\cdot, t), \psi(\cdot) >_{L^2} + < u_{1M}(\cdot, t), \psi(\cdot) >_{H^1} = < S_M f_1(\cdot, t), \psi(\cdot) >_{L^2}$$

$\forall \psi \in H^1(T_N)$ and for a.e. $t \in (0, 2\pi)$. Taking the limit of both sides of this last equality as $M \to \infty$ through a subsequence, we see from (1.37)-(1.39) that (1.18)(ii) does indeed hold for $u_1(x, t)$. ∎

Next, we establish the following lemma, which is useful in showing that the solution u_1 in Lemma 1.3 is unique.

Lemma 1.4. *Suppose $\exists u_1 \in \widetilde{H}^1(\widetilde{\Omega})$ that satisfies (1.18)(i) and (ii) where $f_1 \in L^2(\widetilde{\Omega})$. Then*

$$2^{-1} \int_{T_N} |u_1(x, t)|^2 \, dx + \int_0^t < u_1(\cdot, \tau), u_1(\cdot, \tau) >_{H^1} d\tau$$

$$(1.40) \qquad = \int_0^t [\int_{T_N} f_1(x, \tau) u_1(x, \tau) dx] d\tau \quad \text{for a.e. } t \in (0, 2\pi).$$

Proof of Lemma 1.4.   Since by assumption $u_1 \in \widetilde{H}^1(\widetilde{\Omega})$, we have from (1.17)(ii) that

$$u_1(x, t) \in H^1(T_N) \text{ for } t \in (0, 2\pi) \backslash E \text{ where } meas(E) = 0.$$

Consequently, we have from (1.18)(ii) that

$$< \frac{\partial u_1}{\partial t}(\cdot, t), u_1(\cdot, t) >_{L^2} + < u_1(\cdot, t), u_1(\cdot, t) >_{H^1} = < f_1(\cdot, t), u_1(\cdot, t) >_{L^2}$$

for $t \in (0, 2\pi) \backslash E_1$ where $meas(E_1) = 0$. Hence, using (1.17)(ii) and (iii), we see that

$$\int_r^s [< \frac{\partial u_1}{\partial t}(\cdot, t), u_1(\cdot, t) >_{L^2} + < u_1(\cdot, t), u_1(\cdot, t) >_{H^1}] dt$$

$$(1.41)$$

$$= \int_r^s < f_1(\cdot, t), u_1(\cdot, t) >_{L^2} dt$$

for $0 < r < s < 2\pi$. Now, it is easy to see from the fact that $u_1 \in \widetilde{H}^1(\widetilde{\Omega})$ that there exists $E_2 \subset (0, 2\pi)$ with $meas(E_2) = 0$ such that

$$\int_r^s < \frac{\partial u_1}{\partial t}(\cdot, t), u_1(\cdot, t) >_{L^2} dt = 2^{-1} \int_{T_N} |u_1(x, s)|^2 dx - 2^{-1} \int_{T_N} |u_1(x, r)|^2 dx$$

for $r, s \in (0, 2\pi) \backslash E_2$. Combining this fact with (1.41) and using (1.18)(i), we see that that the equality in (1.40) does indeed hold for $t \in (0, 2\pi) \backslash E_2$, and the proof of Lemma 1.4 is complete. ∎

Lemma 1.5.   *Under the assumptions of Lemma 1.3, the $u_1(x,t) \in \widetilde{H}^1(\widetilde{\Omega})$ which is a weak solution of the initial value problem (1.14) is unique. Furthermore, $u_1(x,t)$ is defined by the equality given in (1.35), i.e., $u_1 = u_1^{(1)} - u_1^{(2)}$ where $u_1^{(1)}$ is given in (1.28), and $u_1^{(2)}$ in (1.32).*

Proof of Lemma 1.5.   Suppose both $u_1, v_1 \in \widetilde{H}^1(\widetilde{\Omega})$ satisfy (1.18)(i) and (ii). Then on setting $w_1 = u_1 - v_1$, we see that $w_1 \in \widetilde{H}^1(\widetilde{\Omega})$ and satisfies

$$\lim_{t \to 0} \|w_1(\cdot, t)\|_{L^2} = 0 \quad t \in (0, 2\pi) \backslash E$$

where $meas(E) = 0$ and also that

$$< \frac{\partial w_1}{\partial t}(\cdot, t), \psi(\cdot) >_{L^2} + < w_1(\cdot, t), \psi(\cdot) >_{H^1} = 0 \ \forall \psi \in H^1(T_N)$$

for a.e. $t \in (0, 2\pi)$. It follows from these two facts and (1.40) in Lemma 1.4 that

$$2^{-1} \int_{T_N} |w_1(x, t)|^2 dx + \int_0^t < w_1(\cdot, \tau), w_1(\cdot, \tau) >_{H^1} d\tau = 0 \ \text{ for a.e. } t \in (0, 2\pi).$$

But by (1.5), the second integral in this last equality is nonnegative. Therefore, the first integal equals zero a.e. in $(0, 2\pi)$. However, $w_1 \in L^2(\widetilde{\Omega})$. Hence,

$$\int_{\widetilde{\Omega}} |w_1(x, t)|^2 dx dt = 0.$$

We conclude that $w_1 = 0$ a.e. in $\widetilde{\Omega}$, which tells us that u_1 is indeed unique. That u_1 is given by the equality in (1.35) follows from the proof given of Lemma 1.3. The proof to Lemma 1.5 is complete. ∎

Next, given $v_1 \in L^2(\widetilde{\Omega})$, we set

$$\mathcal{B}_1 v_1 = \sum_{m \in \Lambda_N} \sum_{k=-\infty}^{\infty} \frac{\widehat{v}_1(m, k)}{ik + \lambda_m} e^{ikt} e^{im \cdot x},$$

(1.45)

$$\mathcal{B}_2 v_1 = \sum_{m \in \Lambda_N} \sum_{k=-\infty}^{\infty} \frac{\widehat{v}_1(m, k)}{ik + \lambda_m} e^{-\lambda_m t} e^{im \cdot x}$$

where $\widehat{v}_1(m, k)$ is defined by (1.11) and meets (1.27) and

(1.46) $$\lambda_m = |m|^2 + 1.$$

It is clear from (1.10) and (1.27) that

$$\mathcal{B}_j : L^2(\tilde{\Omega}) \to L^2(\tilde{\Omega}) \ \text{ for j=1,2.}$$

The following lemma also holds:

Lemma 1.6. $\mathcal{B}_j$ *is both a continuous and a compact map of* $L^2(\tilde{\Omega})$ *into* $L^2(\tilde{\Omega})$ *for j=1,2.*

(For the definition of a compact map, we refer the reader to [Ev, p. 503].)

Proof of Lemma 1.6.   To prove the lemma, suppose $\{v_{1n}\}_{n=1}^{\infty}$ is a sequence in $L^2(\tilde{\Omega})$ with the property that

$$\lim_{n\to\infty} \|v_{1n} - v_1\|_{L^2(\tilde{\Omega})} = 0.$$

Then using the ideas set forth in the proof of Lemma 1.3, we see that

$$(1.47) \qquad \lim_{n\to\infty} \sum_{m\in\Lambda_N} \sum_{k=-\infty}^{\infty} |\widehat{v_{1n}}(m,k) - \widehat{v_1}(m,k)|^2 = 0.$$

Now, it follows from (1.45) and Parsevaal's theorem that

$$(1.48) \ \ \|\mathcal{B}_1(v_{1n} - v_1)\|_{L^2(\tilde{\Omega})}^2 = (2\pi)^{N+1} \sum_{m\in\Lambda_N} \sum_{k=-\infty}^{\infty} \frac{|\widehat{v_{1n}}(m,k) - \widehat{v_1}(m,k)|^2}{|ik + \lambda_m|^2}.$$

But then we see from (1.47) that

$$(1.49) \qquad \lim_{n\to\infty} \|\mathcal{B}_1(v_{1n} - v_1)\|_{L^2(\tilde{\Omega})}^2 = 0.$$

This establishes the continuity of the map $\mathcal{B}_1$.

To show the continuity of the map $\mathcal{B}_2$, we observe from (1.45) that
(1.50)

$$\|\mathcal{B}_2[v_{1n}(\cdot,t) - v_1(\cdot,t)]\|_{L^2}^2 = (2\pi)^N \sum_{m\in\Lambda_N} e^{-2\lambda_m t} \left| \sum_{k=-\infty}^{\infty} \frac{\widehat{v_{1n}}(m,k) - \widehat{v_1}(m,k)}{ik + \lambda_m} \right|^2$$

for almost every $t \in (0, 2\pi)$. Integrating both sides of this last equality from 0 to 2π, we see from (1.10), (1.26), and Schwarz's inequality that there is a constant C such that

$$\|\mathcal{B}_2(v_{1n} - v_1)\|_{L^2(\tilde{\Omega})}^2 \le C \sum_{m\in\Lambda} \sum_{k=-\infty}^{\infty} |\widehat{v_{1n}}(m,k) - \widehat{v_1}(m,k)|^2 \ \ \forall n.$$

We conclude once again from (1.47) that

$$(1.51) \qquad \lim_{n\to\infty} \|\mathcal{B}_2(v_{1n} - v_1)\|_{L^2(\tilde{\Omega})}^2 = 0.$$

This establishes the continuity of the map $\mathcal{B}_2$.

To show that $\mathcal{B}_1$ is a compact map, suppose $\{v_{1n}\}_{n=1}^{\infty}$ is a sequence in $L^2(\tilde{\Omega})$ with the property that $\left\{\|v_{1n}\|_{L^2(\tilde{\Omega})}\right\}_{n=1}^{\infty}$ is a uniformly bounded sequence. Then it follows that for some subsequence $\{v_{1n}\}_{n=1}^{\infty}$ converges weakly in $L^2(\tilde{\Omega})$ to a function v_1 in $L^2(\tilde{\Omega})$ [Ev, p.640]. Hence, we see that the compactness of $\mathcal{B}_1$ will follow (where we are now assuming the full sequence converges weakly) if (i) and (ii) below imply that the limit in (1.49) holds where

$$(i) \sum_{m \in \Lambda} \sum_{k=-\infty}^{\infty} |\widehat{v}_{1n}(m,k) - \widehat{v}_1(m,k)|^2 \leq C \ \ \forall n,$$

$$(ii) \qquad \qquad \lim_{n \to \infty} \widehat{v}_{1n}(m,k) = \widehat{v}_1(m,k) \quad \forall(m,k).$$

We now establish that the limit in (1.49) holds given (i) and (ii). Let $\varepsilon > 0$ be given. Choose $M > 1$ so large so that the following inequality holds:

$$\frac{(2\pi)^{N+1}C}{(M+1)^2} < \frac{\varepsilon}{2}.$$

Then we see from (i), (1.46), and (1.48) that

$$\|\mathcal{B}_1(v_{1n} - v_1)\|_{L^2(\tilde{\Omega})}^2 \leq (2\pi)^{N+1} \sum_{|m| \leq M} \sum_{k=-M}^{M} \frac{|\widehat{v}_{1n}(m,k) - \widehat{v}_1(m,k)|^2}{|ik + \lambda_m|^2} + \frac{\varepsilon}{2} + \frac{\varepsilon}{2}.$$

Consequently, using (ii), we obtain that

$$\limsup_{n \to \infty} \|\mathcal{B}_1(v_{1n} - v_1)\|_{L^2(\tilde{\Omega})}^2 \leq \varepsilon.$$

Since ε is arbitrary, this establishes the fact that the limit in (1.49) is valid and that $\mathcal{B}_1$ is indeed a compact map.

Using the same ideas that we have just used for $\mathcal{B}_1$, we can show that $\mathcal{B}_2$ is a compact map; what we have to do is to show that (i) and (ii) above imply that the limit in (1.51) holds. We now do this.

From (1.50), we see that

(1.52)

$$\|\mathcal{B}_2(v_{1n} - v_1)\|_{L^2(\tilde{\Omega})}^2 \leq (2\pi)^N \sum_{m \in \Lambda_N} \frac{1}{2\lambda_m} \left| \sum_{k=-\infty}^{\infty} \frac{\widehat{v}_{1n}(m,k) - \widehat{v}_1(m,k)}{ik + \lambda_m} \right|^2$$

where λ_m is given by (1.46). Let $\varepsilon > 0$ be given. Choose M so large so that both of the following inequalities hold:

$$\frac{(2\pi)^N C(\pi+1)}{2(M^2+1)^2} < \frac{\varepsilon}{2}; \qquad \frac{(2\pi)^N C}{M} < \frac{\varepsilon}{2}.$$

Then we obtain from Schwarz's inequality, (i), (1.26), and (1.52) that

$$\|\mathcal{B}_2(v_{1n} - v_1)\|^2_{L^2(\tilde{\Omega})} \leq \sum_{|m| \leq M} \frac{1}{2\lambda_m} \left| \sum_{k=-M}^{M} \frac{\widehat{v}_{1n}(m,k) - \widehat{v}_1(m,k)}{ik + \lambda_m} \right|^2 + \frac{\varepsilon}{2} + \frac{\varepsilon}{2}.$$

Consequently, using (ii), we obtain that

$$\limsup_{n \to \infty} \|\mathcal{B}_2(v_{1n} - v_1)\|^2_{L^2(\tilde{\Omega})} \leq \varepsilon.$$

Since ε is arbitrary, this establishes the fact that the limit in (1.51) is valid and that $\mathcal{B}_2$ is indeed a compact map. ∎

Given $v = (v_1, ..., v_J) \in [L^2(\tilde{\Omega})]^J$, we define $\mathcal{B}_n(v)$ for $n = 1, 2$ as follows:

$$(1.53) \qquad \mathcal{B}_n(v) = (\mathcal{B}_n(v_1), ... \mathcal{B}_n(v_J)).$$

It is clear that $\mathcal{B}_n(v) : [L^2(\tilde{\Omega})]^J \to [L^2(\tilde{\Omega})]^J$. As an immediate corollary to Lemma 1.6, we have the following:

Lemma 1.7. *With $\mathcal{B}_n(v)$ defined by (1.53), $\mathcal{B}_n(v)$ is both a continuous and a compact map of $[L^2(\tilde{\Omega})]^J$ into $[L^2(\tilde{\Omega})]^J$ for n=1,2.*

Assuming that $f_j(x, t, s)$ is a Caratheodory function as in (1.2) satisfying the additional condition

$$(1.54) \qquad \exists \, K > 0 \text{ s.t. } |f_j(x, t, s)| \leq K \quad \forall s \in \mathbf{R}^J \text{ and a.e. } (x, t) \in \tilde{\Omega},$$

for $j = 1, ..., J$, we shall study weak solutions of the following nonlinear reaction-diffusion system with periodic boundary conditions:

$$(1.55) \qquad \begin{cases} \dfrac{\partial u_j}{\partial t} - \Delta u_j + u_j = f_j(x, t, u_1, ..., u_J) & \text{in } T_N \times (0, 2\pi), \\[2mm] u_j(x, 0) = 0 \end{cases}$$

for $j = 1, ..., J$.

In particular, we say that $u = (u_1, ..., u_J)$ is a weak solution of (1.55) under the assumption that $f_j(x, t, s)$ satisfies (1.54) provided the following holds:

$u(x, t) \in [\tilde{H}^1(\tilde{\Omega})]^J$ and the following two additional conditions are met (where $E \subset (0, 2\pi)$ is a set of measure zero):

$$(1.56) \qquad (i) \lim_{t \to 0} \|u_j(\cdot, t)\|_{L^2} = 0 \quad t \in (0, 2\pi) \backslash E;$$

.

$$(ii) < \frac{\partial u_j}{\partial t}(\cdot, t), \psi(\cdot) >_{L^2} + < u_j(\cdot, t), \psi(\cdot) >_{H^1} = < f_j(\cdot, t, u), \psi(\cdot) >_{L^2}$$

$\forall \psi \in H^1(T_N)$, for a.e. $t \in (0, 2\pi)$, and for $j = 1, ..., J$.

The following lemma prevails:

Lemma 1.8. *Suppose that $f_j(x,t,s)$ is a Caratheodory function that meets (1.54) for $j = 1, ..., J$. Then there exists $u(x,t) \in [\widetilde{H}^1(\tilde{\Omega})]^J$, which is a weak solution of the reaction-diffusion system with periodic boundary conditions given by (1.55), i.e., both (1.56)(i) and (ii) are valid.*

Proof of Lemma 1.8. To establish Lemma 1.8, we first observe that because of (1.54),

$$f(x,t,\cdot) = (f_1(x,t,\cdot), ..., f_J(x,t,\cdot))$$

is a continuous map of $[L^2(\tilde{\Omega})]^J$ into $[L^2(\tilde{\Omega})]^J$.

Consequently, it follows from Lemma 6 that $\mathcal{B}_n f(x,t,\cdot)$ is a continuous map of $[L^2(\tilde{\Omega})]^J$ into $[L^2(\tilde{\Omega})]^J$ for $n = 1, 2$, where $\mathcal{B}_n$ is defined in (1.45) for $n = 1, 2$. Also, (1.54) and Lemma 1.7 imply that $\mathcal{B}_n f(x,t,\cdot)$ is a compact map of $[L^2(\tilde{\Omega})]^J$ into $[L^2(\tilde{\Omega})]^J$ for $n = 1, 2$. Hence $A(v)$ is a continuous and compact map of $[L^2(\tilde{\Omega})]^J$ into $[L^2(\tilde{\Omega})]^J$, where

$$(1.57) \qquad A(v) = \mathcal{B}_1 f(x,t,v) - \mathcal{B}_2 f(x,t,v) \quad \text{for } v \in [L^2(\tilde{\Omega})]^J.$$

Next, we observe from Lemma 1.5 that given $v \in [L^2(\tilde{\Omega})]^J$, $u = A(v)$ is in $[\widetilde{H}^1(\tilde{\Omega})]^J$ and u_j, the j-th component of u, satisfies (1.56)(i) as well as $(1,56)(ii)$ with $f_j(\cdot,t,u)$ replaced by $f_j(\cdot,t,v)$. Hence, if it can be shown that there is a $u \in [L^2(\tilde{\Omega})]^J$ that is a fixed point of the map $A(u)$, i.e., $u = A(u)$, then Lemma 1.8 will be established.

So to complete the proof of Lemma 1.8, it remains to show that the continuous and compact map $A(u)$ defined in (1.57) has a fixed point. To do this, we invoke Schaefer's fixed point theorem, [Ev, p. 504], and see that for $A(u)$ to have a fixed point, it is sufficient to show that the following set is a bounded set in $[L^2(\tilde{\Omega})]^J$:

$$F = \left\{ u \in [L^2(\tilde{\Omega})]^J : u = \sigma A(u) \text{ for some } 0 < \sigma \leq 1 \right\}.$$

To show that the elements of F are uniformly bounded, we observe from (1.10), (1.26), (1.48), and (1.50) that

$$\|\mathcal{B}_n v_j\|_{[L^2(\tilde{\Omega})]^J} \leq (\pi + 1)^{\frac{1}{2}} \|v_j\|_{[L^2(\tilde{\Omega})]^J} \quad \forall v_j \in L^2(\tilde{\Omega}) \quad \text{for } n = 1, 2.$$

Therefore, from (1.54),

$$\|\mathcal{B}_n f(x,t,v)\|_{[L^2(\tilde{\Omega})]^J} \leq (\pi + 1)^{\frac{1}{2}} J^{\frac{1}{2}} K(2\pi)^{(N+1)/2}.$$

So it follows from (1.57) and and this last inequality that if $u \in F$, then

$$\|u\|_{[L^2(\tilde{\Omega})]^J} \leq 2(\pi + 1)^{\frac{1}{2}} J^{\frac{1}{2}} K(2\pi)^{(N+1)/2}.$$

Hence, we see that the elements of F are uniformly bounded in $[L^2(\tilde{\Omega})]^J$, and the proof of Lemma 1.8 is complete. ■

Proof of Theorem 1.1.   Without loss in generality, we assume from the start that $T = 2\pi$. Next, we set

$$(1.58) \qquad g_j(x,t,s) = s_j + f_j(x,t,s) \quad \text{for} \quad j = 1, \dots J,$$

and instead of dealing with periodic solutions of the IVP (1.1), we deal with periodic solutions of the following IVP:

$$(1.59) \qquad \begin{cases} \dfrac{\partial u_j}{\partial t} - \Delta u_j + u_j = g_j(x,t,u_1,\dots,u_J) & \text{in } T_N \times (0, 2\pi), \\[2mm] \hspace{4cm} u_j(x,0) = 0 \end{cases}$$

for $j = 1, \dots, J$.

It is clear from (1.58) that g_j satisfies (1.2) and (1.3). It also satisfies (1.4) with C_1 replaced by $C_1 + 1$. It is also clear from (1.58) that if we solve (1.59), then we also have solved (1.1) when $T = 2\pi$. So to prove the theorem, we see from (1.6) that we have to show the following:

$$(1.60)$$
$$(i)\exists \ u \in [L^2(0,2\pi; H^1(T_N))]^J \cap [L^\infty \ (0,2\pi; L^2(T_N))]^J$$
$$\text{with both } u_j g_j(x,t,u) \text{ and } g_j(x,t,u) \in L^1(\tilde{\Omega})$$

$$(ii) \int_0^T \int_{T_N} (-u_j \tfrac{\partial \theta}{\partial t} - u_j \Delta \theta + u_j \theta) = \int_{\tilde{\Omega}} g_j(x,t,u)\theta \ \ \forall \theta \in \tilde{C}_c^\infty(\tilde{\Omega}),$$
$$\text{for } j = 1, \dots J$$

$$(iii) \ \lim_{t \to 0} \ \|u_j(\cdot,t)\|_{L^2} = 0 \quad t \in (0,2\pi)\backslash E \text{ where } E \subset (0,2\pi)$$
$$\text{and } meas(E) = 0 \quad \text{for } j = 1, \dots J.$$

To show that (1.60) holds, the first step we take is to set for $n \geq 1$,

$$g_{jn}(x,t,s) = g_j(x,t,s) \quad \text{if } |g_j(x,t,s)| \leq n$$
$$= n \hspace{2.3cm} \text{if } g_j(x,t,s) \geq n$$
$$= -n \hspace{2cm} \text{if } g_j(x,t,s) \leq -n$$

for $j = 1, \dots J$. Then $g_{jn}(x,t,s)$ satisfies (1.54), and we can invoke Lemma 1.8 to obtain

$$(1.61) \qquad u_j^{(n)}(x,t) \in \tilde{H}^1(\tilde{\Omega}) \text{ and } u_j^{(n)} \text{ satisfies } (1.56)(i) \ \text{ for } j = 1, \dots J.$$

Also,

$$(1.62) \qquad u_j^{(n)}(x,t) \text{ and } g_{jn}(x,t,s) \text{ satisfy } (1.56)(ii) \ \forall n \ \text{ and } j = 1, \dots J.$$

Next, we show that the following holds:
$$(1.63)$$
$$\exists C > 0 \text{ s.t. } (i) \left\| u_j^{(n)} \right\|_{L^2(\tilde{\Omega})} \leq C \ \text{ and } (ii) \left\| u_j^{(n)} g_{jn}(\cdot,\cdot,u^{(n)}) \right\|_{L^1(\tilde{\Omega})} \leq C$$

for $j = 1, \dots, J$ and $\forall n$ where $u^{(n)} = (u_1^{(n)}, \dots, u_J^{(n)})$.

To accomplish this, we use (1.61) and (1.62) in conjunction with Lemma 1.4 to obtain

(1.64)
$$2^{-1} \int_{T_N} \left| u_j^{(n)}(x,t) \right|^2 dx + \int_0^t < u_j^{(n)}(\cdot,\tau), u_j^{(n)}(\cdot,\tau) >_{H^1} d\tau$$

$$= \int_0^t [\int_{T_N} g_{jn}(x,\tau,u^{(n)}) u_j^{(n)}(x,\tau) dx] d\tau \quad \text{for a.e. } t \in (0, 2\pi).$$

Observing that for each n and j, $s_j g_{jn}(x,t,s)$ satisfies the inequality in (1.4) with C_1 replaced by $C_1 + 1$ for a.e. $(x,t) \in \tilde{\Omega}$ and $\forall s \in \mathbf{R}^J$, we see from (1.64) that there are two positive constants C_1^* and C_2^* such that

$$(1.65) \quad \int_{T_N} \left| u_j^{(n)}(x,t) \right|^2 dx \le C_1^* \int_0^t [\int_{T_N} \left| u_j^{(n)}(x,\tau) \right|^2 dx] d\tau + C_2^* \quad \forall n,$$

for $j = 1, ..., J$, and for a.e. $t \in (0, 2\pi)$.

From this last inequality, in conjuction with the integral form of Gronwall's inequality [Ev, p. 625], we obtain that there is a positive constant C_1^{**} such that
(1.66)
$$\int_{T_N} \left| u_j^{(n)}(x,t) \right|^2 dx \le C_1^{**} \ \forall n \text{ and for } j = 1, ..., J \text{ and for a.e. } t \in (0, 2\pi).$$

This establishes (1.63)(i) with $C = (2\pi C_1^{**})^{\frac{1}{2}}$.

To establish (1.63)(ii), we see from (1.63)(i) joined with the fact that $s_j g_{jn}(x,t,s)$ satisfies (1.4) that there exists a positive constant C_2^{**} such that

$$(1.67) \quad \int_{B_{jn}} u_j^{(n)}(x,t) g_{jn}(x,t,u^{(n)}) dx dt \le C_2^{**} \ \forall n \text{ and for } j = 1, ..., J,$$

where

$$B_{jn} = \{(x,t) \in \tilde{\Omega} : u_j^{(n)}(x,t) g_{jn}(x,t,u^{(n)}) \ge 0\}.$$

But $\tilde{\Omega} = A_{jn} \cup B_{jn}$ with $A_{jn} \cap B_{jn} = \emptyset$, and also the left-hand side of the equality in (1.64) is nonnegative. In particular, it is nonnegative when $t = 2\pi$. So it follows from (1.67) that

$$-\int_{A_{jn}} u_j^{(n)}(x,t) g_{jn}(x,t,u^{(n)}) dx dt \le C_2^{**}.$$

Consequently, $\int_{\tilde{\Omega}} \left| u_j^{(n)}(x,t) g_{jn}(x,t,u^{(n)}) \right| dx dt \le 2 C_2^{**} \quad \forall n$ and for $j = 1, ..., J$, and (1.63)(ii) is established.

Next, we make use of the fact that $g_j(x,t,s)$ meets the condition in (1.3) for $j = 1, ..., J$, to infer that the following holds:

(1.68)
$$\text{for } R \ge 1, \exists \alpha_R(x,t) \in L^1(\tilde{\Omega}) \text{ s.t. if } \left| u_j^{(n)}(x,t) \right| \le R,$$

$$\text{then } \left| g_{jn}(x,t,u^{(n)}) \right| \le \alpha_R(x,t) \ \forall n \text{ for } j = 1, ..., J.$$

We see, in particular, from this last inequality that

$$\left| g_{jn}(x,t,u^{(n)}) \right| \leq |\alpha_2(x,t)| + \left| u_j^{(n)}(x,t) g_{jn}(x,t,u^{(n)}) \right|$$

$\forall n$, for $j = 1, ..., J$ and for $(x,t) \in \tilde{\Omega}$. So we conclude from (1.63)(ii) that
(1.69)

$$\exists C^\# \text{ such that } \int_{\tilde{\Omega}} \left| g_{jn}(x,t,u^{(n)}) \right| dx dt \leq C^\# \quad \forall n \quad \text{and for } j = 1, ..., J.$$

We also see from (1.68) we can obtain the following:
(1.70)

$$\{g_{jn}(x,t,u^{(n)})\}_{n=1}^\infty \text{ is absolutely equi-integrable in } L^1(\tilde{\Omega}) \text{ for } j = 1, ..., J.$$

What is meant by this last statement is $\forall \varepsilon > 0$, $\exists \delta > 0$ s.t. given $E \subset \tilde{\Omega}$, with $meas(E) < \delta$, then $\int_E \left| g_{jn}(x,t,u^{(n)}) \right| dx dt < \varepsilon$ $\forall n$ and for $j = 1, ..., J$.

To establish (1.70) with C as in (1.63)(ii),

$$(1.71) \qquad\qquad \text{select } R > 1 \text{ such that } CR^{-1} < \frac{\varepsilon}{2}.$$

Next, with $\alpha_R(x,t)$ as in (1.68), choose $\delta > 0$, such that $E \subset \tilde{\Omega}$, and

$$(1.72) \qquad\qquad meas(E) < \delta \Longrightarrow \int_E |\alpha_R(x,t)| \, dx dt < \frac{\varepsilon}{2}.$$

From (1.68), we see that

$$|g_{jn}(x,t,u_n)| \leq |\alpha_R(x,t)| + |u_{jn}(x,t) g_{jn}(x,t,u_n)| \, R^{-1} \, \forall n.$$

The statement in (1.70) follows easily from this last inequality joined with (1.63)(ii), (1.71), and (1.72).

Next, with C as in (1.63)(ii), we see from (1.64) that

$$(1.73) \qquad \int_0^{2\pi} < u_j^{(n)}(\cdot,t), u_j^{(n)}(\cdot,t) >_{H^1} dt \leq C \quad \forall n \text{ and for } j = 1, ..., J.$$

It then follows from (1.5), this last inequality, and standard Hilbert space arguments (see [Ev, pp. 638-640]) that

$$\exists u_j \in L^2(0,2\pi; H^1(T_N)) \text{ s.t. } (i) \lim_{n\to\infty} \int_{\tilde{\Omega}} u_j^{(n)} w \, dx dt = \int_{\tilde{\Omega}} u_j w \, dx dt,$$
(1.74)

$$(ii) \lim_{n\to\infty} \int_0^{2\pi} <u_j^{(n)}(\cdot,t), v(\cdot,t) >_{H^1} dt = \int_0^{2\pi} <u_j(\cdot,t), v(\cdot,t) >_{H^1} dt,$$

$\forall w \in L^2(\tilde{\Omega})$, $\forall v \in L^2(0,2\pi; H^1(T_N))$, and for $j = 1, ..., J$, where we have used the full sequence rather than a subsequence for convenience of notation.

With u_j as in (1.74), we will demonstrate that

$$(1.75) \qquad\qquad \lim_{n\to\infty} \left\| u_j^{(n)} - u_j \right\|_{L^2(\tilde{\Omega})} = 0 \quad \text{for } j = 1, ..., J.$$

To establish (1.75), we use (1.10) and designate by $\hat{u}_j^{(n)}(m,k)$ and $\hat{u}_j(m,k)$ the corresponding Fourier coefficients of $u_j^{(n)}$ and u_j, as in (1.11).

Since both $u_j^{(n)}$ and u_j are in $L^2(\tilde{\Omega})$, it follows from Parsevaal's theorem that

(1.76) $$\left\| u_j^{(n)} - u_j \right\|_{L^2(\tilde{\Omega})}^2 = (2\pi)^{N+1} \sum_{m \in \Lambda_N} \sum_{k=-\infty}^{\infty} \left| \widehat{u}_j^{(n)}(m,k) - \widehat{u}_j(m,k) \right|^2.$$

Also, we have from (1.74)(i) that

(1.77) $$\lim_{n \to \infty} \widehat{u}_j^{(n)}(m,k) = \widehat{u}_j(m,k) \quad \forall(m,k).$$

Next, from (1.61) and (1.17)(ii), we see that $u_j^{(n)} \in H^1(T_N)$ for a.e. t and for every n. Hence, we obtain from Lemma 1.2 that
(1.78)
$$<u_j^{(n)}(\cdot,t), u_j^{(n)}(\cdot,t)>_{H^1} = (2\pi)^{-N} \sum_{m \in \Lambda_N} \lambda_m \left| \int_{T_N} u_j^{(n)}(x,t)e^{-im \cdot x} dx \right|^2$$

$\forall n$, for $j = 1, ..., J$, and for a.e. $t \in (0, 2\pi)$ where

(1.79) $$\lambda_m = |m|^2 + 1.$$

Also, we see from (1.73) that

$$\int_0^{2\pi} \left| \int_{T_N} u_j^{(n)}(x,t)e^{-im \cdot x} dx \right|^2 dt = (2\pi)^{2N+1} \sum_{k=-\infty}^{\infty} \left| \widehat{u}_j^{(n)}(m,k) \right|^2 \quad \forall m \epsilon \Lambda_N.$$

Consequently, we obtain from (1.73) and (178) that

$$(2\pi)^{N+1} \sum_{m \in \Lambda_N} \lambda_m \sum_{k=-\infty}^{\infty} \left| \widehat{u}_j^{(n)}(m,k)) \right|^2 \leq C \quad \forall n \text{ and for } j = 1, ..., J.$$

But this last inequality in conjunction with (1.77) enables us to conclude

(1.80) $$(2\pi)^{N+1} \sum_{m \in \Lambda_N} \lambda_m \sum_{k=-\infty}^{\infty} \left| \widehat{u}_j^{(n)}(m,k) - \widehat{u}_j(m,k) \right|^2 \leq 4C$$

$\forall n$ and for $j = 1, ..., J$.

Continuing, we use (1.62) and (1.56)(ii) to obtain that

$$\int_r^t e^{-ik\tau} \{ \int_{T_N} e^{-im \cdot x} [\partial u_j^{(n)}(x,\tau)/\partial \tau + \lambda_m u_j^{(n)}(x,\tau)] dx \} d\tau$$
(1.81)
$$= \int_r^t e^{-ik\tau} [\int_{T_N} g_{jn}(x,t,u^{(n)})e^{-im \cdot x} dx] d\tau \quad \forall n, \forall m, \forall k, \text{ for } j = 1, ..., J$$

and for $0 < r < t < 2\pi$.

Using the fact that $u_j^{(n)} \in \widetilde{H}^1(\tilde{\Omega})$, we next see that there exists $E_1 \subset (0, 2\pi)$ with $meas(E_1) = 0$ such that

$$\int_{T_N} e^{-im \cdot x} dx \int_r^t e^{-ik\tau} \partial u_j^{(n)}(x,\tau)/\partial \tau \, d\tau = \int_{T_N} e^{-im \cdot x} u_j^{(n)}(x,\tau)e^{-ik\tau} |_{\tau=r}^{\tau=t} \, dx$$

$$+ \int_r^t \int_{T_N} ike^{-ik\tau} e^{-im\cdot x} u_j^{(n)}(x,\tau) \, dx d\tau \quad \forall k \text{ and } \forall m,$$

with $0 < r < t < 2\pi$ and $r,t \in (0,2\pi)\backslash E_1$. Hence, we obtain from this last equality joined with (1.81) that

(1.82)
$$\int_r^t \int_{T_N} (ik+\lambda_m)e^{-ik\tau} e^{-im\cdot x} u_j^{(n)}(x,\tau) dx d\tau =$$

$$- \int_{T_N} e^{-im\cdot x} u_j^{(n)}(x,\tau) e^{-ik\tau} \mid_{\tau=r}^{\tau=t} dx$$

$$+ \int_r^t e^{-ik\tau} [\int_{T_N} g_{jn}(x,t,u^{(n)}) e^{-im\cdot x} dx] d\tau$$

$\forall k$ and $\forall m$ with $0 < r < t < 2\pi$ and $r,t \in (0,2\pi)\backslash E_1$.

From (1.66), we see for a.e. r and t in $(0,2\pi)$ with $r < t$, that the absolute value of the first expression on the right-hand side of (1.82) is majorized by $2(2\pi)^{N/2}(C_1^{**})^{\frac{1}{2}}$. Likewise, we see from (1.69) that the absolute value of the second expression on the right-hand side of (1.82) is majorized by $C^{\#}$. Hence, we obtain from (1.82) that there exists $E_2 \subset (0,2\pi)$ with $meas(E_2) = 0$ such that
(1.83)
$$\left| (ik+\lambda_m) \int_r^t \int_{t_n} e^{-ik\tau} e^{-im\cdot x} u_j^{(n)}(x,\tau) dx d\tau \right| \le 2(2\pi)^{N/2}(C_1^{**})^{\frac{1}{2}} + C^{\#}$$

$\forall k$ and $\forall m$ and for $0 < r < t < 2\pi$ with $r,t \in (0,2\pi)\backslash E_2$. Taking the limit in (1.83) as $r \to 0$ and as $t \to 2\pi$ with $r,t \in (0,2\pi)\backslash E_2$, we conclude that

$$\left| \widehat{u}_j^{(n)}(m,k) \right| \le [2(2\pi)^{N/2}(C_1^{**})^{\frac{1}{2}} + C^{\#}]/|ik+\lambda_m| \quad \forall k, \forall m, \forall n,$$

and for $j = 1, ..., J$.

This last inequality joined with the limit in (1.77), in turn, gives us

(1.84)
$$\left| \widehat{u}_j^{(n)}(m,k) - \widehat{u}_j(m,k) \right|^2 \le C^{\Diamond}/(k^2+\lambda_m^2)$$

$\forall k, \forall m, \forall n$, and for $j = 1, ..., J$ where $C^{\Diamond} = 4[2(2\pi)^{N/2}(C_1^{**})^{\frac{1}{2}} + C^{\#}]^2$.

To see that the limit in (1.75) actually prevails, from the equality in (1.76), we see it is sufficient to show that

(1.85)
$$\lim_{n\to\infty} \limsup_{M\to\infty} \sum_{|m|\le M} \sum_{k=-\infty}^{\infty} \left| \widehat{u}_j^{(n)}(m,k) - \widehat{u}_j(m,k) \right|^2 = 0$$

for $j = 1, ..., J$.

To establish this last fact, given $\varepsilon > 0$, we choose M_0 to be a positive integer so large that

$$M_0^{-2} 4C < \frac{\varepsilon}{2}$$

where C is the constant in (1.80). Also, we observe from (1.79) that

$$M_0 \le |m| \implies 1 \le \frac{\lambda_m}{M_0^2}.$$

Hence, it follows from (1.80) that

$$(1.86) \qquad \sum_{|m|>M_0} \sum_{k=-\infty}^{\infty} \left| \widehat{u}_j^{(n)}(m,k) - \widehat{u}_j(m,k) \right|^2 \le M_0^{-2} 4C < \frac{\varepsilon}{2}$$

$\forall n$ and for $j = 1, ..., J$.

Next, let the number of integral lattice points in the following set

$$\{m : |m| \le M_0\}$$

be designated by $M_0^{\#}$, and choose K_0 to be a positive integer so large that

$$M_0^{\#} K_0^{-1} C^{\diamond} \le \frac{\varepsilon}{4},$$

where $C^{\diamond}$ is the constant in (1.84). Then, using the fact that $\sum_{k=K_0+1}^{\infty} \frac{1}{k^2} \le K_0^{-1}$, we see from (1.79) and (1.84) that

$$\sum_{|m|\le M_0} \sum_{k=K_0+1}^{\infty} \left| \widehat{u}_j^{(n)}(m,k) - \widehat{u}_j(m,k) \right|^2 \le \sum_{|m|\le M_0} K_0^{-1} C^{\diamond} \le M_0^{\#} K_0^{-1} C^{\diamond} \le \frac{\varepsilon}{4}$$

$\forall n$ and for $j = 1, ..., J$. Similarly,

$$\sum_{|m|\le M_0} \sum_{k=-\infty}^{-(K_0+1)} \left| \widehat{u}_j^{(n)}(m,k) - \widehat{u}_j(m,k) \right|^2 \le \frac{\varepsilon}{4}$$

$\forall n$ and for $j = 1, ..., J$, and we obtain from these last two inequalities that

$$\sum_{|m|\le M_0} \sum_{k=-\infty}^{\infty} \left| \widehat{u}_j^{(n)}(m,k) - \widehat{u}_j(m,k) \right|^2$$

$$\le \sum_{|m|\le M_0} \sum_{k=-K_0}^{K_0} \left| \widehat{u}_j^{(n)}(m,k) - \widehat{u}_j(m,k) \right|^2 + \frac{\varepsilon}{2}$$

$\forall n$ and for $j = 1, ..., J$.

From (1.77), (1.86), and this last inequality, we therefore obtain that

$$\lim_{n\to\infty} \limsup_{M\to\infty} \sum_{|m|\le M} \sum_{k=-\infty}^{\infty} \left| \widehat{u}_j^{(n)}(m,k) - \widehat{u}_j(m,k) \right|^2 \le \varepsilon.$$

But ε is an arbitrary positive number. Hence, the left-hand side of this last inequality is zero, and the assertion in (1.85) is established. Consequently, the limit in (1.75) also holds, and this is the key fact in the proof of the theorem.

Continuing, from (1.64) and (1.75), we see that the following three facts also hold:

$\exists E_3 \subset (0, 2\pi)$ with $meas(E_3) = 0$ such that

$$(1.87) \qquad \lim_{n \to \infty} \int_{T_N} \left| u_j^{(n)}(x,t) - u_j(x,t) \right|^2 dx = 0 \quad \text{for } t \in (0, 2\pi) \backslash E_3;$$

$$(1.88) \qquad \lim_{n \to \infty} u_j^{(n)}(x,t) = u_j(x,t) \quad \text{for a.e. } (x,t) \in \tilde{\Omega};$$

$$(1.89) \quad \int_{T_N} \left| u_j^{(n)}(x,t) \right|^2 dx \leq 2 \int_0^t [\int_{T_N} u_j^{(n)}(x,\tau) g_{jn}(x,\tau, u^{(n)}(x,\tau)) dx] d\tau$$

for $t \in (0, 2\pi) \backslash E_3$, for $j = 1, ..., J$ and $\forall n$, where we have used the full sequence rather than a subsequence for convenience of notation.

We next show that the limit in (1.60)(iii) holds, i.e.,

$$(1.90) \qquad \lim_{t \to 0} \int_{T_N} |u_j(x,t)|^2 dx = 0 \quad t \in (0, 2\pi) \backslash E_3$$

for $j = 1, ..., J$.

To do this, we observe that the inequality in (1.4) holds for $s_j g_{jn}(x,t,s)$ with C_1 replaced by $C_1 + 1$, and therefore, from (1.89), we obtain

$$\int_{T_N} \left| u_j^{(n)}(x,t) \right|^2 dx \leq 2 \int_0^t \int_{T_N} [(C_1 + 1) \left| u_j^{(n)}(x,t) \right|^2$$
$$+ C_2(x,t) \left| u_j^{(n)}(x,t) \right| + C_3(x,t)] dx d\tau$$

for $t \in (0, 2\pi) \backslash E_3$, $\forall n$, and for $j = 1, ..., J$, where C_1 is a positive constant, $C_2(x,t) \in L^2(\tilde{\Omega})$, and $C_3(x,t) \in L^1(\tilde{\Omega})$. Also, C_2 and C_3 are nonnegative functions. Passing to the limit on both sides of the above inequality as $n \to \infty$ and making use of (175) and (1.90), we see that

(1.91)
$$\int_{T_N} |u_j(x,t)|^2 dx \leq 2 \int_0^t \int_{T_N} [(C_1 + 1) |u_j(x,t)|^2$$

$$+ C_2(x,t) |u_j(x,t)| + C_3(x,t)] dx d\tau$$

for $t \in (0, 2\pi) \backslash E_3$. But $u_j \in L^2(\tilde{\Omega})$. Consequently, the limit as $t \to 0$ of the integral on the right-hand side of the inequality in (1.91) is zero. Hence the limit of the integral on the left-hand side of the inequality in (1.91) is zero if $t \to 0$ through the values in $(0, 2\pi) \backslash E_3$, and (1.90) is completely established.

Next, using the fact that $u_j \in L^2(\tilde{\Omega})$ for $j = 1, ..., J$, we designate the value of the integral when $t = 2\pi$ on the right-hand of the inequality in (1.91) by the positive constant K_1^*. Hence,

$$\int_{T_N} |u_j(x,t)|^2 dx \leq 2K_1^* \quad for \ t \in (0, 2\pi) \backslash E_3.$$

Since $meas(E_3) = 0$, we obtain that

$$(1.92) \qquad\qquad u_j \in L^\infty(0, 2\pi; L^2(T_N))$$

for $j = 1, ..., J$.

We would like to show that the rest of the items in (1.60)(i) also prevail. To do this, we use (1.63)(ii), (1.69), (1.88), and Fatou's Lemma [Ev, p. 648] to obtain that

$$(1.93) \quad (i) \int_{\tilde{\Omega}} |u_j g_j(x,t,u)| \, dxdt \le C \text{ and } (ii) \int_{\tilde{\Omega}} |g_j(x,t,u)| \, dxdt \le C^\sharp,$$

for $j = 1, ..., J$ where C and $C^\sharp$ are positive constants. From (1.74), we see that $u_j \in L^2(0, 2\pi; H^1(T_N))$ for $j = 1, ..., J$. So (1.92) along with the inequalities in (1.93) show that all the items in (1.60)(i) do indeed prevail.

Since the limits in (1.90) are the same as those in (1.60)(iii), to complete the proof to the theorem, it remains to show that the equalities in (1.60)(ii) hold. We now do this.

From (1.61) and (1.62), we infer that $u_j^{(n)} \epsilon \tilde{H}^1(\tilde{\Omega})$ and

$$\int_0^{2\pi} < \frac{\partial u_j^{(n)}}{\partial t}(\cdot,t), \theta(\cdot,t) >_{L^2} dt + \int_0^{2\pi} <u_j^{(n)}(\cdot,t), \theta(\cdot,t) >_{H^1} dt$$

$$(1.94) \qquad = \int_0^{2\pi} < g_{jn}(\cdot,t,u^{(n)}), \theta(\cdot,t) >_{L^2} dt \quad \forall \theta \in \tilde{C}_c^\infty(\tilde{\Omega}),$$

$\forall n$, and for $j = 1, ..., J$. Now the first integral on the left-hand side of the equality in (1.94) is equal to

$$\int_{T_N} [\int_0^{2\pi} \frac{\partial u_j^{(n)}}{\partial t}(x,t)\theta(x,t)dt]dx.$$

But $\theta \in \tilde{C}_c^\infty(\tilde{\Omega})$ and $u_j^{(n)} \in \tilde{H}^1(\tilde{\Omega})$. Hence this latter integral is equal to

$$-\int_0^{2\pi} \int_{T_N} u_j^{(n)}(x,t)\frac{\partial \theta}{\partial t}(x,t)dxdt.$$

We record this fact as

$$(1.95) \qquad \int_0^{2\pi} < \frac{\partial u_j^{(n)}}{\partial t}(\cdot,t), \theta(\cdot,t) >_{L^2} dt = -\int_0^{2\pi} \int_{T_N} u_j^{(n)} \frac{\partial \theta}{\partial t} dxdt.$$

Also, from the fact that $\theta \in \tilde{C}_c^\infty(\tilde{\Omega})$ and $u_j^{(n)} \in \tilde{H}^1(\tilde{\Omega})$, we see that

$$\int_0^{2\pi} \int_{T_N} \frac{\partial}{\partial x_k} u_j^{(n)}(x,t)\frac{\partial}{\partial x_k}\theta(x,t)dxdt = -\int_0^{2\pi} \int_{T_N} u_j^{(n)}(x,t)\frac{\partial^2}{\partial x_k^2}\theta(x,t)dxdt$$

for $k = 1, ..., N$. Hence, we obtain from (1.5) that

$$(1.96) \qquad \int_0^{2\pi} <u_j^{(n)}(\cdot,t), \theta(\cdot,t) >_{H^1} dt = \int_0^{2\pi} \int_{T_N} [-u_j^{(n)}\Delta\theta + u_j^{(n)}\theta]dxdt.$$

From (1.95) and (1.96) joined with (1.94), we consequently have that

(1.97)
$$\int_0^{2\pi} \int_{T_N} [-u_j^{(n)} \frac{\partial \theta}{\partial t} - u_j^{(n)} \Delta \theta + u_j^{(n)} \theta] dx dt$$

$$= \int_0^{2\pi} \int_{T_N} g_{jn}(x, t, u_j^{(n)}) \theta(x, t) dx dt \quad \forall \theta \in \tilde{C}_c^\infty(\tilde{\Omega}),$$

$\forall n$, and for $j = 1, ..., J$.

Now the limit in (1.88) implies that

$$\lim_{n \to \infty} g_{jn}(x, t, u_j^{(n)}) = g_j(x, t, u) \quad \text{for a.e. } (x, t) \in \tilde{\Omega}.$$

From the observations that $\theta \in \tilde{C}_c^\infty(\tilde{\Omega})$, that

$$g_{jn}(x, t, u_j^{(n)}) \text{ is absolutely equi-integrable in } L^1(\tilde{\Omega})$$

(see (1.70)), and that $g_j(x, t, u) \in L^1(\tilde{\Omega})$ (see (1.93)(ii)), we obtain from this last limit and Egoroff's Theorem [Ev, p. 647] that the limit of the integral on the right-hand side of the equality in (1.97) as $n \to \infty$ is equal to

$$\int_0^{2\pi} \int_{T_N} g_j(x, t, u) \theta(x, t) dx dt \quad \forall \theta \in \tilde{C}_c^\infty(\tilde{\Omega})$$

and for $j = 1, ..., J$.

On the other hand, from (1.75), we see that the limit of the integral on the left-hand side of the equality in (1.97) as $n \to \infty$ is

$$\int_0^{2\pi} \int_{T_N} [-u_j(x, t) \frac{\partial \theta}{\partial t}(x, t) - u_j(x, t) \Delta \theta(x, t) + u_j(x, t) \theta(x, t)] dx dt.$$

Hence, we obtain that

$$\int_0^{2\pi} \int_{T_N} [-u_j(x, t) \frac{\partial \theta}{\partial t}(x, t) - u_j(x, t) \Delta \theta(x, t) + u_j(x, t) \theta(x, t)] dx dt$$

$$= \int_0^{2\pi} \int_{T_N} g_j(x, t, u) \theta(x, t) dx dt \quad \forall \theta \in \tilde{C}_c^\infty(\tilde{\Omega}),$$

and for $j = 1, ..., J$.

But this last equality is the same as the equality in (1.60)(ii). Therefore (1.60)(ii) is indeed true, and the proof of the theorem is complete. ∎

If we replace (1.4) by the two-sided condition

(1.98) $$|f_j(x, t, s)| \leq C_1 |s_j| + C_2$$

for a.e. $(x, t) \in \tilde{\Omega}$ and $\forall s \in \mathbf{R}$, where C_1 and C_2 are positive constants and $j = 1, ..., J$, then we can obtain an improvement to the conclusion of Theorem 1.1. In particular, we can show that the solution u to the reaction-diffusion problem (1.1) under assumption (1.98) is in $[\tilde{H}^1(\tilde{\Omega})]^J$, which is defined in (1.16) above with $T = 2\pi$. Also, we can obtain a weak solution to the problem (1.1). In other words, we can show that

$u(x, t) \in [\widetilde{H}^1(\tilde{\Omega})]^J$ and the following two additional conditions are met (where $E \subset (0, 2\pi)$ is a set of measure zero):

(1.99) (i) $\lim\limits_{t \to 0} \|u_j(\cdot, t)\|_{L^2} = 0$ $t \in (0, 2\pi) \backslash E;$

.

(ii) $< \dfrac{\partial u_j}{\partial t}(\cdot, t), \psi(\cdot) >_{L^2} + < \nabla u_j(\cdot, t), \nabla \psi(\cdot) >_{L^2} = < f_j(\cdot, t, u), \psi(\cdot) >_{L^2}$

$\forall \psi \in H^1(T_N)$, for a.e. $t \in (0, 2\pi)$, and for $j = 1, ..., J$.

We do all this in the following theorem, which we shall prove:

Theorem 1.9.   *Let $T_N \times (0, T) = \tilde{\Omega}$. With $s = (s_1, ..., s_J)$, assume that $f_j(x, t, s)$ satisfies the conditions stated in (1.2) and (1.98) for $j = 1, ..., J$. Then there exists*

$$u \in [\widetilde{H}^1(\tilde{\Omega})]^J \cap [L^\infty (0, T; L^2(T_N))]^J$$

with $u_j f_j(x, t, u) \in L^1(\tilde{\Omega})$ and $f_j(x, t, u) \in L^2(\tilde{\Omega})$ for $j = 1, ..., J$, such that $u = (u_1, ..., u_J)$ satisfies the conditions in (1.99) and is, therefore, a weak solution of the reaction-diffusion system (1.1).

Proof of Theorem 1.9. With no loss in generality, we once again assume that $T = 2\pi$ and observe that $f_j(x, t, s)$ clearly satisfies the conditons in the hypothesis of Theorem 1.1. So, by that theorem, we have the existence of a

(1.100) $\qquad u \in [L^2 (0, 2\pi; H^1(T_N))]^J \cap [L^\infty (0, T; L^2(T_N))]^J,$

which is also a generalized periodic solution of the reaction-diffusion system (1.1).

What we have to do is show that this $u \in [\widetilde{H}^1(\tilde{\Omega})]^J$ and that it meets condition (ii) in (1.99). We already know that it meets condition (i) in (1.99).

As in the proof of Theorem 1.1, we set

(1.101) $\qquad g_j(x, t, s) = s_j + f_j(x, t, s)$ for $j = 1, ...J,$

and

(1.102) $\qquad \begin{aligned} g_{jn}(x, t, s) &= g_j(x, t, s) && \text{if } |g_j(x, t, s)| \le n \\ &= n && \text{if } g_j(x, t, s) \ge n \\ &= -n && \text{if } g_j(x, t, s) \le -n. \end{aligned}$

Then from (1.61) and (1.62) in the proof of Theorem 1.1, we obtain

(1.103) $\quad u_j^{(n)}(x, t) \in \widetilde{H}^1(\tilde{\Omega})$ and $u_j^{(n)}$ satisfies (1.56)(i) for $j = 1, ...J.$

Also,

(1.104) $\quad u_j^{(n)}(x, t)$ and $g_{jn}(x, t, u^{(n)})$ satisfy (1.56)(ii) $\forall n$ and $j = 1, ...J.$

Next, we observe from (1.100) that, in particular, $u \in [L^2(\tilde{\Omega})]^J$. Hence, it follows from (1.98) that

$$(1.105) \qquad f_j(x,t,u) \in L^2(\tilde{\Omega}) \text{ for } j = 1, ...J,$$

and also from (1.101) that

$$(1.106) \qquad g_j(x,t,u) \in L^2(\tilde{\Omega}) \text{ for } j = 1, ...J.$$

So, we invoke Lemma 1.3 and obtain the existence of

$$(1.107) \qquad v = (v_1, ...v_J) \in [\widetilde{H}^1(\tilde{\Omega})]^J$$

such that

$$(1.108) \qquad v_j \text{ and } g_j(x,t,u) \text{ satisfy } (1.18) \text{ (i) and (ii) for } j = 1, ...J.$$

As a consequence of (1.104) and this last fact, we have that

$$< \frac{\partial u_j^{(n)}}{\partial t}(\cdot,t) - \frac{\partial v_j}{\partial t}(\cdot,t), \psi(\cdot) >_{L^2} + < u_j^{(n)}(\cdot,t) - v_j(\cdot,t), \psi(\cdot) >_{H^1}$$

$$= < g_{jn}(\cdot,t,u^{(n)}) - g_j(x,t,u), \psi(\cdot) >_{L^2}$$

$\forall \psi \in H^1(T_N)$, for a.e. $t \in (0, 2\pi)$, and for $j = 1, ..., J$.

But then, invoking Lemma 1.4, we obtain

$$2^{-1} \int_{T_N} \left| u_j^{(n)}(x,t) - v_j(x,t) \right|^2 dx$$
$$+ \int_0^t < u_j^{(n)}(x,\tau) - v_j(x,\tau), u_j^{(n)}(x,\tau) - v_j(x,\tau) >_{H^1} d\tau$$

$$(1.109)$$
$$= \int_0^t \left\{ \int_{T_N} \left[g_{jn}(x,t,u^{(n)}) - g_j(x,t,u) \right] \left[u_j^{(n)}(x,\tau) - v_j(x,\tau) \right] dx \right\} d\tau$$

for a.e. $t \in (0, 2\pi)$ and for $j = 1, ..., J$.

Next, we recall certain facts from the proof of Theorem 1.1, namely from (1.75) that the following holds:

$$(i) \lim_{n \to \infty} \left\| u_j^{(n)} - u_j \right\|_{L^2(\tilde{\Omega})} ;$$

$$(1.110)$$

$$(ii) \lim_{n \to \infty} u_j^{(n)}(x,t) = u_j(x,t) \text{ for } a.e. (x,t) \in \tilde{\Omega}.$$

We observe from (1.98) and (1.101) that

$$\left| g_{jn}(x,t,u^{(n)}) \right|^2 \leq \left[\left| u_j^{(n)} \right| + C_1 \left| u_j^{(n)} \right| + C_2 \right]^2$$

$\forall n$. Hence it follows from (1.110)(i) that

$\{ \left| g_{jn}(x,t,u^{(n)}) \right|^2 \}_{n=1}^\infty$ is absolutely equi-integrable in $L^1(\tilde{\Omega})$ for $j = 1, ..., J$.

Also, we obtain from (1.110)(ii) that

$$\lim_{n \to \infty} g_{jn}(x,t,u^{(n)}) = g_j(x,t,u) \text{ for } a.e.(x,t) \in \tilde{\Omega}.$$

From these last two facts joined with Egoroff's theorem and Schwarz's inequality, we see that the integral on the right-hand side of the equality in (1.109) goes to zero as $n \to \infty$. We state this as

$$\lim_{n \to \infty} \int_0^t \int_{T_N} \left[g_{jn}(x, t, u^{(n)}) - g_j(x, t, u) \right] \left[u_j^{(n)}(x, \tau) - v_j(x, \tau) \right] dx d\tau = 0$$

for $t \in (0, 2\pi)$ and for $j = 1, ..., J$.

But then it follows from (1.109) and this last limit that

$$\lim_{n \to \infty} \int_0^t \int_{T_N} \left[u_j^{(n)}(x, \tau) - v_j(x, \tau) \right]^2 dx d\tau = 0$$

for *a.e.* $t \in (0, 2\pi)$ and for $j = 1, ..., J$.

We conclude from (1.110)(i) and this last fact that

(1.111) $u_j(x, t) = v_j(x, t)$ for   *a.e.* $(x, t) \in \dot{\tilde{\Omega}}$ and for $j = 1, ..., J$.

From this last equality and (1.107), we see that indeed

$$u = (u_1, ... u_J) \in [\widetilde{H}^1(\tilde{\Omega})]^J.$$

Also, from (1.108) joined with (1.111), we obtain that

$$< \frac{\partial u_j}{\partial t}(\cdot, t), \psi(\cdot) >_{L^2} + < u_j(\cdot, t), \psi(\cdot) >_{H^1} = < g_j(\cdot, t, u), \psi(\cdot) >_{L^2}$$

$\forall \psi \in H^1(T_N)$, for a.e. $t \in (0, 2\pi)$, and for $j = 1, ..., J$.

But using (1.05) and (1.101), we see this last equality can be replaced with

$$< \frac{\partial u_j}{\partial t}(\cdot, t), \psi(\cdot) >_{L^2} + < \nabla u_j(\cdot, t), \nabla \psi(\cdot) >_{L^2} = < f_j(\cdot, t, u), \psi(\cdot) >_{L^2}$$

This establishes (1.99)(ii) and completes the proof of the theorem. ∎

Exercises.

   1. Prove that $H^1(T_2)$, which is defined in (1.5), can be generated also by the functions in $C^2(T_N)$ where

$$C^2(T_2) = \{ \phi \in C^2(\mathbf{R}^2) : \phi(x) \text{ is periodic of period } 2\pi \text{ in each variable} \}.$$

   2. Given $\{f_n\}_{n=1}^\infty$, a sequence of functions absolutely equi-integrable in $L^1(T_2)$ with the property that

$$\lim_{n \to \infty} f_n(x) = 0 \text{ for a.e. } x \in T_N,$$

prove that $\lim_{n \to \infty} \int_{T_2} |f_n(x)| dx = 0$.

   3. Using the notation of this section, given $\tilde{\Omega} = T_2 \times (0, 2\pi)$ and $f \in [\tilde{C}^2(\tilde{\Omega})]^2$, find $u \in [\widetilde{H}^1(\tilde{\Omega})]^2 \cap [L^\infty(0, 2\pi; L^2(T_2))]^2$, which satisfies the conditions in (1.99) where

$$f_1(x,t) \;=\; \cos t \cos x_1 + \sin 2t \sin 3x_2,$$

$$f_2(x,t) \;=\; (\sum_{k=1}^{\infty} \frac{\cos kt}{k^5})(\sum_{n=1}^{\infty} \frac{\cos nx_1}{n^8}).$$

4. Using the notation of this section, given $\tilde{\Omega} = T_2 \times (0, 2\pi)$ and $f(x,t,s) \in [C(\tilde{\Omega} \times \mathbf{R}^2)]^2$ and meets the conditions in (1.98), find

$$u \in [\tilde{H}^1(\tilde{\Omega})]^2 \cap [L^\infty(0, 2\pi; L^2(T_2))]^2,$$

which satisfies the conditions in (1.99) where

$$f_1(x,t,u) \;=\; -u_1 + \sin t \cos x_1 + \cos 2t \sin x_2$$
$$f_2(x,t,u) \;=\; u_2/2 + 1 + \cos x_1.$$

2. Quasilinear Ellipticity on the N-Torus

In this section, we shall study the existence of solutions to quasilinear elliptic partial differential equations under periodic boundary conditions. Once again, we will see the power of Fourier analysis in handling these difficult problems.

The material presented here comes mainly from the author's Transaction paper, [Sh12], and makes use of an ingenious (according to the referee of said paper) Galerkin-type argument. It serves quite well as an introduction to the general subject of quasilinear ellipticity.

We will use the notation introduced in the previous section of this chapter except that now u will represent a function and not a vector function as it did previously.

In particular, we will operate in N-dimensional Euclidean space $\mathbf{R}^N$, $N \geq 2$, and as before, let T_N be the N-dimensional torus

$$T_N = \{x : -\pi \leq x_n < \pi, \; n = 1, ..., N\}.$$

$g(x) \in C^\infty(T_N)$ will mean that $g(x) \in C^\infty(\mathbf{R}^N)$ and is periodic of period 2π in each of the variables x_n for $n = 1, ..., N$.

We introduce the Hilbert space $H^1(T_N)$ as follows:

$H^1(T_N)$ is the closure of the set of functions in $C^\infty(T_N)$ under the norm generated by the following real inner product

$$< g, h >_{H^1} = \int_{T_N} [gh + Dg \cdot Dh] dx \quad \text{for } g, h \in C^\infty(T_N),$$

where $Du = (D_1u, ..., D_N u)$ and $D_n u = \partial u/\partial x_n$. (In this section, we will use the notation $Du = \nabla u$.)

So if $u, v \in H^1(T_N)$, then

(2.1) $< u, v >_{H^1} = \int_{T_N} [uv + Du \cdot Dv] dx$ for $u, v \in H^1(T_N)$.

Also, for each n, u has a weak partial derivative on T_N with respect to x_n, that there is a function $w \in L^2(T_N)$ such that

$$\int_{T_N} u D_n \phi \, dx = - \int_{T_N} w \phi \, dx \quad \forall \phi \in C^\infty(T_N),$$

for $n = 1, ..., N$. We refer to w as $D_n u$.

This last equality could also have been written as

$$< u, D_n \phi >_{L^2} = - < w, \phi >_{L^2}$$

We shall consider second order, quasilinear elliptic operators Q, operating on $H^1(T_N)$ of the form

(2.2) $Qu = - \sum_{i,j=1}^{N} D_i [a^{ij}(x, u) D_j u] + \sum_{j=1}^{N} b^j(x, u, Du) D_j u.$

The coefficients of Q, namely the functions $a^{ij}(x, s)$ and $b^j(x, s, p)$ are assumed to be defined for $(x, s) \in T_N \times \mathbf{R}$ and $(x, s, p) \in T_N \times \mathbf{R} \times \mathbf{R}^N$, respectively. Furthermore, we shall suppose the following throughout this section.

(Q_1) The coefficients $a^{ij}(x, s)$ and $b^j(x, s, p)$ satisfy the Caratheodory conditions: For each fixed $s \in \mathbf{R}$ and $p \in \mathbf{R}^N$, the functions $a^{ij}(x, s)$ and $b^j(x, s, p)$ are measureable; for a.e. $x \in T_N$, the functions $a^{ij}(x, s)$ and $b^j(x, s, p)$ are respectively continuous in $\mathbf{R}$ and $\mathbf{R} \times \mathbf{R}^N$, $i, j = 1, ..., N$.

(Q_2) $\exists$ a nonnegative function $a(x) \in L^2(T_N)$ and an $\eta > 0$ such that

$$\left| a^{ij}(x, s) \right| \leq a(x) + \eta |s|$$

for $s \in \mathbf{R}$ and a.e. $x \in T_N$, $i, j = 1, ..., N$.

(Q_3) The principal part of Q is symmetric, that is, $a^{ij}(x, s) = a^{ji}(x, s)$ for $s \in \mathbf{R}$ and a.e. $x \in T_N$, $i, j = 1, ..., N$.

(Q_4) Q is uniformly elliptic almost everywhere in T_N, that is, there is a constant $\eta_0 > 0$ such that

$$\sum_{i,j=1}^{N} a^{ij}(x, s) \xi_i \xi_j \geq \eta_0 |\xi|^2$$

for $s \in \mathbf{R}$, a.e. $x \in T_N$, and $\xi \in \mathbf{R}^N$ ($|\xi|^2 = \xi_1^2 + \cdots + \xi_N^2$).

(Q_5) There is a nonnegative function $b(x) \in L^2(T_N)$ and positive constants η_1 and η_2 such that

$$\left| b^j(x, s, p) \right| \leq b(x) + \eta_1 |s| + \eta_2 |p|$$

for $s \in \mathbf{R}$, a.e. $x \in T_N$, and $p \in \mathbf{R}^N$, $j = 1, ..., N$.

(Q_6) For every $u \in H^1(T_N)$, the vector function

$$\mathbf{b}(x, u, Du) = \left[b^1(x, u, Du), ..., b^N(x, u, Du) \right]$$

is weakly solenoidal, i.e.,

$$(2.3) \qquad \int_{T_N} \left[\sum_{j=1}^{N} b^j(x, u, Du) D_j v \right] dx = 0$$

for $u, v \in H^1(T_N)$.

(Q_7) If $\{u_n\}_{n=1}^\infty$ is a sequence of functions in $L^2(T_N)$, which tends strongly to $u \in L^2(T_N)$ and $\{\mathbf{w}_n\}_{n=1}^\infty$ is a sequence of vector-valued functions in $[L^2(T_N)]^N$, which tends weakly to $\mathbf{w} \in [L^2(T_N)]^N$, then $\{\mathbf{b}(x, u_n, \mathbf{w}_n)\}_{n=1}^\infty$ is a sequence of vector-valued functions in $[L^2(T_N)]^N$, which tends weakly to $\mathbf{b}(x, u, \mathbf{w}) \in [L^2(T_N)]^N$.

By strong convergence in (Q_7), we mean convergence in norm.

It is clear that there are many examples of $a^{ij}(x, s)$ that satisfy $(Q_1) - (Q_4)$.

Before proceeding, we give an example of a vector $\mathbf{b}(x, s, p)$ that meets (Q_5), (Q_6), and (Q_7). Define the j-th component of $\mathbf{b}(x, s, p)$ as follows:

$$b^1(x, s, p) = p_2 \sin x_1$$

$$b^2(x, s, p) = -p_1 \sin x_1 - s \cos x_1$$

$$b^j(x, s, p) = 0 \qquad \text{for } j = 3, ..., N.$$

Clearly, condition (Q_5) is met. If $u \in H^1(T_N)$ and $\phi \in C^\infty(T_N)$, then

$$\int_{T_N} [D_2 (u \sin x_1) D_1 \phi - D_1 (u \sin x_1) D_2 \phi] dx$$

$$= \int_{T_N} u \sin x_1 (D_1 D_2 \phi - D_2 D_1 \phi) dx = 0.$$

So (2.3) holds for ϕ, and therefore, it is easy to see from the definition of $H^1(T_N)$ that condition (Q_6) is met.

To see that condition (Q_7) holds, let $\{u_n\}_{n=1}^\infty$ be a sequence of functions in $L^2(T_N)$, which tend strongly to $u \in L^2(T_N)$, and $\{\mathbf{w}_n\}_{n=1}^\infty$ be a sequence of vector-valued functions in $[L^2(T_N)]^N$, which tend weakly to $\mathbf{w}$ in $[L^2(T_N)]^N$. Then,

$$b^1(x, u_n, \mathbf{w}_n) = w_{n2} \sin x_1$$

$$b^2(x, u_n, \mathbf{w}_n) = -w_{n1} \sin x_1 - u_n \cos x_1$$

$$b^j(x, u_n, \mathbf{w}_n) = 0 \qquad \text{for } j = 3, ..., N.$$

It is clear that

$$\lim_{n \to \infty} \int_{T_N} b^j(x, u_n, \mathbf{w}_n) v \, dx = \int_{T_N} b^j(x, u, \mathbf{w}) v \, dx \quad \forall v \in L^2(T_N),$$

for $j = 1, ..., N$. Consequently, condition (Q_7) is met. Therefore, $\mathbf{b}(x, s, p)$ has met all the asserted conditions, and our example is complete.

We shall study equations of the form

(2.4) $$Qu = f(x)$$

and

(2.5) $$Qu = f(x, u).$$

For the former, we shall suppose $f(x) \in L^2(T_N)$, and for the latter, we shall suppose $f(x, s)$ meets the following two conditions:

$(f - 1)$ $f(x, s)$ meets the same Caratheodory conditions that $a^{ij}(x, s)$ meets in (Q_1) above;

$(f - 2)$ For each $r > 0$, there is a $\zeta_r \in L^2(T_N)$ such that

$$|f(x, s)| \leq \zeta_r(x) \quad \text{for } |s| \leq r \text{ and for a.e. } x \in T_N.$$

To establish the first theorem in the paper, we shall also need the following one-sided condition on $f(x, s)$.

$(f - 3)$ Given $\varepsilon > 0$, $\exists$ a nonnegative function $c_\varepsilon(x) \in L^2(T_N)$, and a constant $s_0(\varepsilon)$ such that

$$sf(x, s) \leq \varepsilon s^2 + c_\varepsilon(x)|s| \quad \text{for } |s| \geq s_0(\varepsilon) \text{ and for a.e. } x \in T_N,$$

we note that $(f - 3)$ is a generalization of the notion

$$\limsup_{|s| \to \infty} f(x, s)/s \leq 0 \quad \text{uniformly for } x \in T_N.$$

We set

(2.6) $$\mathcal{F}_\pm(x) = \limsup_{s \to \pm\infty} f(x, s)/s,$$

and will establish the following theorem:

Theorem 2.1. *Assume* $(Q_1) - (Q_7), (f - 1), (f - 2),$ *and* $(f - 3)$. *Suppose*

$$(i) \int_{T_N} \mathcal{F}_+(x)\, dx < 0 \text{ and } (ii) \int_{T_N} \mathcal{F}_-(x)\, dx < 0,$$

where $\mathcal{F}_+(x)$ *and* $\mathcal{F}_-(x)$ *are defined in (2.6). Then, there exists a distribution solution* $u \in H^1(T_N)$ *of* $Qu = f(x, u)$ *on* T_N *with* $f(x, u) \in L^1(T_N)$ *and* $f(x, u)u \in L^1(T_N)$.

We shall prove two more theorems about the quasilinear elliptic operator Qu in this section. The statement of these theorems can be found immediately after the proof of Theorem 2.1, which will be given next.

To be quite explicit, what we mean by $u \in H^1(T_N)$ is a distribution solution of $Qu = f(x, u)$ *on* T_N with $f(x, u) \in L^1(T_N)$ is the following:
(2.7)

$$\int_{T_N} \left[\sum_{i,j=1}^N a^{ij}(x, u) D_j u D_i \phi + \sum_{j=1}^N \phi b^j(x, u, Du) D_j u \right] dx = \int_{T_N} f(x, u)\, \phi dx$$

$\forall \phi \in C^\infty(T_N)$.

Because $Q1 = 0 \cdot 1$, we see that 0 acts like an eigenvalue of Q. Also, since Q is a generalization of $-\Delta$, 0 acts like the first eigenvalue of Q. So Theorem 2.1 is referred to in the literature as a result at resonance (see [DG, p.4]). Theorem 2.1 is a generalization to the quasilinear case (with a different proof) of the result given by De Figueiredo and Gossez in [DG, p. 10].

Theorem 2.1 is in a certain sense a best possible result, i.e., if we replace (i) above by (i') below and keep (ii), the theorem is false. Likewise, if we keep (i) and replace (ii) above by (ii'') below, the theorem is false.

(i') $\int_{T_N} \mathcal{F}_+(x)\, dx = 0$.

(ii'') $\int_{T_N} \mathcal{F}_-(x)\, dx = 0$.

To see that this is the case for (i'), we shall suppose Theorem 2.1 holds if (i) is replaced by (i') and arrive at a contradiction. Set

$$\beta(s) = \begin{cases} 0, & s \geq 0 \\ -s, & s < 0 \end{cases}$$

and

$$f(x, s) = 1 + \beta(s) \quad \text{for} \ \ x \in T_N \text{ and } s \in \mathbf{R}.$$

Then, $f(x, s)$ meets $(f-1), (f-2)$, and $(f-3)$. Furthermore, it follows from (2.6) that

$$\mathcal{F}_+(x) = 0 \text{ and } \mathcal{F}_-(x) = -1 \text{ for } x \in T_N.$$

Therefore, condition (i') is met. Likewise, condition (ii) is met. So if Theorem 2.1 were true with (i) replaced by (i'), a distribution solution $u \in H^1(T_N)$ of $Qu = f(x, u)$ on T_N would exist, i.e., we would have that (2.7) holds.

On setting $\phi = 1$ in (2.7), and observing that $D_i(1) = 0$ for $i = 1, ..., N$, we obtain from (2.7) that

(2.8)
$$\int_{T_N} \left[\sum_{j=1}^N b^j(x, u, Du) D_j u \right] dx = \int_{T_N} f(x, u)\, dx.$$

Now $f(x, u) = 1 + \beta(u(x))$ for $x \in T_N$. So

(2.9)
$$f(x, u) \geq 1 \quad \text{a.e. in } T_N.$$

Also, we observe from (Q_6) that the left-hand side of (2.8) is zero. Therefore, we obtain from (2.8) and (2.9) that

$$0 \geq (2\pi)^N$$

is a contradiction. We conclude that Theorem 2.1 does not hold when (i) is replaced by (i'). A similar example shows that (i) cannot be replaced by (i''). Theorem 2.1 is indeed a best possible result.

If we take

$$f(x, s) = 1 + x_1 - s \quad \text{for} \ \ x \in T_N \text{ and } s \in \mathbf{R},$$

then it follows that $f(x, s)$ satisfies $(f-1) - (f-3)$. Also, $\mathcal{F}_+(x) = -1$ and $\mathcal{F}_-(x) = -1$ for $x \in T_N$. So this $f(x, s)$ is an example of a function that will work for Theorem 2.1.

We shall use a Galerkin technique to establish Theorem 2.1. In order to accomplish this, we observe that there is a sequence $\{\psi_k\}_{k=1}^{\infty}$ of real-valued functions in $C^{\infty}(T_N)$ with the properties stated in (2.10) and (2.11) below.

$$(a) \ \ \psi_1 = (2\pi)^{-N/2},$$

(2.10)

$$(b) \ \ \int_{T_N} \psi_k \psi_l dx = \delta_{kl}$$

where δ_{kl} is the Kronecker -delta $k, l = 1, \dots$.

Given $\psi \in C^{\infty}(T_N)$ and $\varepsilon > 0$, $\exists$ constants $c_1, \dots, c_n$ such that

$$(a) \ |\psi(x) - \textstyle\sum_{k=1}^n c_k \psi_k(x)| \leq \varepsilon,$$

(2.11)

$$(b) \ |D_j \psi(x) - \textstyle\sum_{k=1}^n c_k D_j \psi_k(x)| \leq \varepsilon,$$

uniformly for $x \in T_N$ and $j = 1, \dots, N$.

To show that such a sequence with the properties enumerated in (2.10) and (2.11) does indeed exist, we proceed as follows: First of all, we see from Corollary 2.3 in Chapter 1 that in dimension 2

$$\{\cos m_1 x_1 \cos m_2 x_2\}_{m_1=0, m_2=0}^{\infty,\infty} \cup \{\cos m_1 x_1 \sin m_2 x_2\}_{m_1=0, m_2=1}^{\infty,\infty}$$

$$\cup \{\sin m_1 x_1 \cos m_2 x_2\}_{m_1=1, m_2=0}^{\infty,\infty} \cup \{\sin m_1 x_1 \sin m_2 x_2\}_{m_1=1, m_2=1}^{\infty,\infty}$$

properly normalized gives rise to a CONS (a complete orthonormal system) in $L^2(T_2)$. Likewise, in dimension 3, we see that

$$\{\cos m_1 x_1 \cos m_2 x_2 \cos m_3 x_3\}_{m_1=0, m_2=0, m_3=0}^{\infty,\infty,\infty}$$

$$\cup \{\cos m_1 x_1 \cos m_2 x_2 \sin m_3 x_3\}_{m_1=0, m_2=0, m_3=1}^{\infty,\infty,\infty}$$

$$\vdots$$

$$\cup \{\sin m_1 x_1 \sin m_2 x_2 \cos m_3 x_3\}_{m_1=1, m_2=1, m_3=0}^{\infty,\infty,\infty}$$

$$\cup \{\sin m_1 x_1 \sin m_2 x_2 \sin m_3 x_3\}_{m_1=1, m_2=1, m_3=1}^{\infty,\infty,\infty}$$

(where we have in all a union of eight sequences) properly normalized gives rise to a CONS in $L^2(T_3)$.

We see that we can proceed like this in any dimension N and obtain a union of 2^N sequences of sines and cosines that will give rise to a CONS in $L^2(T_N)$. So we have established the existence of a real-valued sequence $\{\psi_k\}_{k=1}^{\infty}$ in $C^{\infty}(T_N)$ which is also a CONS in $L^2(T_N)$, and therefore meets the conditions in (2.10).

Next, we observe that any real-valued trigonometric polynomial of the form

$$\sum_{|m|\leq M} a_m e^{im\cdot x} \quad \text{where } \bar{a}_m = a_{-m}$$

can be written as a finite real linear combination of the elements in the sequence $\{\psi_k\}_{k=1}^{\infty}$. For example, for $N = 2$,

$$
\begin{aligned}
e^{im\cdot x} + e^{-im\cdot x} &= 2\cos(m_1 x_1 + m_2 x_2) \\
&= 2\cos m_1 x_1 \cos m_2 x_2 - 2\sin m_1 x_1 \sin m_2 x_2.
\end{aligned}
$$

So it is easy to see from Theorem 2.1 in Chapter 1 that conditions (2.11)(a) and (2.11)(b) do indeed hold.

The first lemma we prove is where $\{\psi_k\}_{k=1}^{\infty}$ is the sequence in (2.10) and (2.11).

Lemma 2.2. *Let $F(x)$ be a nonnegative function in $L^1(T_N)$, and let $f(x,s)$ satisfy $(f-1)$ and $(f-2)$. Suppose that*

$$|f(x,s)| \leq F(x) \quad \text{for } s \in \mathbf{R} \text{ and a.e. } x \in T_N.$$

Suppose also that Q satisfies $(Q_1) - (Q_6)$. Then if n is a given positive integer, there is a function $u = \gamma_1\psi_1 + \cdots + \gamma_n\psi_n$ such that
(2.12)
$$\int_{T_N}\left[\sum_{i,j=1}^{N} a^{ij}(x,u)D_j u D_i \psi_k + \sum_{j=1}^{N} \psi_k b^j(x,u,Du)D_j u + u\psi_k n^{-1}\right]dx$$

$$= \int_{T_N} f(x,u)\psi_k dx$$

for $k = 1, ..., n$.

Proof of Lemma 2.2. This makes use of the Galerkin argument cited in the introduction.

For each $\alpha = (\alpha_1, ..., \alpha_n) \in \mathbf{R}^n$, we introduce an $n \times n$ matrix $A(\alpha)$ with components $A_{kl}(\alpha)$ given as follows:

$$A_{kl}(\alpha) = \sum_{i,j=1}^{N} < D_i\psi_k, a^{ij}(\cdot, \sum_{q=1}^{n}\alpha_q\psi_q)D_j\psi_l >_{L^2}$$

(2.13)
$$+ < \psi_k, \sum_{j=1}^{N} b^j(\cdot, \sum_{q=1}^{n}\alpha_q\psi_q, D\sum_{q=1}^{n}\alpha_q\psi_q)D_j\psi_l >_{L^2}$$

$$+ < \psi_k, \psi_l >_{L^2} n^{-1}.$$

We observe from $(Q_1), (Q_2)$, and (Q_5) that

(2.14)
$$A_{kl}(\alpha) \in C(\mathbf{R}^n) \text{ for } k, l = 1, ..., n.$$

Next, we observe from (2.13) that for $\beta = (\beta_1, ..., \beta_n)$

$$\sum_{l=1}^n A_{kl}(\alpha)\beta_l =$$

(2.15)

$$\sum_{l=1}^n \sum_{i,j=1}^N < D_i\psi_k, a^{ij}(\cdot, \sum_{q=1}^n \alpha_q\psi_q)\beta_l D_j\psi_l >_{L^2}$$

$$+ \sum_{l=1}^n < \psi_k, \sum_{j=1}^N b^j(\cdot, \sum_{q=1}^n \alpha_q\psi_q, D\sum_{q=1}^n \alpha_q\psi_q)\beta_l D_j\psi_l >_{L^2}$$

$$+ \sum_{l=1}^n < \psi_k, \beta_l\psi_l >_{L^2} n^{-1}.$$

Now it follows from (Q_6) and the fact that $D_j\psi^2 = 2\psi D_j\psi$ that

(2.16)
$$\sum_{j=1}^n < \psi, b^j(\cdot, \phi, D\phi)D_j\psi >_{L^2} = 0$$

for $\phi, \psi \in C^\infty(T_N)$. Consequently, we see from (2.15) that the quadratic form $\beta \cdot A(\alpha)\beta = \sum_{k,l=1}^n A_{kl}(\alpha)\beta_k\beta_l$ is such that

(2.17) $\beta \cdot A(\alpha)\beta = \sum_{i,j=1}^N < D_i\psi, a^{ij}(\cdot, \sum_{q=1}^n \alpha_q\psi_q)D_j\psi >_{L^2} + < \psi, \psi >_{L^2} n^{-1}$

where $\psi = \beta_1\psi_1 + \cdots + \beta_n\psi_n$.

It follows from (Q_4) that the right-hand side of (2.17) majorizes

$$\eta_0 \int_{T_N} |D\psi|^2 dx + < \psi, \psi >_{L^2} n^{-1},$$

where $\psi = \beta_1\psi_1 + \cdots + \beta_n\psi_n$ and $|D\psi|^2 = |D\psi_1|^2 + \cdots + |D\psi_n|^2$. We conclude therefore from (2.17) and (2.10)(b) that

$$\beta \cdot A(\alpha)\beta \geq |\beta|^2 n^{-1}.$$

Since β is arbitrary in $\mathbf{R}^n$, it follows from this last inequality that for each $\alpha \in \mathbf{R}^n$, the inverse matrix $[A(\alpha)]^{-1}$ exists, and furthermore

(2.18)
$$\left\| [A(\alpha)]^{-1} \right\|_{\mathcal{M}} \leq n \quad \forall \alpha \in \mathbf{R}^n,$$

where $\|\cdot\|_{\mathcal{M}}$ designates the usual $n \times n$ matrix norm.

Next, for each $\alpha \in \mathbf{R}^n$, we set

(2.19)
$$S_k(\alpha) = < \psi_k, f(\cdot, \sum_{q=1}^n \alpha_q\psi_q) >_{L^2}$$

for $k = 1, ..., n$. We observe from $(f-1)$ and $(f-2)$ that $S_k(\alpha) \in C(\mathbf{R}^n)$. Also, we see from the hypothesis of the lemma that there is a constant Γ_1 such that

$$|S_k(\alpha)| \leq \Gamma_1$$

$\forall \alpha \in \mathbf{R}^n$ and $k = 1, ..., n$. We set

(2.20)
$$S(\alpha) = (S_1(\alpha), ..., S_n(\alpha)),$$

and conclude that

(2.21)

S is a continuous map of $\mathbf{R}^n$ into the closed ball of $\mathbf{R}^n$ with center 0 and radius $n^{1/2}\Gamma_1$.

Next, we set $G(\alpha) = [A(\alpha)]^{-1}S(\alpha)$ and observe from (2.14), (2.18), and (2.21) that $G(\alpha) \in [C(\mathbf{R}^n)]^n$ and maps $\mathbf{R}^n$ into the closed ball of $\mathbf{R}^n$ with center 0 and radius $n^{3/2}\Gamma_1$. Consequently, G is a continuous map of this last mentioned closed ball into itself. We invoke the Brouwer fixed point theorem (see [Ev, p. 441]) and conclude that there exists $\gamma = (\gamma_1, ..., \gamma_n)$ such that $[A(\gamma)]^{-1}S(\gamma) = \gamma$, i.e.,

$$A(\gamma)\gamma = S(\gamma).$$

We set $u = \gamma_1\psi_1 + \cdots + \gamma_n\psi_n$ and obtain from (2.15), (2.19), and (2.20) with $\alpha = \beta = \gamma$ that

$$\sum_{l=1}^n \sum_{i,j=1}^N < D_i\psi_k, a^{ij}(\cdot, \sum_{q=1}^n \gamma_q\psi_q)\gamma_l D_j\psi_l >_{L^2}$$

$$+ \sum_{l=1}^n < \psi_k, \sum_{j=1}^N b^j(\cdot, \sum_{q=1}^n \gamma_q\psi_q, D\sum_{q=1}^n \gamma_q\psi_q)\gamma_l D_j\psi_l >_{L^2}$$

$$+ \sum_{l=1}^n < \psi_k, \gamma_l\psi_l >_{L^2} n^{-1} = < \psi_k, f(\cdot, \sum_{q=1}^n \gamma_q\psi_q) >_{L^2}$$

for $k = 1, ..., n$. But this is (2.12) in the statement of the lemma, and the proof of the lemma is complete. ∎

The next lemma we prove is the following:

Lemma 2.3. *Let n be a given positive integer and $f(x,s)$ satisfy (f-1) and (f-2). Suppose that Q satisfies $(Q_1)-(Q_6)$ and that there is a nonnegative function $F(x)$ in $L^2(T_N)$ such that*

(2.22) $$sf(x,s) \leq F(x)|s| + s^2/2n$$

for a.e. $x \in T_N$ and $\forall s \in \mathbf{R}$. Then there is a function $u = \gamma_1\psi_1 + \cdots + \gamma_n\psi_n$ such that (2.12) in Lemma 2.2 holds.

Proof of Lemma 2.3. For each positive integer M, set

$$f^M(x,s) = \begin{cases} f(x,M) & s \geq M \\ f(x,s) & -M \leq s \leq M \\ f(x,-M) & s \leq -M. \end{cases}$$

It follows from $(f-2)$ that there is a $\zeta_M(x) \in L^2(T_N)$ such that

$$\left| f^M(x,s) \right| \leq \zeta_M(x) \quad \text{for a.e. } x \in T_N \quad \text{and } \forall s \in \mathbf{R}.$$

But then we obtain from Lemma 2.2 that there exists $\{\gamma_i^M\}_{i=1}^n$ such that

(2.23) $$u^M = \gamma_1^M \psi_1 + \cdots + \gamma_n^M \psi_n$$

satisfies (2.12) with f replaced by f^M, i.e.,

$$\sum_{i,j=1}^n < D_i\psi_k, a^{ij}\left(\cdot, u^M\right) D_j u^M >_{L^2} + <\psi_k, u^M >_{L^2} n^{-1}$$

(2.24) $$+\sum_{j=1}^n <\psi_k, b^j\left(\cdot, u^M, Du^M\right) D_j u^M >_{L^2}$$

$$=<\psi_k, f^M\left(\cdot, u^M\right) >_{L^2}$$

for $k = 1, ..., n$.

Now it follows from the definition of $f^M(x, s)$ and from (2.22) that

$$sf^M(x, s) \le F(x)|s| + s^2/2n$$

for a.e. $x \in T_N$ and $\forall s \in \mathbf{R}$. Consequently,

(2.25) $$u^M(x) f^M\left[x, u^M(x)\right] \le F(x)\left|u^M(x)\right| + u^M(x)^2/2n$$

for a.e. $x \in T_N$ and $M = 1, 2, ...$

Next, we multiply both sides of (2.24) by γ_n^M and sum over $k = 1, ..., n$. Using (Q_4), (2.16), and (2.23), we then obtain

$$< u^M, u^M >_{L^2} n^{-1} \le < u^M, f^M\left(\cdot, u^M\right) >_{L^2}$$

for each positive integer M. But then it follows from (2.25) that

$$< u^M, u^M >_{L^2}(2n)^{-1} \le < \left|u^M\right|, F >_{L^2}.$$

By hypothesis $F \in L^2(T_N)$, we obtain from this last inequality in conjunction with Schwarz's inequality that

$$< u^M, u^M >_{L^2} \le 4n^2 < F, F >_{L^2}.$$

We next employ (2.10)(b) and (2.23) in conjunction with this last fact to obtain

$$(\gamma_1^M)^2 + \cdots + (\gamma_n^M)^2 \le 4n^2 < F, F >_{L^2}$$

for every positive integer M.

Since n is a fixed positive integer, we infer from this last inequality that there is a subsequence of $\{\gamma_k^M\}_{k=1}^\infty$, which converges for each $k = 1, ..., n$. For ease of notation, we shall suppose this subsequence is the full sequence and record this fact as

(2.26) $$\lim_{M\to\infty} \gamma_k^M = \gamma_k, \quad k = 1, ..., n.$$

We set $u = \gamma_1\psi_1 + \cdots + \gamma_n\psi_n$ and obtain from (2.23) and (2.26) that
(2.27)
$$\lim_{M\to\infty} u^M(x) = u(x) \quad \text{uniformly for } x \in T_N,$$

$$\lim_{M\to\infty} D_j u^M(x) = D_j u(x) \quad \text{uniformly for } x \in T_N \text{ and } j = 1, ..., N.$$

From (Q_1) and this last fact, we see that

$$(a) \quad \lim_{M \to \infty} a^{ij} \left(x, u^M \left(x \right) \right) = a^{ij} \left(x, u \left(x \right) \right),$$

(2.28)

$$(b) \quad \lim_{M \to \infty} b^j \left(x, u^M \left(x \right), Du^M \left(x \right) \right) = b^j \left(x, u \left(x \right), Du \left(x \right) \right)$$

for a.e. $x \in T_N$ and $i, j = 1, ..., N$.

From (Q_2), in conjunction with (2.27) and (2.28), we obtain that

$$(2.29) \qquad \lim_{M \to \infty} < D_i \psi_k, a^{ij} \left(\cdot, u^M \right) D_j u^M >_{L^2} = < D_i \psi_k, a^{ij} \left(\cdot, u \right) D_j u >_{L^2}$$

for $k = 1, ..., n$.

Likewise, we see from (Q_5) that

(2.30)

$$\lim_{M \to \infty} < \psi_k, b^j \left(\cdot, u^M, Du^M \right) D_j u^M >_{L^2} = < \psi_k, b^j \left(\cdot, u, Du \right) D_j u >_{L^2}$$

for $k = 1, ..., n$.

Next, we observe from (2.23) and from (2.26) that the sequence

$$\left\{ u^M \left(x \right) \right\}_{M=1}^{\infty}$$

is uniformly bounded on T_N and in $C^\infty \left(T_N \right)$ for each M. We consequently obtain from the definition of $f^M \left(x, s \right)$ that there is an M_0 such that for

$$f^M \left(x, s \right) = f \left(x, s \right) \qquad \text{for} \quad M \geq M_0$$

for $x \in T_N$. But then it follows from $(f-1)$, $(f-2)$, and (2.27) that

$$\lim_{M \to \infty} < \psi_k, f^M \left(\cdot, u^M \right) >_{L^2} = < \psi_k, f \left(\cdot, u \right) >_{L^2}$$

for $k = 1, ..., n$.

We conclude from (2.24), and the limits in (2.27), (2.29), and (2.30) joined with this last limit that

$$\sum_{i,j=1}^n < D_i \psi_k, a^{ij} \left(\cdot, u \right) D_j u >_{L^2} + < \psi_k, u >_{L^2} n^{-1}$$

$$+ \sum_{j=1}^n < \psi_k, b^j \left(\cdot, u, Du \right) D_j u >_{L^2}$$

$$= < \psi_k, f \left(\cdot, u \right) >_{L^2},$$

for $k = 1, ..., n$. But this establishes (2.12) for $u = \gamma_1 \psi_1 + \cdots + \gamma_n \psi_n$, and the proof of the lemma is complete. ∎

Lemma 2.4. *Suppose $f(x,s)$ satisfies (f-1) and (f-2), and there is a non-negative function $F \in L^2 \left(T_N \right)$ such that*

$$(2.31) \qquad sf \left(x, s \right) \leq F \left(x \right) |s| + s^2 \quad \text{for a.e. } x \in T_N \quad \text{and} \quad \forall s \in \mathbf{R}.$$

Suppose $\mathcal{Q}$ satisfies (Q_1)-(Q_6), and also for every positive integer n, there is $u^n = \gamma_1^n \psi_1 + \cdots + \gamma_n^n \psi_n$,   which satisfies

$$\sum_{i,j=1}^n < D_i \psi_k, a^{ij}(\cdot, u^n) D_j u^n >_{L^2} + < \psi_k, u^n >_{L^2} n^{-1}$$

(2.32)
$$+ \sum_{j=1}^n < \psi_k, b^j(\cdot, u^n, Du^n) D_j u^n >_{L^2}$$

$$= < \psi_k, f(\cdot, u^n) >_{L^2},$$

for $k=1,...,n$. Suppose furthermore there is a constant K such that

(2.33)
$$\|u^n\|_{L^2} \leq K \qquad for \ n = 1, 2, ...$$

Then there is a constant $K^\star$ such that

(2.34)
$$< |f(\cdot, u^n)|, |u^n| >_{L^2} \leq K^\star \qquad for \ n = 1, 2, ...$$

Proof of Lemma 2.4.   Multiplying both sides of (2.32) by γ_k^n and summing over $k=1,...,n$, we obtain from the hypothesis of the lemma and (2.16) that

$$\sum_{i,j=1}^n < D_i u^n, a^{ij}(\cdot, u^n) D_j u^n >_{L^2} + < u^n, u^n >_{L^2} n^{-1} = < u^n, f(\cdot, u^n) >_{L^2}.$$

Consequently, we have from (Q_4) and this last fact that

(2.35)
$$0 \leq < u^n, f(\cdot, u^n) >_{L^2}.$$

Next, we set

(2.36)
$$A_n = \{x \in T_N : u^n(x) f(x, u^n(x)) \geq 0\},$$
$$B_n = \{x \in T_N : u^n(x) f(x, u^n(x)) < 0\},$$

and observe from (2.31) that

$$\int_{A_n} u^n(x) f(x, u^n(x)) \, dx \leq \|F\|_{L^2} \|u^n\|_{L^2} + \|u^n\|_{L^2}^2$$

for $n = 1, 2, ...$ Consequently, we have from (2.33) that there is a constant $K_1 > 0$ such that

(2.37)
$$\int_{A_n} u^n(x) f(x, u^n(x)) \, dx \leq K_1 \qquad for \ n = 1, 2, ...$$

Also, it follows from (2.35) and (2.36) that

$$-\int_{B_n} u^n(x) f(x, u^n(x)) \, dx \leq \int_{A_n} u^n(x) f(x, u^n(x)) \, dx$$

Therefore, we obtain from (2.37) and this last fact

$$\int_{B_n} |u^n(x)| |f(x, u^n(x))| \, dx \leq K_1.$$

But this fact in conjunction with (2.36) and (2.37) gives

$$\int_{T_N} |u^n(x)| \, |f(x, u^n(x))| \, dx \leq 2K_1$$

for $n = 1, 2, \ldots$. This last inequality is (2.34) with $K^\star = 2K_1$. ∎

Lemma 2.5. *Suppose the conditions in the hypothesis of Lemma 2.4 hold. Then the sequence*

$$\{f(x, u^n(x))\}_{n=1}^{\infty}$$

is absolutely equi-integrable.

Proof of Lemma 2.5. The precise definition of absolutely equi-integrable was given earlier in this chapter (1.70).

To prove the lemma, let $\varepsilon > 0$ be given. Then choose $r > 0$ so that

(2.38) $K^\star/r < \varepsilon/2$

where $K^\star$ is the constant in the conclusion of Lemma 2.4. Next, using $(f-2)$, choose $\zeta_r \in L^2(T_N)$ so that

(2.39) $|f(x, s)| \leq \zeta_r(x)$ for a.e. $x \in T_N$ and $|s| \leq r$.

Also, set

$$A_n = \{x \in T_N : |u^n(x)| \leq r\},$$

(2.40)

$$B_n = \{x \in T_N : |u^n(x)| > r\},$$

and choose $\delta > 0$, so that

(2.41) $meas(E) < \delta \Rightarrow \int_E \zeta_r(x) \, dx < \varepsilon/2.$

Now, suppose $meas(E) < \delta$ with δ as in (2.41). Then it follows from Lemma 2.4 and (2.39)-(2.41) that

$$\int_E |f(x, u^n(x))| \, dx \leq \int_{E \cap A_n} \zeta_r(x) \, dx + \frac{1}{r} \int_{E \cap B_n} |u^n(x) f(x, u^n(x))| \, dx$$

$$\leq \varepsilon/2 + K^\star/r$$

for $n = 1, 2, \ldots$. From (2.38), we see that the right-hand side of this last inequality is $< \varepsilon$. Hence $\{f(x, u^n(x))\}_{n=1}^{\infty}$ is absolutely equi-integrable. ∎

Next, with $<\cdot, \cdot>_{H^1}$ defined in (2.1), we establish the following lemma:

Lemma 2.6. *Suppose $\{v_n\}_{n=1}^{\infty}$ is a sequence in $H^1(T_N)$ with*

$$\|v_n\|_{H^1} \leq K \qquad \forall n,$$

where K is a positive constant. Then there exists a subsequence $\left\{v_{n_j}\right\}_{j=1}^{\infty}$ and $v \in H^1\left(T_N\right)$ such that

$$(2.42) \qquad \lim_{j \to \infty} \left\|v_{n_j} - v\right\|_{L^2} = 0.$$

Proof of Lemma 2.6.   Since $\|v_n\|_{H^1} \leq K \quad \forall n$, it follows from standard Hilbert space theory (see [Ev, p. 639]) that there exists $v \in H^1\left(T_N\right)$ and a subsequence $\left\{v_{n_j}\right\}_{j=1}^{\infty}$ such that

$$(2.43) \qquad \lim_{j \to \infty} <v_{n_j},\ w>_{H^1} = <v,\ w>_{H^1} \quad \forall w \in H^1\left(T_N\right).$$

From Parsevaal's theorem, which is Corollary 2.5 in Chapter 1, we see that

$$(2.44) \qquad (2\pi)^{-N} \left\|v_{n_j} - v\right\|_{L^2}^2 = \lim_{R \to \infty} \sum_{|m| \leq R} \left|\widehat{v}_{n_j}(m) - \widehat{v}(m)\right|^2 \quad \forall n_j,$$

where $\widehat{v}(m)$ is the Fourier coefficient of v introduced in Chapter 1 above (1.1).

Also, it is easy to see from the uniform boundedness condition in the hypothesis of the lemma and from Parsevaal's theorem that there is a constant K_1 such that

$$(2.45) \qquad \lim_{R \to \infty} \sum_{|m| \leq R} |m|^2 \left|\widehat{v}_{n_j}(m) - \widehat{v}(m)\right|^2 \leq K_1 \quad \forall n_j.$$

We see from (2.44) that the proof of the lemma will be established once we show that the following fact holds: Given $\varepsilon > 0$,

$$(2.46) \qquad \limsup_{j \to \infty} \lim_{R \to \infty} \sum_{|m| \leq R} \left|\widehat{v}_{n_j}(m) - \widehat{v}(m)\right|^2 \leq \varepsilon.$$

To establish this last inequality, we choose R_0 so large that with K_1 as in (2.45)

$$K_1/R_0^2 \leq \varepsilon.$$

Then,

$$
\begin{aligned}
\lim_{R \to \infty} \sum_{|m| \leq R} \left|\widehat{v}_{n_j}(m) - \widehat{v}(m)\right|^2 &\leq \sum_{|m| \leq R_0} \left|\widehat{v}_{n_j}(m) - \widehat{v}(m)\right|^2 \\
&\quad + \frac{1}{R_0^2} \sum_{|m| > R_0} |m|^2 \left|\widehat{v}_{n_j}(m) - \widehat{v}(m)\right|^2 \\
&\leq \sum_{|m| \leq R_0} \left|\widehat{v}_{n_j}(m) - \widehat{v}(m)\right|^2 + K_1/R_0^2 \\
&\leq \sum_{|m| \leq R_0} \left|\widehat{v}_{n_j}(m) - \widehat{v}(m)\right|^2 + \varepsilon
\end{aligned}
$$

$\forall n_j$. From (2.43), we see that $\lim_{j\to\infty} \sum_{|m|\le R_0} |\widehat{v}_{n_j}(m) - \widehat{v}(m)|^2 = 0$. Therefore, we obtain from this last set of inequalities that

$$\limsup_{j\to\infty} \lim_{R\to\infty} \sum_{|m|\le R} |\widehat{v}_{n_j}(m) - \widehat{v}(m)|^2 \le \varepsilon,$$

which is (2.46). ■

We shall also need the following lemma:

Lemma 2.7. *Suppose $v \in H^1(T_N)$ and*

$$D_j v(x) = 0 \text{ for a.e. } x \in T_N,$$

for j=1,...,N. Then $v(x) = $ constant for a.e. $x\in T_N$.

Proof of Lemma 2.7. Using the notation of Chapter 1, as in the proof of Lemma 2.6 above, we see that if $\phi \in C^\infty(T_N)$, then it is easy to see that

$$\widehat{D_j\phi}(m) = im_j\widehat{\phi}(m) \quad \forall m \in \Lambda_N$$

for j=1,...,N where $\widehat{\phi}(m)$ is the Fourier coefficient of ϕ. It follows from the definition of $H^1(T_N)$ that similarly

$$\widehat{D_j v}(m) = im_j\widehat{v}(m) \quad \forall m \in \Lambda_N$$

for j=1,...,N. But then we obtain from the hypothesis of the lemma that

$$\widehat{v}(m) = 0 \ \forall m \in \Lambda_N\setminus\{0\}.$$

Consequently, it follows from Corollary 3.4 in Chapter 1 that

$$v(x) = \widehat{v}(0) \quad \text{for a.e. } x \in T_N,$$

which proves the lemma. ■

Proof of Theorem 2.1. Since f satisfies $(f-1) - (f-3)$, we see that for every $\varepsilon > 0$, there exists a nonnegative function $F_\varepsilon(x) \in L^2(T_N)$ such that

(2.47) $$sf(x,s) \le \varepsilon s^2 + F_\varepsilon(x)|s|$$

for a.e. $x \in T_N$ and $\forall s \in \mathbf{R}$. Consequently, it follows from Lemma 2.3 that there is a sequence $\{u^n\}_{n=1}^\infty$ with the following properties:

(2.48) $$u^n = \gamma_1^n\psi_1 + \cdots + \gamma_n^n\psi_n;$$

$$\int_{T_N}[\sum_{i,j=1}^N a^{ij}(x,u^n)D_j u^n D_i\psi_k$$

(2.49) $$+ \sum_{j=1}^N \psi_k b^j(x,u^n,Du^n)D_j u^n + u^n\psi_k n^{-1}]dx$$

$$= \int_{T_N} f(x,u^n)\psi_k dx$$

with $k = 1, ..., n$ where $\{\psi_k\}_{k=1}^{\infty}$ is the orthonormal sequence in (2.10)(a),(b).

We claim there is a constant K such that

$$(2.50) \qquad\qquad \|u^n\|_{H^1} \leq K \quad \forall n.$$

Suppose that (2.50) is false. Then there is a subsequence of $\{u^n\}_{n=1}^{\infty}$, which, for ease of notation, we take to be the full sequence with the following properties:

$$(2.51a) \qquad \lim_{n \to \infty} \|u^n\|_{H^1} = \infty,$$

$$(2.51b) \qquad \text{with } v^n = \frac{u^n}{\|u^n\|_{H^1}}, \ \exists v \in H^1 \text{ such that } \lim_{n \to \infty} \|v^n - v\|_{L^2} = 0,$$

$$(2.51c) \quad \lim_{n \to \infty} v^n(x) = v(x) \quad \text{for a.e. } x \in T_N,$$

$$(2.51d) \quad \lim_{n \to \infty} <w, D_j v^n>_{L^2} = <w, D_j v>_{L^2} \ \forall w \in L^2(T_N) \quad j = 1, ..., N,$$

where we have also made use of Lemma 2.6.

Next, we observe that $D_j(u^n)^2 = 2u^n D_j u^n$ and consequently obtain from (Q_6) that

$$\int_{T_N} u^n \sum_{j=1}^{N} b^j(x, u^n, Du^n) D_j u^n dx = 0.$$

Using this fact in conjunction with (2.48), (2.49), and (2.51b), gives us that

$$(2.52) \qquad \begin{aligned} &\int_{T_N} \left[\sum_{i,j=1}^{N} a^{ij}(x, u^n) D_j v^n D_i v^n + (v^n)^2 n^{-1} \right] dx \\ &= \|u^n\|_{H^1}^{-2} \int_{T_N} f(x, u^n) u^n dx. \end{aligned}$$

Now it follows from (Q_4) that

$$\sum_{i,j=1}^{N} a^{ij}(x, u^n) D_j v^n D_i v^n \geq \eta_0 |Dv^n(x)|^2$$

for a.e. $x \in T_N$, where η_0 is a positive constant. Consequently, we obtain from (2.47) and (2.52) that for every $\varepsilon > 0$, there is an $F_\varepsilon \in L^2(T_N)$ such that

$$(2.53) \qquad \eta_0 \||Dv^n\|_{L^2}^2 \leq \varepsilon \|v^n\|_{L^2}^2 + \|u^n\|_{H^1}^{-1} \|v^n\|_{L^2} \|F_\varepsilon\|_{L^2}.$$

From (2.51b) we see that $\|v^n\|_{H^1} = 1$ and furthermore that $\|v^n\|_{L^2} \leq 1$. We conclude from this last fact in conjunction with (2.51a) and (2.53) that

$$\limsup_{n \to \infty} \||Dv^n\|_{L^2}^2 \leq \varepsilon \eta_0^{-1},$$

and therefore, since $\varepsilon > 0$ is arbitrary, that

$$(2.54) \qquad\qquad \lim_{n\to\infty} \sup \, |||Dv^n|||^2_{L^2} = 0.$$

Since $|D_j v^n(x)| \le |Dv^n(x)|$, we obtain from this last *limsup* and (2.51d) that

$$\int_{T_N} \phi D_j v \, dx = 0 \quad \forall \phi \in C^\infty(T_N)$$

for $j = 1, ..., N$. Hence,

$$(2.55) \qquad D_j v(x) = 0 \ \text{ for a.e. } x \in T_N \text{ and for } j = 1, ..., N.$$

But then it follows from Lemma 2.7 that v is equal to a constant almost everywhere in T_N.

To calculate this constant, we observe from (2.1) that

$$\|v^n\|^2_{H^1} = |||v^n|||^2_{L^2} + |||Dv^n|||^2_{L^2}.$$

Also, from (2.51b), we see that

$$(a) \quad \|v^n\|^2_{H^1} = 1,$$

$$(b) \quad \lim_{n\to\infty} \|v^n\|^2_{L^2} = \|v\|^2_{L^2}.$$

We conclude from these last three equalities in conjunction with the limit in (2.54) that

$$\|v\|^2_{L^2} = 1.$$

But v is equal to a constant almost everywhere in T_N. Therefore, $v = (2\pi)^{-N/2}$ almost everywhere in T_N, or $v = -(2\pi)^{-N/2}$ almost everywhere in T_N.

We shall suppose

$$(2.56) \qquad\qquad v = (2\pi)^{-N/2} \quad \text{a.e. in } T_N$$

and arrive at a contradiction. A similar line of reasoning prevails in case the other alternative holds, and we leave the details of this part to the reader.

To arrive at a contradiction, we see from (Q_4) and (2.52) that

$$(2.57) \qquad\qquad 0 \le \int_{T_N} f(x, u^n) u^n dx \ \ \forall n.$$

Also, from (2.47) with $\varepsilon = 1$, we see that

$$(2.58) \qquad \begin{aligned} &(v^n)^2 + F_1(x) v^n \|u^n\|^{-1}_{H^1} \\ &\qquad - f(x, u^n) u^n \|u^n\|^{-2}_{H^1} \ge 0 \ \ \forall n. \end{aligned}$$

We set

$$(2.59) \qquad g_n = \text{ the left-hand side of the inequality in (2.58)}$$

and obtain from (2.57) that

$$\int_{T_N} g_n dx \le \|v^n\|^2_{L^2} + \|u^n\|^{-1}_{H^1} \int_{T_N} |F_1(x) v^n| \, dx.$$

We conclude from (2.51a,b) and (2.56) in conjunction with this last inequality that

$$\lim_{n\to\infty} \inf \int_{T_N} g_n dx \leq 1.$$

This last inequality joined with (2.58), (2.59), and Fatou's lemma [Ru2, p. 24] gives

(2.60) $$\int_{T_N} \lim_{n\to\infty} \inf g_n(x)\, dx \leq 1.$$

From (2.51b), we see that $u^n(x) = v^n(x) \|u^n\|_{H^1}$. Therefore, we have that

$$\lim_{n\to\infty} v^n(x) = (2\pi)^{-N/2} \quad \text{and} \quad \lim_{n\to\infty} u^n(x) = \infty \quad \text{a.e. in } T_N.$$

Also, we have that

$$f(x, u^n)\, u^n \|u^n\|_{H^1}^{-2} = (v^n)^2\, f(x, u^n) / u^n.$$

Consequently, from (2.6), (2.58), and (2.59), we know that

$$\lim_{n\to\infty} \inf g_n(x) \geq (2\pi)^{-N} - (2\pi)^{-N} \mathcal{F}_+(x) \quad \text{a.e. in } T_N.$$

But then we obtain from (2.60) that

$$\int_{T_N} \mathcal{F}_+(x)\, dx \geq 0.$$

This last inequality contradicts condition (i) in the hypothesis of the theorem. We have arrived at a contradiction. Therefore, (2.51a) is false, and (2.50) is indeed true.

From the fact that (2.50) holds, it follows from Lemma 2.6 that there is a subsequence of $\{u^n\}_{n=1}^{\infty}$, which, for ease of notation, we take to be the full sequence with the following properties:

(2.61a) $\quad \exists u \in H^1(T_N)$ such that $\lim_{n\to\infty} \|u^n - u\|_{L^2} = 0,$

(2.61b) $\quad \lim_{n\to\infty} u^n(x) = u(x) \quad$ a.e. in $T_N,$

(2.61c) $\quad \lim_{n\to\infty} <w, D_j u^n>_{L^2} = <w, D_j u>_{L^2} \forall w \in L^2(T_N) \quad j = 1, ..., N.$

From (2.61b) and (Q_1), we see that

$$\lim_{n\to\infty} a^{ij}[x, u^n(x)] = a^{ij}[x, u(x)] \quad \text{a.e. in } T_N.$$

Also, it is easy to see from (Q_2) and (2.61a) that the sequence

$$\left\{ |a^{ij}[x, u^n(x)]|^2 \right\}_{n=1}^{\infty} \quad \text{is absolutely equi-integrable in } L^1(T_N).$$

It consequently follows from Egoroff's theorem, that

$$\lim_{n\to\infty} \|a^{ij}(\cdot, u^n) - a^{ij}(\cdot, u)\|_{L^2} = 0.$$

But this fact in conjunction with (2.50) and Schwarz's inequality implies that for fixed k,

$$\lim_{n\to\infty} \sum_{i,j=1}^{N} < D_i\psi_k, \left[a^{ij}(\cdot, u^n) - a^{ij}(\cdot, u)\right] D_j u^n >_{L^2} = 0.$$

From (2.61c), we also see that

$$\lim_{n\to\infty} \sum_{i,j=1}^{N} \int_{T_N} a^{ij}(x, u) D_i\psi_k D_j u^n dx = \sum_{i,j=1}^{N} \cdot \int_{T_N} a^{ij}(x, u) D_i\psi_k D_j u \, dx.$$

So we conclude from these last two limits that
(2.62)

$$\lim_{n\to\infty} \sum_{i,j=1}^{N} < D_i\psi_k, a^{ij}(\cdot, u^n) D_j u^n >_{L^2} = \sum_{i,j=1}^{N} \cdot \int_{T_N} a^{ij}(x, u) D_i\psi_k D_j u \, dx$$

for $k = 1, 2, \ldots$.

Next, from (Q_6) and (2.3), we obtain that

$$\int_{T_N} \left[\sum_{j=1}^{N} b^j(x, u^n, Du^n) D_j(u^n\psi_k) \right] dx = 0,$$

and consequently that
(2.63)

$$\sum_{j=1}^{N} \int_{T_N} b^j(x, u^n, Du^n)\psi_k D_j u^n dx = -\sum_{j=1}^{N} \int_{T_N} b^j(x, u^n, Du^n)u^n D_j\psi_k dx.$$

On the other hand, from (Q_5), (2.50), and (2.61a), we see that

$$\lim_{n\to\infty} \sum_{j=1}^{N} \int_{T_N} |u^n - u| \left| b^j(x, u^n, Du^n) \right| dx = 0,$$

and consequently from (2.63), we obtain that

(2.64)
$$\lim_{n\to\infty} \sum_{j=1}^{N} \int_{T_N} b^j(x, u^n, Du^n)\psi_k D_j u^n dx$$
$$= -\lim_{n\to\infty} \sum_{j=1}^{N} \int_{T_N} b^j(x, u^n, Du^n)u D_j\psi_k dx.$$

However, from (2.61c), we see that $\{Du^n\}_{n=1}^{\infty}$ is a sequence of vector-valued functions in $\left[L^2(T_N)\right]^N$, which tends weakly to $Du \in \left[L^2(T_N)\right]^N$. Also, from (2.61a), we see that $\{u^n\}_{n=1}^{\infty}$ is a sequence of functions in $L^2(T_N)$, which tends strongly to u in $L^2(T_N)$. Therefore, it follows from (Q_7) that

$$\lim_{n\to\infty} \sum_{j=1}^{N} < u, b^j(\cdot, u^n, Du^n) D_j\psi_k >_{L^2}$$

$$= \sum_{j=1}^{N} < u, b^j(\cdot, u, Du) D_j\psi_k >_{L^2}.$$

But since $u \in H^1(T_N)$, we obtain as earlier in (2.63) that

$$\sum_{j=1}^N < u, b^j(\cdot, u, Du) D_j \psi_k >_{L^2}$$

$$= -\sum_{j=1}^N \int_{T_N} b^j(x, u, Du) \psi_k D_j u dx.$$

We consequently conclude from these last two equalities joined with (2.64) that

(2.65)
$$\lim_{n \to \infty} \sum_{j=1}^N < \psi_k, b^j(\cdot, u^n, Du^n) D_j u^n >_{L^2}$$

$$= \sum_{j=1}^N \int_{T_N} b^j(x, u, Du) \psi_k D_j u dx$$

for $k = 1, 2$.

Next, we see from (2.50) and Lemma 2.5 that the sequence $\{f(x, u^n(x))\}_{n=1}^\infty$ is absolutely equi-integrable. Also, we see from $(f-1)$ and (2.61b) that

$$\lim_{n \to \infty} f(x, u^n(x)) = f(x, u(x)) \text{ for a.e. } x \in T_N.$$

It consequently follows from (2.50), Lemma 2.4, and Fatou's lemma that

(2.66a)
$$uf(x, u) \in L^1(T_N),$$

and then from $(f-2)$ that

(2.66b)
$$f(x, u) \in L^1(T_N).$$

Also, we obtain from the absolute equi-integrability above joined with Egoroff's theorem that

(2.67)
$$\lim_{n \to \infty} \int_{T_N} \psi_k(x) f(x, u^n) \, dx = \int_{T_N} \psi_k(x) f(x, u) \, dx$$

for $k = 1, 2, ...$

From (2.49), (2.62), (2.65), and (2.67), we obtain

(2.68)
$$\sum_{i,j=1}^N \cdot \int_{T_N} a^{ij}(x, u) D_i \psi_k D_j u \, dx$$

$$+ \sum_{j=1}^N \int_{T_N} b^j(x, u, Du) \psi_k D_j u dx$$

$$= \int_{T_N} \psi_k(x) f(x, u) \, dx$$

for $k = 1, ..., n$.

From (Q_2), we see that $a^{ij}(\cdot, u) D_j u \in L^1(T_N)$. From (Q_5), we see that $b^j(\cdot, u, Du) D_j u \in L^1(T_N)$. From (2.66b), we have that $f(\cdot, u) \in L^1(T_N)$. Also, it follows from (2.11) that given $\phi \in C^\infty(T_N)$, there is a sequence $\{\phi_n\}_{n=1}^\infty$ and there are real constants $\{c_q^n\}_{q=1}^n$ such that

$$\phi_n = c_1^n \psi_1 + \cdots + c_1^n \psi_n$$

and

$$\lim_{n \to \infty} \phi_n(x) = \phi(x),$$

$$\lim_{n \to \infty} D_j \phi_n(x) = D_j \phi(x)$$

uniformly for $x \in T_N$, $j = 1, ..., N$. We conclude first that (2.68) holds with ψ_k replaced by ϕ_n, and next on passing to the limit as $n \to \infty$ that

$$\sum_{i,j=1}^{N} \cdot \int_{T_N} a^{ij}(x,u) D_i \phi D_j u \, dx$$

$$+ \sum_{j=1}^{N} \int_{T_N} b^j(x,u,Du) \phi D_j u dx$$

$$= \int_{T_N} \phi f(x,u) \, dx$$

$\forall \phi \in C^\infty(T_N)$.

But this last equality is the same as (2.7). So indeed u is a distribution solution of $Qu = f(x,u)$ on T_N. This fact joined with (2.66a) and (2.66b) completes the proof of the theorem. ∎

We intend to prove two more theorems in this section. The first of the two deals with $Qu = f(x,u)$ where $f(x,u) = g(u) - h(x)$. The second theorem studies $Qu = f(x)$.

Theorem 2.8.   *Assume $(Q_1) - (Q_7)$, where $g \in C(\mathbf{R}) \cap L^\infty(\mathbf{R})$, $h \in L^2(T_N)$ and that the limits $\lim_{s \to \infty} g(s) = g(\infty)$ and $\lim_{s \to -\infty} g(s) = g(-\infty)$ exist. Suppose also that*

(2.69) $g(\infty) < g(s) < g(-\infty)$ $\forall s \in \mathbf{R}$.

Then a necessary and sufficient condition that a distribution solution $u \in H^1(T_N)$ of $Qu = g(u) - h(x)$ exists is that

(2.70) $(2\pi)^N g(\infty) < \displaystyle\int_{T_N} h(x) \, dx < (2\pi)^N g(-\infty)$.

Theorem 2.8 is also referred to in the literature as a result at resonance and can be found in [Sh12, p. 570]. It is motivated by the work of Landesman and Lazer [LL, p. 611], but the proof given here is different. More elliptic resonance results can be found in [BN, Chapter IV] and in [KW].

An example of a $g \in C(\mathbf{R}) \cap L^\infty(\mathbf{R})$ that meets the condition in (2.69) is

$$g(s) = -Arc\tan s \quad \forall s \in \mathbf{R}.$$

Theorem 2.9.   *Assume* $(Q_1) - (Q_7)$ *and let* $f \in L^2(T_N)$. *Then a necessary and sufficient condition that a distribution solution* $u \in H^1(T_N)$ *of* $Qu = f(x)$ *exists is that*

$$\int_{T_N} f(x)\, dx = 0.$$

This result first appeared in [Sh13, p. 204].

Proof of Theorem 2.8.   We first establish the necessary condition of the theorem. Suppose then that

$$\sum_{i,j=1}^{N} \cdot \int_{T_N} a^{ij}(x,u)\, D_i\phi D_j u\ dx$$

$$+ \sum_{j=1}^{N} \int_{T_N} b^j(x,u,Du)\phi D_j u dx$$

$$= \int_{T_N} \phi\, [g(u) - h(x)]\, dx$$

$\forall \phi \in C^\infty(T_N)$, where $u \in H^1(T_N)$ and that (2.69) also holds. We choose $\phi = 1$ in this last equality and observe that the left-hand side then becomes zero. Consequently,

$$\int_{T_N} g(u)\, dx = \int_{T_N} h(x)\, dx.$$

But because of (2.69),

$$(2\pi)^N g(\infty) < \int_{T_N} g(u)\, dx < (2\pi)^N g(-\infty).$$

We conclude that

$$(2\pi)^N g(\infty) < \int_{T_N} h(x)\, dx < (2\pi)^N g(-\infty),$$

and the necessary condition of the theorem is established.

To establish the sufficiency condition, we observe that

$$|g(s) - h(x)| \le \|g\|_{L^\infty} + |h(x)|$$

$\forall s \in \mathbf{R}$ and $\forall x \in T_N$. Since the right-hand side of this last equality is in $L^2(T_N)$, we see that the conditions in the hypothesis of Lemma 2.2 are met. Hence, there exists a sequence of functions $\{u^n\}_{n=1}^{\infty}$ with

$$u^n = \gamma_1^n \psi_1 + \cdots + \gamma_n^n \psi_n,$$

where the $\{\gamma_k^n\}_{k=1}^{n}$ are real constants, such that

$$\sum_{i,j=1}^{N} \int_{T_N} a^{ij}(x,u^n)\, D_i\psi_k D_j u^n\ dx\ + n^{-1} \int_{T_N} \psi_k u^n\ dx$$

(2.71) $$+ \sum_{j=1}^{N} \int_{T_N} b^j(x,u^n,Du^n)\psi_k D_j u^n dx$$

$$= \int_{T_N} \psi_k\, [g(u^n) - h(x)]\, dx$$

for $k = 1, ..., n$.

We claim, as in the proof of Theorem 2.1, that there is a constant K such that

$$(2.72) \qquad \|u^n\|_{H^1} \leq K \quad \forall n.$$

Suppose that (2.72) is false. Then there is a subsequence of $\{u^n\}_{n=1}^{\infty}$, which, for ease of notation, we take to be the full sequence such that the properties enumerated in (2.51a)-(2.51d) hold.

Now, as in the proof of Theorem 2.1:

$$\int_{T_N} u^n \sum_{j=1}^N b^j (x, u^n, Du^n) D_j u^n dx = 0.$$

So we obtain from (2.71) that

$$(2.73) \qquad \begin{aligned} &\sum_{i,j=1}^N \int_{T_N} a^{ij}(x, u^n) D_i v^n D_j v^n \; dx \; + n^{-1} \int_{T_N} (v^n)^2 \; dx \\ &\qquad = \|u^n\|_{H^1}^{-1} \int_{T_N} v^n [g(u^n) - h(x)] dx \end{aligned}$$

where $v^n = u^n / \|u^n\|_{H^1}$.

From (Q_4), we obtain that

$$\eta_0 \||Dv^n\||_{L^2}^2 \leq \sum_{i,j=1}^N \int_{T_N} a^{ij}(x, u^n) D_i v^n D_j v^n \; dx.$$

Also, from (2.51b), we see that $\|v^n\|_{L^2} \leq 1$. So we conclude from (2.73) that there exists a constant K_1 such that

$$\eta_0 \||Dv^n\||_{L^2}^2 \leq K_1 \|u^n\|_{H^1}^{-1} \quad \forall n,$$

where $\eta_0 > 0$.

Using (2.51a) in conjuction with this last inequality gives

$$(2.74) \qquad \lim_{n \to \infty} \||Dv^n\||_{L^2}^2 = 0.$$

From (2.51b), we see that

$$1 = \|v^n\|_{L^2}^2 + \sum_{j=1}^N \|D_j v^n\|_{L^2}^2 \quad \forall n.$$

So from (2.74), we obtain that $\lim_{n \to \infty} \|v^n\|_{L^2}^2 = 1$, and hence from (2.51b) that

$$(2.75) \qquad \|v\|_{L^2}^2 = 1.$$

From (2.51d) joined with (2.74), we also obtain that

$$< w, D_j v >_{L^2} = 0 \quad \forall w \in L^2(T_N) \text{ and } j = 1, ..., N.$$

Therefore, $D_j v(x) = 0$ almost everywhere in T_N for $j = 1, ..., N$, and we see from Lemma 2.7 that

$$v = \text{ constant almost everywhere in } T_N.$$

It follows therefore from (2.75) that $v = (2\pi)^{-N/2}$ almost everywhere in T_N or $v = -(2\pi)^{-N/2}$ almost everywhere in T_N.

We shall suppose

$$(2.76) \qquad v = (2\pi)^{-N/2} \quad \text{a.e. in } T_N,$$

and arrive at a contradiction. As the reader will easily see, a similar line of reasoning gives a contradiction in case the other alternative holds.

Suppose then that (2.76) holds. We observe from (Q_4) and (2.73)

$$\|u^n\|_{H^1}^{-1} \int_{T_N} v^n \left[g(u^n) - h(x) \right] dx \geq 0$$

and consequently that

$$(2.77) \qquad \int_{T_N} v^n h(x)\, dx. \leq \int_{T_N} v^n g(u^n)\, dx.$$

Since $u^n = v^n \|u^n\|_{H^1}$, we see from (2.51a), (2.51c), and (2.76) that

$$(2.78) \qquad \lim_{n\to\infty} v^n(x)\, g(u^n(x)) = (2\pi)^{-N/2} g(\infty) \quad \text{a.e. in } T_N.$$

Also, since $g \in L^\infty(\mathbf{R})$, it follows from (2.51b) that the sequence $\{v^n g(u^n)\}_{n=1}^\infty$ is absolutely equi-integrable on T_N. Therefore, from Egoroff's theorem and (2.78), we obtain

$$(2.79) \qquad \lim_{n\to\infty} \int_{T_N} v^n g(u^n)\, dx = (2\pi)^{N/2} g(\infty).$$

On the other hand, from (2.51b) and the fact that $h \in L^2(T_N)$, we see that·

$$\lim_{n\to\infty} \int_{T_N} |(v^n - v)\, h|\, dx = 0.$$

Consequently, from (2.73), we have

$$\lim_{n\to\infty} \int_{T_N} v^n h(x)\, dx = (2\pi)^{-N/2} \int_{T_N} h(x)\, dx.$$

We conclude from (2.77) and (2.79) that

$$\int_{T_N} h(x)\, dx \leq (2\pi)^N g(\infty).$$

But this contradicts the first inequality in (2.70), which is strict. We conclude (2.51a) is false, and hence that the inequality in (2.72) is indeed true.

From the fact that (2.72) holds, it follows from Lemma 2.6 that there is a subsequence of $\{u^n\}_{n=1}^\infty$, which for ease of notation we take to be the full sequence, and a $u \in H^1(T_N)$ with the properties enumerated in (2.61a)-(2.61c).

Using (Q_1), (Q_2), (2.61a), (2.61b), and (2.72) exactly as in the proof of Theorem 2.1, we obtain that

(2.80)
$$\lim_{n \to \infty} \sum_{i,j=1}^{N} < D_i \psi_k, a^{ij}(\cdot, u^n) D_j u^n >_{L^2} = \sum_{i,j=1}^{N} \cdot \int_{T_N} a^{ij}(x, u) D_i \psi_k D_j u \, dx$$

for $k = 1, 2, \ldots$.

Similarly, we obtain from the proof of Theorem 2.1 that

(2.81)
$$\lim_{n \to \infty} \sum_{j=1}^{N} \int_{T_N} b^j(x, u^n, Du^n) \psi_k D_j u^n dx$$

$$= - \cdot \sum_{j=1}^{N} \int_{T_N} b^j(x, u^n, Du^n) u D_j \psi_k dx.$$

Next, from (2.61b) and the fact that $g \in C(\mathbf{R})$, we see that

$$\lim_{n \to \infty} g(u^n(x)) = g(u(x)) \qquad \text{a.e. in } T_N.$$

Since g is also in $L^\infty(\mathbf{R})$, we obtain from this last equality and the Lebesgue dominated convergence theorem that

$$\lim_{n \to \infty} < \psi_k, g(u^n) >_{L^2} = < \psi_k, g(u) >_{L^2} \qquad \text{for} \quad k = 1, 2, \ldots.$$

From this last fact in conjunction with (2.80), (2.81), (2.71), and (2.72), we finally obtain that

(2.82)
$$\sum_{i,j=1}^{N} \int_{T_N} a^{ij}(x, u) D_i \psi_k D_j u \, dx$$

$$+ \sum_{j=1}^{N} \int_{T_N} b^j(x, u, Du) \psi_k D_j u dx$$

$$= \int_{T_N} \psi_k [g(u) - h(x)] \, dx$$

for $k = 1, \ldots, n$.

From (Q_2), we see that $a^{ij}(\cdot, u) D_j u \in L^1(T_N)$. From (Q_5), we see that $b^j(\cdot, u, Du) D_j u \in L^1(T_N)$. From the hypothesis of the theorem, we have that $g(u) - h(x) \in L^1(T_N)$. Also, it follows from (2.11) that given $\phi \in C^\infty(T_N)$, there is a sequence $\{\phi_n\}_{n=1}^\infty$ and there are real constants $\{c_q^n\}_{q=1}^n$ such that

$$\phi_n = c_1^n \psi_1 + \cdots + c_1^n \psi_n$$

and

$$\lim_{n \to \infty} \phi_n(x) = \phi(x),$$

$$\lim_{n \to \infty} D_j \phi_n(x) = D_j \phi(x)$$

uniformly for $x \in T_N$, $j = 1, ..., N$. We conclude first that (2.82) holds with ψ_k replaced by ϕ_n, and next on passing to the limit as $n \to \infty$ that

$$\sum_{i,j=1}^{N} \cdot \int_{T_N} a^{ij}(x, u) D_i \phi D_j u \, dx$$

$$+ \sum_{j=1}^{N} \int_{T_N} b^j(x, u, Du) \phi D_j u dx$$

$$= \int_{T_N} \phi [g(u) - h(x)] \, dx$$

$\forall \phi \in C^\infty(T_N)$. So indeed $u \in H^1(T_N)$ is a distribution solution of

$$Qu = g(u) - h(x)$$

on T_N. ∎

Proof of Theorem 2.9. We prove the necessary condition of the theorem first. So suppose that $u \in H^1(T_N)$ is a distribution solution of $Qu = f(x)$ where $f \in L^2(T_N)$. Then

$$\sum_{i,j=1}^{N} \cdot \int_{T_N} a^{ij}(x, u) D_i \phi D_j u \, dx$$

$$+ \sum_{j=1}^{N} \int_{T_N} b^j(x, u, Du) \phi D_j u dx$$

$$= \int_{T_N} \phi f(x) \, dx$$

$\forall \phi \in C^\infty(T_N)$. We take $\phi = 1$ in this last equality and observe that $D_i 1 = 0$. So we are left with

$$\sum_{j=1}^{N} \int_{T_N} b^j(x, u, Du) D_j u dx = \int_{T_N} f(x) \, dx.$$

But from (Q_6) and (2.3), we see that the left-hand side of this last equality is zero. We conclude that

$$\int_{T_N} f(x) \, dx = 0,$$

and the necessary condition of the theorem is established.

To establish the sufficiency condition of the theorem, we introduce the new Hilbert space $H_0^1(T_N)$ defined as follows:

$$H_0^1(T_N) = \left\{ v \in H^1(T_N) : \int_{T_n} v(x) \, dx = 0 \right\}.$$

We observe that if $\phi \in C^\infty(T_N) \cap H_0^1(T_N)$, then

$$\widehat{D_j \phi}(m) = im_j \widehat{\phi}(m) \quad \forall m \in \Lambda_N$$

for $j = 1, ..., N$ where $\widehat{\phi}(m)$ is the Fourier coefficient of ϕ introduced in Chapter 1. Also, the fact that $\phi \in H_0^1(T_N)$ implies that

$$\widehat{\phi}(0) = 0.$$

From Parsevaal's Theorem, (Corollary 2.5 in Chapter 1), we obtain from the above that

$$(2\pi)^N \sum_{|m|\geq 1} |m|^2 \left|\widehat{\phi}(m)\right|^2 = \sum_{j=1}^{N} \|D_j\phi\|_{L^2}^2$$

and that

$$(2\pi)^N \sum_{|m|\geq 1} \left|\widehat{\phi}(m)\right|^2 = \|\phi\|_{L^2}^2 .$$

Consequently, $\phi \in C^\infty(T_N) \cap H_0^1(T_N)$ implies that

$$\|\phi\|_{L^2}^2 \leq \sum_{j=1}^{N} \|D_j\phi\|_{L^2}^2 .$$

Since every $v \in H_0^1(T_N)$ is the limit of a sequence of such ϕ in in the H^1-*norm*, we see that a similar inequality holds for $v \in H_0^1(T_N)$. We record this fact as

$$(2.83) \qquad v \in H_0^1(T_N) \Rightarrow \|v\|_{L^2}^2 \leq \sum_{j=1}^{N} \|D_j v\|_{L^2}^2 .$$

Continuing with the proof of the theorem, since the conditions in the hypothesis of Lemma 2.2 are met, there is a sequence $\{u^n\}_{n=1}^{\infty}$ with the following properties:

$$u^n = \gamma_1^n \psi_1 + \cdots + \gamma_n^n \psi_n,$$

$$\int_{T_N} [\sum_{i,j=1}^{N} a^{ij}(x, u^n) D_j u^n D_i \psi_k$$

$$(2.84) \qquad + \sum_{j=1}^{N} \psi_k b^j(x, u^n, Du^n) D_j u^n + u^n \psi_k n^{-1}] dx$$

$$= \int_{T_N} f(x) \psi_k dx,$$

with $k = 1, ..., n$ where $\{\psi_k\}_{k=1}^{\infty}$ is the orthonormal sequence in (2.10)(a),(b).

Now, from (2.10)(a), $\psi_1 = (2\pi)^{-N/2}$. Putting this value in (2.84) and observing from (Q_6) that

$$\int_{T_N} [\sum_{j=1}^{N} b^j(x, u^n, Du^n) D_j u^n] dx = 0,$$

we obtain from the hypothesis of the theorem that

$$(2\pi)^{-N/2} \int_{T_N} u^n n^{-1} dx = 0.$$

Consequently, $u^n \in H_0^1(T_N)$ and is of the form

$$(2.85) \qquad u^n = \gamma_2^n \psi_2 + \cdots + \gamma_n^n \psi_n$$

$\forall n \geq 2$.

Observing that $D_j (u^n)^2 = 2u^n D_j u^n$, we see from (Q_6) that

$$\int_{T_N} [\sum_{j=1}^{N} u^n b^j(x, u^n, Du^n) D_j u^n] dx = 0.$$

So on multiplying both sides of (2.84) by γ_k^n and summing on k from 2 thru n, we obtain

$$\int_{T_N} [\sum_{i,j=1}^{N} a^{ij}(x, u^n) D_j u^n D_i u^n + (u^n)^2 n^{-1}] dx = \int_{T_N} f(x) u^n dx.$$

We conclude from (Q_4) and this last fact that

$$\eta_0 \sum_{j=1}^{N} \|D_j u^n\|_{L^2}^2 \leq \|u^n\|_{L^2} \|f\|_{L^2}$$

where η_0 is a positive constant.

Since $u^n \in H_0^1(T_N)$, we obtain from (2.83) and this last inequality that

$$\eta_0 \left(\sum_{j=1}^{N} \|D_j u^n\|_{L^2}^2 \right)^{1/2} \leq \|f\|_{L^2}$$

$\forall n \geq 2$. Using (2.83) once again, we see that this in turn implies that there is a constant K such that

(2.86) $$\|u^n\|_{H^1} \leq K \quad \forall n \geq 2.$$

From the fact that this last inequality holds, it follws from Lemma 2.6 that there is a subsequence of $\{u^n\}_{n=2}^{\infty}$, which, for ease of notation, we take to be the full sequence and a $u \in H_0^1(T_N)$ such that (2.61a), (2.61b), and (2.61c) hold.

Proceeding exactly as in the proof of Theorem 2.1, we see from (2.62) that

$$\lim_{n \to \infty} \sum_{i,j=1}^{N} < D_i \psi_k, a^{ij}(\cdot, u^n) D_j u^n >_{L^2}$$

$$= \sum_{i,j=1}^{N} \cdot \int_{T_N} a^{ij}(x, u) D_i \psi_k D_j u \, dx$$

for $k = 2, 3, \ldots$.

In a similar manner, we obtain from (2.65) that

$$\lim_{n \to \infty} \sum_{j=1}^{N} < \psi_k, b^j(\cdot, u^n, Du^n) D_j u^n >_{L^2}$$

$$= \sum_{j=1}^{N} \int_{T_N} b^j(x, u, Du) \psi_k D_j u \, dx$$

for $k = 2, 3, \ldots$.

So, we obtain from these last two limits joined with (2.84) that

$$\int_{T_N}[\sum_{i,j=1}^{N} a^{ij}(x,u)D_j u D_i \psi_k$$

$$(2.87) \qquad\qquad + \sum_{j=1}^{N} \psi_k b^j(x,u,Du)D_j u]dx$$

$$= \int_{T_N} f(x)\psi_k dx,$$

for $k = 2, 3, ...$

From (Q_2), we see that $a^{ij}(\cdot, u) D_j u \in L^1(T_N)$. From (Q_5), we see that $b^j(\cdot, u, Du) D_j u \in L^1(T_N)$. From the hypothesis of the theorem, we have that $f(x) \in L^2(T_N)$ and $\int_{T_N} f(x) dx = 0$. Also, it follows from (2.11) that given $\phi \in C^\infty(T_N)$, there is a sequence $\{\phi_n\}_{n=1}^{\infty}$ and there are real constants $\{c_q^n\}_{q=1}^{n}$ such that

$$\phi_n = c_1^n \psi_1 + c_2^n \psi_2 + \cdots + c_1^n \psi_n$$

and

$$\lim_{n\to\infty} \phi_n(x) = \phi(x),$$

$$\lim_{n\to\infty} D_j \phi_n(x) = D_j \phi(x)$$

uniformly for $x \in T_N$, $j = 1, ..., N$. We conclude first that (2.87) holds with ψ_k replaced by ϕ_n, and next on passing to the limit as $n \to \infty$ that

$$\sum_{i,j=1}^{N} \cdot \int_{T_N} a^{ij}(x,u) D_i\phi D_j u\, dx$$

$$+ \sum_{j=1}^{N} \int_{T_N} b^j(x,u,Du)\phi D_j u dx$$

$$= \int_{T_N} \phi f(x)\, dx$$

$\forall \phi \in C^\infty(T_N)$.

So indeed $u \in H^1(T_N)$ is a distribution solution of

$$Qu = f(x)$$

on T_N. ∎

In the quasilinear elliptic resonance results that we have discussed in this section, i.e., Theorem 2.1 and Theorem 2.8, the second order coefficients depended on x and u, and were of the form $a^{ij}(x,u)$. For results where the coefficients are of the form $a^{ij}(x,u,Du)$, we refer the reader to the author's research monograph published by the American Mathematical Socciety [Sh, 14].

Exercises.

1. Give an example to show that Theorem 2.1 is false if (ii) in the theorem is replaced by (ii'') where

$$(ii'') \quad \int_{T_N} \mathcal{F}_-(x)\, dx = 0.$$

2. Prove that if A is a real $n \times n$ matrix with the property that

$$\beta \cdot A\beta \geq K \left| \beta \right|^2 \qquad \forall \beta \ s \ \mathbf{R}^n,$$

where K is a positive constant, then A^{-1} exists and $\left\| A^{-1} \right\|_{\mathcal{M}} \leq K^{-1}$ where $\left\| \cdot \right\|_{\mathcal{M}}$ designates the usual $n \times n$ matrix norm.

3. Give an example of a solenoidal two-vector

$$\mathbf{b}(x,s,p) = \left(b^1(x,s,p), b^2(x,s,p) \right)$$

on T_2 that meets (Q_5), (Q_6), and (Q_7) where b^1 depends upon s but b^2 does not depend upon s.

4. Given that $f(x,s)$ meets $(f-1)$ and $(f-2)$, suppose there exists a $u \in L^2(T_2)$ such that

$$uf(x,u) \in L^1(T_2).$$

Prove that $f(x,u) \in L^1(T_2)$.

3. Further Results and Comments

1. In an infinite strip in the upper half-plane, a uniqueness theorem for solutions of the heat equation holds, which is similar to the uniqueness theorem for harmonic functions in the unit disk described in §4 of Chapter 3. With $\ Str_{\alpha,\beta} = \mathbf{R} \times (\alpha, \beta)$ where α, β are nonnegative real numbers, the following theorem holds [Sh19]:

Theorem. *Let* $u \in C^2(Str_{0,\beta})$ *with* $\beta > 0$, *and suppose*

$$\frac{\partial u}{\partial t}(x,t) = \frac{\partial^2 u}{\partial x^2}(x,t) \qquad \forall (x,t) \in \ Str_{0,\beta}.$$

Suppose also that $u \in L^\infty(Str_{t_0,\beta})$ $\forall t_0$ *such that* $0 < t_0 < \beta$. *Suppose furthermore*

$$(i) \lim_{t \to 0} u(x,t) = 0 \qquad \forall x \in \mathbf{R},$$

$$(ii) \sup_{x \in \mathbf{R}} |u(x,t)| = o(t^{-1}) \text{ as } t \to 0.$$

Then $u(x,t)$ *is identically zero for* $(x,t) \in Str_{0,\beta}$.

This theorem is false if (i) is replaced with (i') or if (ii) is replaced with (ii') where

$$(i') \lim_{t \to 0} u(x,t) = 0 \qquad \forall x \in \mathbf{R} \backslash \{0\},$$

$$(ii') \sup_{x \in \mathbf{R}} |u(x,t)| = O(t^{-1}) \text{ as } t \to 0.$$

The counter-example for (i') is $k(x,t) = t^{-\frac{1}{2}} e^{-x^2/4t}$.

The counter-example for (ii') is $\partial k\left(x,t\right)/\partial x = -\frac{x}{2t^{3/2}}e^{-x^2/4t}$. To see that $\partial k\left(x,t\right)/\partial x$ is indeed a counter-example for (ii'), observe

$$\lim_{t\to 0}\left|t\frac{\partial k}{\partial x}\left(t^{\frac{1}{2}},t\right)\right| = 2^{-1}e^{-1/4}.$$

2. Using Galerkin techniques and harmonic analysis, we can also obtain results about time-periodic quasilinear parabolic boundary-value problems. In particular, let $\Omega \subset \mathbf{R}^N$ be a bounded open set, $N \geq 2$, and define $\widetilde{\Omega} = \Omega \times T_1$ where $T_1 = [-\pi, \pi)$. Define

$$\mathcal{A} = \{v\left(x,t\right) \in C^\infty\left(\Omega \times \mathbf{R}\right) : v \text{ satisfies } (\mathcal{A}_1) \text{ and } (\mathcal{A}_2)\},$$

where

$$(\mathcal{A}_1)\qquad v\left(x,t\right) = v\left(x,t+2\pi\right) \ \forall x \in \Omega \text{ and } \forall t \in \mathbf{R},$$

$$(\mathcal{A}_2)\ \exists E, \text{ a compact subset of } \Omega, \text{ such that } v\left(x,t\right) = 0 \\ \forall x \in \Omega\backslash E \ \text{ and } \forall t \in \mathbf{R}.$$

Introduce, for $u,v \in \mathcal{A}$, the following inner product:

$$< u,v >_{\widetilde{H}} = < D_t u, D_t v >_{L^2\left(\widetilde{\Omega}\right)} + \sum_{j=1}^{N} < D_j u, D_j v >_{L^2\left(\widetilde{\Omega}\right)}.$$

The Hilbert space we obtain from $\mathcal{A}$ by completing $\mathcal{A}$ by means of Cauchy sequences generated from the above inner product will be called $\widetilde{H}$.

Let $Q\left(u\right) = -\sum_{j=1}^{N} D_j\{1 + [1 + \nabla u \cdot \nabla u]^{-\frac{1}{2}} \}D_j u$ and introduce the two-form for $u, v \in \widetilde{H}$:

$$Q\left(u,v\right) = \sum_{j=1}^{N} \int_{\widetilde{\Omega}} [1 + (1 + \nabla u \cdot \nabla u)^{-\frac{1}{2}}] D_j u D_j v\, dx\, dt.$$

With $\|u\|_{L^2} = \|u\|_{L^2\left(\widetilde{\Omega}\right)}$, define

$$\lambda_1^* = \liminf_{\|u\|_{L^2}\to\infty} Q\left(u,u\right)/\|u\|_{L^2}^2 \quad \text{for } u \in \widetilde{H}.$$

Consider the problem

$$(3.1)\qquad\qquad \frac{\partial u}{\partial t} + Q\left(u\right) = g\left(x,t,u\right) + h\left(x,t\right)$$

where $u \in \widetilde{H}$, $h\left(x,t\right) \in L^2\left(\widetilde{\Omega}\right)$, and $g\left(x,t,s\right)$ satisfies the following three conditions:

$$(g-1)\qquad\qquad g \ \in \ C\left(\widetilde{\Omega}\times\mathbf{R}\right)$$
$$(g-2)\left|g\left(x,t,s\right)\right| \ \leq \ c_1\left|s\right| + \left|h_1\left(x,t\right)\right|$$
$$(g-3)\,sg\left(x,t,s\right) \ \leq \ \Gamma\left|s\right|^2 + \left|s\right|\left|h_2\left(x,t\right)\right|$$

where c_1 is a positive constant, $h_1, h_2 \in L^2\left(\widetilde{\Omega}\right)$, and $\Gamma < \lambda_1^*$.

We say u is a weak solution of the time-periodic quasilinear parabolic boundary-value problem (3.1) provided $u \in \widetilde{H}$ and

$$\int_{\widetilde{\Omega}} v D_t u \, dx \, dt + Q(u,v) = \int_{\widetilde{\Omega}} [g(x,t,u) + h(x,t)] v \, dx \, dt \quad \forall v \in \widetilde{H}.$$

The following result about these matters was obtained by Lefton and Shapiro in [LefS].

Theorem. *With* $\widetilde{\Omega}$ *as above,* g *satisfying* $(g-1), (g-2), (g-3)$, *and* $h \in L^2\left(\widetilde{\Omega}\right)$, *there exists* $u \in \widetilde{H}$ *that is a weak solution of (3.1).*

This theorem is a nonresonant result because the inequality $\Gamma < \lambda_1^*$ is strict. Resonant results also are presented in [LefS], and other quasilinear operator besides the $Q(u)$ above are considered.

More results about time-periodic quasilinear parabolic differential equations are presented in [Sh14, Ch. II].

3. When Fourier series methods are combined with recent developments in the Calculus of Variations, interesting one-sided resonant results can be obtained for the nonlinear Schrodinger differential equation in $\mathbf{R}^N$, $N \geq 1$,

$$(3.2) \qquad -\Delta u + q(x) u = \lambda_1 u - \alpha u^- + g(x,u) + h.$$

Here, $q \in C\left(\mathbf{R}^N\right), q \geq 0$, and $q(x) \to \infty$ as $|x| \to \infty$. Also, λ_1 is the first eigenvalue associated with the Schrodinger operator, $\alpha > 0$, $u^-(x) = -\min(u(x), 0), h \in L^2\left(\mathbf{R}^N\right)$, and $g(x,t) \in C\left(\mathbf{R}^N \times \mathbf{R}\right)$.

Let $C_q^1\left(R^N\right) = \left\{ u \in C^1\left(R^N\right) : \int_{R^N} [|\nabla u|^2 + (1+q)u^2] dx < \infty \right\}$ and introduce in this space the inner product

$$<u,v>_{1,q} = \int_{R^N} [\nabla u \cdot \nabla v + (1+q)uv] dx.$$

Close $C_q^1\left(R^N\right)$ using this inner product and the method of Cauchy sequences. Call the resulting Hilbert space H_q^1.

For the function g appearing in (3.2), the following assumptions are made:

$(g-1) \exists b \in L^2\left(R^N\right)$ such that $|g(x,t)| \leq b(x) \ \forall x \in R^N$ and $\forall t \in R$,

$(g-2) \lim_{t\to\infty} g(x,t) = g_+(x) \quad \forall x \in R^N$,

$(g-3) \ g(x,t) < g_+(x) \quad \forall x \in \mathbf{R}^N$ and $\forall t \in R$.

Say u is a weak solution of the Schrodinger equation (3.2), if $u \in H_q^1$ and if

$$\int_{R^N} [\nabla u \cdot \nabla v + quv] dx = \int_{\mathbf{R}^N} \left[\lambda_1 u - \alpha u^- + g(x,u) + h \right] v \, dx \ \forall v \in H_q^1.$$

Next, we introduce the solvability condition

$$(3.3) \qquad \int_{\mathbf{R}^N} [g_+(x)\,\phi_1(x) + h(x)\,\phi_1(x)]dx > 0,$$

where $\phi_1(x)$ is the first eigenfunction associated with the Schrodinger operator.

Combining linking theory in the variational calculus [St, p. 127] with Fourier analytic methods, the following one-sided resonant result is obtained in [Sh20].

Theorem. *Let $g \in C\left(\mathbf{R}^N \times \mathbf{R}\right)$ satisfy $(g-1), (g-2),$ and $(g-3)$, let $\alpha > 0$, and let $h \in L^2\left(\mathbf{R}^N\right)$, $N \geq 1$. Then the solvability condition (3.3) is both a necessary and sufficient condition for obtaining a $u \in H_q^1$, which is a weak solution to the Schrodinger equation (3.2).*

In [Sh20], a more general solvability condition than (3.3) is used for the sufficiency condition in the above theorem, namely the following:

$$\lim_{t\to\infty}\left\{\int_{\mathbf{R}^N} [G(x, t\phi_1(x)) + th(x)\,\phi_1(x)]dx\right\} = +\infty,$$

where $G(x, t) = \int_0^t g(x, s)\, ds$.

CHAPTER 6

The Stationary Navier-Stokes Equations

1. Distribution Solutions

The results in this Chapter are motivated by our two manuscripts on the stationary Navier-Stokes equations entitled, "Generalized and Classical Solutions of the Nonlinear Stationary Navier-Stokes Equations" [Sh10] and "One-sided Conditions for the Navier-Stokes Equations" [Sh15]. The first paper appeared in the Transactions of the American Mathematical Society and the second in the Journal of Differential Equations.

The main results in §1 deal with the situation when the driving force depends on x and $\mathbf{v}$ in a nonlinear manner, i.e., is of the form $\mathbf{f}(x, \mathbf{v})$, and come from the Journal of Differential Equations paper. The main theorem in §2, motivated by the Transactions paper, establishes a regularity result for the stationary Navier-Stokes equations when $\mathbf{f}(x) \in C^{\alpha}(T_N)$, $0 < \alpha < 1$, obtaining in two and three-dimensions a solution pair $(\mathbf{v}, p)$ with $\mathbf{v} \in C^{2+\alpha}(T_N)$ and $p \in C^{1+\alpha}(T_N)$.

As far as we can tell, the nonlinear results of the type mentioned above do not appear in the standard texts dealing with the stationary Navier-Stokes equations, i.e., they are not in the books by Ladyzhenskya [La], Temam [Te], or Galdi [Ga]. Also, as far as we can tell, the proof of the theorem in §2 using the C^{α}-Calderon-Zygmund theory likewise does not appear in the standard texts dealing with the stationary Navier-Stokes equations.

Also, in §1, we establish the basic result for the stationary Navier-Stokes equations under periodic boundary conditions when the driving force is

$$\mathbf{f}(x) \in L^2(T_N).$$

We shall operate in real N-dimensional Euclidean space, $\mathbf{R}^N$, $N \geq 2$, and use the following notation:

$$
\begin{aligned}
x &= (x_1, ..., x_N) \qquad y = (y_1, ..., y_N) \\
\alpha x + \beta y &= (\alpha x_1 + \beta y_1, ..., \alpha x_N + \beta y_N) \\
x \cdot y &= x_1 y_1 + ... + x_N y_N, \qquad |x| = (x \cdot x)^{\frac{1}{2}}.
\end{aligned}
$$

With T_N, the N-dimensional torus

$$T_N = \{x : -\pi \leq x_j < \pi, \ j = 1, ..., N\},$$

we shall say $f_1 \in L^r(T_N), 1 \leq r < \infty$, provided f_1 is a real-valued (unless explicitly stated otherwise) Lebesgue measurable function defined on $\mathbf{R}^N$ of period 2π in each variable such that

$$\int_{T_N} |f_1|^r \, dx < \infty.$$

$\phi = (\phi_1, ..., \phi_N) \in [C^\infty (T_N)]^N$ means that $\phi_j \in C^\infty (\mathbf{R}^N)$ and periodic of period 2π in each variable, $j = 1, ... N$.

$H^1(T_N)$ is the closure of the set of functions in $C^\infty(T_N)$ under the norm generated by the following real inner product:

$$< g_1, h_1 >_{H^1} = \int_{T_N} [g_1 h_1 + \nabla g_1 \cdot \nabla h_1] dx \quad \text{for } g_1, h_1 \in C^\infty(T_N).$$

In particular, we see if $g_1 \in H^1(T_N)$, then g_1 is in the familiar Sobolev space $W^{1,2}(T_N)$. By this, we mean, $g_1 \in L^2(T_N)$ and there are functions $w_1, ..., w_N \in L^2(T_N)$ such that

$$\int_{T_N} g_1 \partial \phi_1 / \partial x^j \, dx = -\int_{T_N} w_j \phi_1 dx \quad \forall \phi_1 \in C^\infty (T_N)$$

for $j = 1, ... N$. We refer to w_j as a weak partial derivative of g_1, and frequently write w_j as $\partial g_1 / \partial x^j$.

In this section, we will establish three resonance-type existence theorems for periodic solutions of the stationary Navier-Stokes equations. So, we deal with the equations

$$-\nu \Delta \mathbf{v}(x) + (\mathbf{v}(x) \cdot \nabla) \mathbf{v}(x) + \nabla p(x) = \mathbf{f}(x, \mathbf{v}(x))$$

(1.1)

$$(\nabla \cdot \mathbf{v})(x) = 0$$

where ν is a positive constant, and $\mathbf{v}$ and $\mathbf{f}$ are vector-valued functions.

$\mathbf{f} = (f_1, ..., f_N) : T_N \times \mathbf{R}^N \to \mathbf{R}^N$ and, throughout this section, $\mathbf{f}$ will meet the following two Caratheodory conditions:

$(f-1)$ For each fixed $\mathbf{s} = (s_1, ..., s_N) \in \mathbf{R}^N$, $f_j(x, s)$ is a real-valued measurable function on T_N, and for almost every $x \in T_N$, $f_j(x, \mathbf{s})$ is continuous on $\mathbf{R}^N$, $j = 1, ... N$.

$(f-2)$ For each $r > 0$, there is a finite-valued nonnegative function $\zeta_r(x) \in L^2(T_N)$ such that

$$|f_j(x, \mathbf{s})| \leq \zeta_r(x) \quad \text{for } |s_j| \leq r, \text{ for a.e. } x \in T_N,$$

and for $s_k \in R$, $k \neq j$, $j, k = 1, ... N$.

We will deal with the pair $(\mathbf{v}, p)$ where $\mathbf{v} \in [L^2(T_N)]^N$ and $p \in L^1(T_N)$. We say such a pair is a distribution solution of the stationary Navier-Stokes equations (1.1) provided the components of $\mathbf{f}(x, \mathbf{v}(x))$ are in $L^1(T_N)$ and

$$-\int_{T_N} [\nu \mathbf{v} \cdot \Delta \phi + \mathbf{v} \cdot (\mathbf{v} \cdot \nabla) \phi + p \nabla \cdot \phi] dx$$
$$= \int_{T_N} [\mathbf{f}(x, \mathbf{v}(x)) \cdot \phi(x)] dx$$

(1.2)

$$\int_{T_N} \mathbf{v} \cdot \nabla \xi dx = 0$$

for all $\phi \in [C^\infty(T_N)]^N$ and $\xi \in C^\infty(T_N)$.

First of all, however, we will present the very basic theorem for the stationary Navier-Stokes equations on the N-torus where the components of $\mathbf{f}$ do not depend on $\mathbf{v}$, i.e.,

$$\mathbf{f}(x) = (f_1(x), ..., f_N(x)).$$

The next three theorems we present will deal with the situation where the components of $\mathbf{f}$ actually do depend on $\mathbf{v}$, i.e.,

$$\mathbf{f}(x, \mathbf{v}(x)) = (f_1(x, \mathbf{v}(x)), ..., f_N(x, \mathbf{v}(x))).$$

The first theorem we prove is the following:

Theorem 1.1.   *Let $\mathbf{f}(x) = (f_1(x), ..., f_N(x))$ be a vector-valued function where $f_j \in L^2(T_N)$ for j=1,...N. Then a necessary and sufficient condition that there is a pair $(\mathbf{v}, p)$ with $v_j \in H^1(T_N)$ for j=1,...N and $p \in L^2(T_N)$ such that $(\mathbf{v}, p)$ is a distribution solution of the stationary Navier-Stokes equations (1.1) is that*

$$(\bigstar) \qquad\qquad \int_{T_N} f_j(x)\, dx = 0 \ \ for\ j\text{=}1,...N.$$

If $\lambda \in \mathbf{R}$ and the pair $(\mathbf{v}, p)$ is a classical solution of (1.1) with $\mathbf{f} = \lambda \mathbf{v}$, then it is easy to see from (1.2) that $\lambda \geq 0$. Also, it is easy to see that the pair $(\mathbf{v}, 0)$ satisfies (1.1) with $\mathbf{f} = 0\mathbf{v}$ whenever $\mathbf{v}$ is a constant vector field. This indicates that, in a way, zero plays the role of the first eigenvalue for the system (1.1). The literature refers to results for elliptic equations that occur at the first eigenvalue as results at resonance. Motivated by these facts, and the papers of Landesman and Lazer [LL], DeFigueiredo and Gossez [DG], and Brezis and Nirenberg [BN], we also present in this section two best possible resonance existence theorems where the components of $\mathbf{f}$ are subjected to one-sided growth conditions. The third resonance theorem is not one-sided, but it does present a necessary and sufficient result.

We say that $\mathbf{f}$ satisfies the one-sided growth condition $(f - 3)$ provided the following holds:

$$(f - 3) \qquad\qquad \limsup_{|s_j| \to \infty} f_j(x, \mathbf{s}) / s_j \leq 0 \quad \text{uniformly for } x \in T_N,$$

and $s_k \in \mathbf{R}$, $k \neq j$, $j, k = 1, ..., N$.

To be specific, by $(f - 3)$, we mean for each fixed j, given $\varepsilon > 0$, there is an $s_0 > 0$, such that

$$(1.3) \qquad\qquad f_j(x, \mathbf{s}) / s_j \leq \varepsilon \quad \text{for} \quad s_0 \leq |s_j|$$

for $x \in T_N$ and $s_k \in \mathbf{R}$, $k \neq j$, $j, k = 1, ..., N$.

We also note for future use that (1.3) is equivalent to

$$f_j(x, \mathbf{s}) \leq \varepsilon s_j \qquad \text{for} \quad s_0 \leq s_j$$

(1.4)

$$\geq \varepsilon s_j \qquad \text{for} \quad -s_0 \geq s_j.$$

Now, Theorem 1.2 will deal with the growth condition $(f-3)$ and a more restrictive growth condition on sets of positive measure. In particular, we define

(1.5)

$$E_j(\mathbf{f}) = \{x \in T_N : \limsup_{|s_j| \to \infty} f_j(x, \mathbf{s})/s_j < 0$$

$$\text{uniformly for } s_k \in \mathbf{R}, k \neq j, \ k = 1, ..., N\}$$

for $j = 1, ..., N$.

Also, for each set of positive integers n and l, we set

$$E_j(\mathbf{f}, n, l) = \{x \in T_N : f_j(x, \mathbf{s})/s_j < -n^{-1} \text{ for } |s_j| > l$$

(1.6)

$$\text{and for } s_k \in \mathbf{R}, k \neq j, \ k = 1, ..., N\}$$

Explicitly, $x_0 \in E_j(\mathbf{f})$ means there is a pair of positive integers n and l such that $x_0 \in E_j(\mathbf{f}, n, l)$. Consequently,

(1.7)
$$E_j(\mathbf{f}) = \bigcup_{n=1}^{\infty} \bigcup_{l=1}^{\infty} E_j(\mathbf{f}, n, l).$$

Also, it is easy to see that if $\mathbf{f}$ meets $(f-1)$, then $E_j(\mathbf{f}, n, l)$ is a measurable set. With $|E_j(\mathbf{f})|$ designating the N-dimensional Lebesgue measure of $E_j(\mathbf{f})$, Theorem 1.2 will have as part of its hypothesis that $|E_j(\mathbf{f})| > 0$ for $j = 1, ..., N$.

We now state Theorem 1.2.

Theorem 1.2. *Let $\mathbf{f} = (f_1, ..., f_N)$ be a vector-valued function satisfying $(f-1), (f-2)$, and $(f-3)$ where $N \geq 2$. Suppose that*

(1.8) $$|E_j(\mathbf{f})| > 0 \qquad for \qquad j = 1, ..., N,$$

where $E_j(\mathbf{f})$ is defined by (1.5). Then there exists a pair $(\mathbf{v}, p)$ with

$$p, \ f_j(x, \mathbf{v}), \ and \ v_j \, f_j(x, \mathbf{v}) \in L^1(T_N) \quad and \quad v_j \in H^1(T_N)$$

for $j = 1, ...N$ such that $(\mathbf{v}, p)$ is a distribution solution of the stationary Navier-Stokes equations (1.1).

An example of an $\mathbf{f}$ that meets the hypothesis of the theorem is $\mathbf{f} = (f_1, ..., f_N)$ where

$$f_j(x, s) = -s_j \eta_j(x) + \eta_j(x)$$

and η_j is a nonnegative function that is in $L^2(T_N)$ with

$$\int_{T_N} \eta_j(x)\, dx > 0 \qquad \text{for } j = 1, ..., N.$$

Clearly, there are many more such functions.

Theorem 1.2 is a best possible result, i.e., it is false if we replace condition (1.8) by the slightly less restrictive condition "$(N-1)$ of the sets $E_j(\mathbf{f})$ are of positive measure and the remaining set is of measure ≥ 0."

To see that Theorem 1.2 is false under such an assumption, we set

$$f_1(x,s) = -1 \quad \text{and} \quad f_j(x,s) = -s_j \quad \text{for} \quad j = 2, ..., N.$$

Then, it is clear that $\mathbf{f} = (f_1, ..., f_N)$ satisfies $(f-1), (f-2)$, and $(f-3)$, that $E_1(\mathbf{f}) = $ the empty set, and that $E_j(\mathbf{f}) = T_N$ for $j = 2, ..., N$. Consequently, $|E_1(\mathbf{f})| = 0$, and $|E_j(\mathbf{f})| = (2\pi)^N$ for $j = 2, ..., N$, and the less restrictive assumption is also met. Suppose there is a pair $(\mathbf{v}, p)$ with $p \in L^1(T_N)$ and $\mathbf{v} \in [H^1(T_N)]^N$ such that the first equation in (1.2) holds for all $\boldsymbol{\phi} \in [C^\infty(T_N)]^N$. Taking $\boldsymbol{\phi} = (1, 0, ..., 0)$ in the first equation in (1.2) gives 0 on the left-hand side of the equal sign and $-(2\pi)^N$ on the right-hand side of the equal sign. Since $0 \neq -(2\pi)^N$, we conclude that no such pair exists, and our assertion concerning the best possibility of Theorem 1.2 is established.

In order to state Theorem 1.3, we need to introduce two further one-sided growth conditions.

We say $\mathbf{f}$ satisfies $(f-4)$ if the following holds:

$(f-4)$ There exists a finite-valued nonnegative function $\zeta \in L^2(T_N)$ such that

$$\begin{aligned} f_j(x,\mathbf{s}) \quad &\leq \quad \zeta(x) \qquad \text{for} \ \ s_j \geq 0 \\ &\geq \quad -\zeta(x) \qquad \text{for} \ \ s_j \leq 0 \end{aligned}$$

for $x \in T_N, s \in \mathbf{R}^N$, and $j = 1, ..., N$.

To introduce the next one-sided condition, we suppose $\mathbf{g}(x, \mathbf{s})$ satisfies $(f-1)$ and $\mathbf{h} = (h_1, ..., h_N)$ is a vector-valued function with $h_j(x)$ finite-valued and in $L^1(T_N)$ for $j = 1, ..., N$. We say $(g, \mathbf{h})$ satisfies $(f-5)$ provided the following holds:

$(f-5)$
$$\begin{aligned} \limsup_{s_j \to \infty} g_j(x,\mathbf{s}) - h_j(x) &\leq 0 \\ \text{and} \\ \liminf_{s_j \to -\infty} g_j(x,\mathbf{s}) - h_j(x) &\geq 0 \end{aligned}$$

uniformly for $x \in T_N, s_k \in \mathbf{R}, k \neq j, k = 1, ..., N$ for $j = 1, ..., N$.

We also set
(1.9)
$$E_j^+(g,\mathbf{h}) = \{x \in T_N : \limsup_{s_j \to \infty} g_j(x,\mathbf{s}) - h_j(x) < 0$$

$$\text{uniformly for } s_k \in \mathbf{R}, k \neq j, \ k = 1, ..., N\};$$

(1.10)
$$E_j^- (g, \mathbf{h}) = \{x \in T_N : \liminf_{s_j \to -\infty} g_j (x, \mathbf{s}) - h_j (x) > 0$$

$$\text{uniformly for } s_k \in \mathbf{R}, k \neq j, \ k = 1, ..., N\};$$

(1.11)
$$E_j^+ (\mathbf{g}, \mathbf{h}, n, l) = \{x \in T_N : g_j (x, \mathbf{s}) - h_j (x) < -n^{-1} \text{ for } s_j > l$$

$$\text{and for } s_k \in \mathbf{R}, k \neq j, \ k = 1, ..., N\};$$

(1.12)
$$E_j^- (\mathbf{g}, \mathbf{h}, n, l) = \{x \in T_N : g_j (x, \mathbf{s}) - h_j (x) > n^{-1} \text{ for } s_j < -l$$

$$\text{and for } s_k \in \mathbf{R}, k \neq j, \ k = 1, ..., N\}.$$

We observe that

(1.13)
$$E_j^+ (g, \mathbf{h}) = \bigcup_{n=1}^{\infty} \bigcup_{l=1}^{\infty} E_j^+ (\mathbf{g}, \mathbf{h}, n, l)$$

$$E_j^- (g, \mathbf{h}) = \bigcup_{n=1}^{\infty} \bigcup_{l=1}^{\infty} E_j^- (\mathbf{g}, \mathbf{h}, n, l)$$

for $j = 1, ..., N$.

We now state Theorem 1.3.

Theorem 1.3. *Let* $\mathbf{f} (x, s) = \mathbf{g} (x, \mathbf{s}) - \mathbf{h} (x)$ *be vector-valued functions with* $\mathbf{h} (x)$ *finite-valued and in* $[L^2 (T_N)]^N, N \geq 2$. *Suppose that* $\mathbf{g}$ *satisfies* $(f - 1), (f - 2),$ *and* $(f - 4)$ *and* $(\mathbf{g}, \mathbf{h})$ *satisfies* $(f - 5)$. *Suppose furthermore that*

(1.14)
$$\left| E_j^+ (\mathbf{g}, \mathbf{h}) \right| > 0 \quad and \quad \left| E_j^- (\mathbf{g}, \mathbf{h}) \right| > 0$$

for $j = 1, ..., N$ *where* $E_j^+ (\mathbf{g}, \mathbf{h})$ *and* $E^- (\mathbf{g}, \mathbf{h})$ *are defined by (1.9) and (1.10), respectively. Then there exists a pair* $(\mathbf{v}, p)$ *with*

$$p, \ f_j (x, \mathbf{v}), \ and \ v_j \ f_j (x, \mathbf{v}) \in L^1 (T_N)$$

and $v_j \in H^1 (T_N)$ *for* $j = 1, ..., N$ *such that* $(\mathbf{v}, p)$ *is a distribution solution of the stationary Navier-Stokes equations (1.1).*

An example of an $\mathbf{f}$ that meets the conditions in the hypothesis of Theorem 1.3 but not Theorem 1.2 is $\mathbf{f} = (f_1, ..., f_N)$ where

$$f_j (x, \mathbf{s}) = -\eta_j (x) s_j / \left(1 + s_j^2 \right)^{1/2} - \eta_j (x) /2$$

and η_j is a nonnegative function that is in $L^2 (T_N)$ with

$$\int_{T_N} \eta_j (x) \, dx > 0 \qquad \text{for } j = 1, ..., N.$$

By modifying the counter-example used for Theorem 1.2, it is easy to show that Theorem 1.3 is also a best possible result. In particular, we now show that Theorem 1.3 is false if we replace (1.14) by the following less restrictive condition:

$$\left|E_j^+(\mathbf{g},\mathbf{h})\right| > 0, \quad j = 1, ..., N$$

(1.15)
$$\left|E_j^-(\mathbf{g},\mathbf{h})\right| > 0, \quad j = 2, ..., N$$

$$\left|E_1^-(\mathbf{g},\mathbf{h})\right| \geq 0.$$

We take $h_j = 0$ for $j = 1, ..., N$, $g_j(x,s) = -s_j$ for $j = 2, ..., N$, and

$$g_1(x,\mathbf{s}) = -1 \qquad \text{for } s_1 \geq 0$$
$$= -\left(1 + s_1^2\right)^{-1} \qquad \text{for } s_1 \leq 0.$$

Then it is clear that $\mathbf{g}$ meets $(f-1), (f-2)$, and $(f-4)$, that $(\mathbf{g},\mathbf{h})$ meets $(f-5)$, and that (1.15) holds where $E_1^-(\mathbf{g},\mathbf{h}) = $ the empty set.

Suppose that there is a pair $(\mathbf{v},p)$ with p in $L^1(T_N)$ and $\mathbf{v}$ in $H^1(T_N)$ that satisfies (1.2) for all $\phi \in C^\infty(T_N)$. Taking $\phi = (1,0,...,0)$ in the first equation in (1.2) gives 0 on the left-hand side of the equal sign. Since $v_1 \in L^2(T_N)$, there is a set $A \subset T_N$ and a constant K such that $|A| \geq \pi^N$ and $|v_1(x)| \leq K$ for $x \in A$. Consequently, after multiplying both sides of the equation by -1, the right-hand side of the equal sign in (1.2) is $\geq \pi^N\left(1 + K^2\right)^{-1}$. But $0 \geq \pi^N\left(1 + K^2\right)^{-1}$ is not true. So we have arrived at a contradiction. We conclude that no such pair $(\mathbf{v},p)$ exists, and our assertion concerning the best possibility of Theorem 1.3 is established.

We will also establish the following result, which is the direct analogue of a familiar Landesman-Lazer result [LL, p. 611]:

Theorem 1.4. *Suppose that* $\mathbf{g}(\mathbf{s}) = (g_1(s_1), ..., g_N(s_N))$ *and* $g_j(s_j) \in C(\mathbf{R})$ *for* $j = 1, ...N$, $N \geq 2$. *Suppose also that the limits*

(1.16) $$\lim_{t \to \infty} g_j(t) = g_j(\infty) \qquad \text{and} \qquad \lim_{t \to -\infty} g_j(t) = g_j(-\infty)$$

exist and are finite, and that

(1.17) $$g_j(\infty) < g_j(t) < g_j(-\infty)$$

for $t \in \mathbf{R}$ *and* $j = 1, ...N$. *Suppose furthermore that the components of* $\mathbf{h}(x)$ *are finite-valued and in* $L^2(T_N)$ *and that* $\mathbf{f}(x,\mathbf{s}) = \mathbf{g}(\mathbf{s}) - \mathbf{h}(x)$. *Then a necessary and sufficient condition that there exists a pair* $(\mathbf{v},p)$ *with*

$$p, \ f_j(x,\mathbf{v}), \ \text{and} \ v_j \ f_j(x,\mathbf{v}) \in L^1(T_N)$$

and $v_j \in H^1(T_N)$ *such that* $(\mathbf{v},p)$ *is a distribution solution of (1.1) is that*

(1.18) $$(2\pi)^N g_j(\infty) < \int_{T_N} h_j(x)\,dx < (2\pi)^N g_j(-\infty)$$

for $j = 1, ...N$.

To establish these four theorems, we will make strong use of the theory of multiple Fourier series. In particular, if $f_1 \in L^1(T_N)$, we set

$$\widehat{f_1}(m) = (2\pi)^{-N} \int_{T_N} f_1(x) e^{-im \cdot x} dx \quad \forall m \in \Lambda_N$$

where Λ_N represents the set of integral lattice points in $\mathbf{R}^N$, and

$$S_R(f_1, x) = \sum_{|m| \leq R} \widehat{f_1}(m) e^{im \cdot x}.$$

If $f_1 \in L^2(T_N)$, it is well-known that

$$\lim_{R \to \infty} \int_{T_N} |S_R(f_1, x) - f_1(x)|^2 dx = 0.$$

Also, Parsevaal's Theorem tells us that

$$\int_{T_N} |f(x)|^2 dx = (2\pi)^N \sum_{m \in \Lambda_N} \left| \widehat{f_1}(m) \right|^2.$$

For $f_1 \in L^1(T_N)$, we will also set for $t > 0$,

$$(1.19) \qquad f_1(x, t) = \sum_{m \in \Lambda_N} \widehat{f_1}(m) e^{im \cdot x - |m|t}.$$

$f_1(x, t)$ is also called the Abel means of f_1, and it is well-known (see Theorem 4.3 of Chapter 1) that

$$(1.20) \qquad \lim_{t \to 0} \int_{T_N} |f_1(x, t) - f_1(x)| dx = 0.$$

Also, for $t > 0$, set

$$(1.21) \qquad H_0(x, t) = \sum_{|m| > 0} |m|^{-2} e^{im \cdot x - |m|t},$$

and for $j = 1, ..., N$, set

$$(1.22) \qquad H_j(x, t) = \sum_{|m| > 0} im_j |m|^{-2} e^{im \cdot x - |m|t}.$$

We also define
(1.23)
$$\Phi(x) = (2\pi)^N [|S_{N-1}|(N-2)]^{-1} |x|^{-(N-2)} \quad \forall x \in T_N \backslash \{0\} \text{ and } N \geq 3$$

$$= (2\pi) \log |x|^{-1} \quad \forall x \in T_2 \backslash \{0\} \text{ and } N = 2,,$$

where $|S_{N-1}|$ is the $(N-1)$-dimensional volume of the $(N-1)$-sphere, i.e., for $N = 3$, $|S_{N-1}| = 4\pi$. In particular, for general $N \geq 2$, $|S_{N-1}| = 2(\pi)^{N/2}/\Gamma(\frac{N}{2})$.

A well-known lemma (see Lemma 1.4 of Chapter 3) concerning the functions just introduced is the following:

Lemma A. *The following facts hold for the functions defined in (1.21), (1.22), and (1.23) where $N \geq 2$:*

(i) $\lim_{t \to 0} H_0(x,t) = H_0(x)$ *exists and is finite and*

$$\lim_{t \to 0} H_j(x,t) = H_j(x) \text{ exists and is finite}$$

$\forall x \in \mathbf{R}^N \setminus \cup_{m \in \Lambda_N} \{2\pi m\}$ *and $j = 1, ..., N$;*

(ii) $\lim_{|x| \to 0} [H_0(x) - \Phi(x)]$ *exists and is finite;*

(iii) $\lim_{|x| \to 0} [H_j(x) + (2\pi)^N x_j / |S_{N-1}| \, |x|^N]$ *exists and is finite for $j = 1, ..., N$;*

(iv) $H_0(x), H_j(x) \in L^1(T_N)$ *for $j = 1, ..., N$;*

(v) $H_0(x) - |x|^2 / 2N$ *is harmonic in $\mathbf{R}^N \setminus \cup_{m \in \Lambda_N} \{2\pi m\}$;*

(vi) $H_j(x)$ *is harmonic in $\mathbf{R}^N \setminus \cup_{m \in \Lambda_N} \{2\pi m\}$ for $j = 1, ..., N$;*

(vii) $\lim_{t \to 0} \int_{T_N} |H_0(x,t) - H_0(x)| \, dx = 0$ *and*

$$lim_{t \to 0} \int_{T_N} |H_j(x,t) - H_j(x)| \, dx = 0$$

for $j = 1, ..., N$.

Two other well-known lemmas about multiple Fourier series are the following:

Lemma B. *If f_1 is a function in $L^r(T_N)$, where $1 < r < \infty$, then*

$$\sum_{|m| > 0} \widehat{f_1}(m) \frac{m_j m_k}{|m|^2} e^{im \cdot x}$$

is also the Fourier series of a function in $L^r(T_N)$ for $j, k = 1, ..., N$.

Lemma C. *If f_1 is a function in $C^\alpha(T_N)$, where $0 < \alpha < 1$, then*

$$\sum_{|m| > 0} \widehat{f_1}(m) \frac{m_j m_k}{|m|^2} e^{im \cdot x}$$

is also the Fourier series of a function in $C^\alpha(T_N)$ for $j, k = 1, ..., N$.

The statement made in Lemma C is (5.5) and (5.6) in the proof of Theorem 5.1 in Chapter 2. Likewise, the statement made in Lemma B is (6.8) in the proof of Theorem 6.1 in Chapter 2.

In order to study the system of stationary Navier-Stokes equations, we will need a result about the system of stationary Stokes equations, which is the following:

$$-\nu \Delta \mathbf{v}(x) + \nabla p(x) = \mathbf{h}(x)$$

(1.24)

$$(\nabla \cdot \mathbf{v})(x) = 0$$

where ν is a positive constant, and $\mathbf{v}$ and $\mathbf{h}$ are vector-valued functions.

We will deal with the pair $(\mathbf{v}, p)$ where $\mathbf{v} \in \left[L^1(T_N)\right]^N$ and $p \in L^1(T_N)$.

If $\mathbf{h} \in \left[L^1 \left(T_N \right) \right]^N$, we say such a pair is a distribution solution of (1.24) provided

$$- \int_{T_N} [\nu \mathbf{v} \cdot \Delta \phi + p \nabla \cdot \phi] dx$$
$$= \int_{T_N} [\mathbf{h} \left(x \right) \cdot \phi \left(x \right)] dx$$

(1.25)

$$\int_{T_N} \mathbf{v} \cdot \nabla \xi dx = 0$$

for all $\phi \in [C^\infty \left(T_N \right)]^N$ and $\xi \in C^\infty \left(T_N \right)$.

In order to obtain distribution solutions of the stationary Stokes equations, we introduce a set of functions using Lemmas A and B above. First from Lemma A above, we observe that there is a function $H_0 \left(x \right) \in L^1 \left(T_N \right)$, which is periodic of period 2π in each variable and is in

$$C^\infty \left(\mathbf{R}^N \setminus \cup_{m \in \Lambda_N} \{ 2\pi m \} \right)$$

with the following Fourier series:

$$\sum_{|m|>0} |m|^{-2} e^{im \cdot x}.$$

Because of (ii) in Lemma A, the singularity of $H_0 \left(x \right)$ at the origin is the same as that of $\Phi \left(x \right)$, which is defined in (1.23) above. So we observe the following is true:

For each $N \geq 2$, there exists r_0 with $1 < r_0 < \infty$, such that $H_0 \in L^{r_0} \left(T_N \right)$.

Because of this fact and Lemma B, we consequently see that there are functions

(1.26) $u_j^k \in L^{r_0} \left(T_N \right), \qquad j, k = 1, ..., N,$

which have the following Fourier coefficients:

$$\widehat{u}_j^k \left(m \right) = \left[\delta_j^k - m_j m_k |m|^{-2} \right] |m|^{-2} \nu^{-1} \quad \text{for} \quad m \neq 0$$
$$= 0 \qquad\qquad\qquad\qquad\qquad \text{for} \quad m = 0.$$

where δ_j^k is the usual Kronecker-δ.

Also, from (1.22) and Lemma A, we set

(1.27) $q_j \left(x, t \right) = -H_j \left(x, t \right)$

and see there are functions

(1.28) $q_j \in L^1 \left(T_N \right), \qquad j = 1, ..., N,$

which have the following Fourier coefficients:

$$\widehat{q}_j \left(m \right) = -im_j / |m|^{-2} \quad \text{for} \quad m \neq 0$$
$$= 0 \qquad\qquad \text{for} \quad m = 0.$$

Using the functions u_j^k and q_j, we next establish the following lemma about distribution solutions of the Stokes equations:

Lemma 1.5. *Suppose $h_j \in L^1(T_N)$ and $\widehat{h}_j(0) = 0$ for $j = 1, ..., N$. Set*

$$v_j(x) = (2\pi)^{-N} \int_{T_N} \left[\sum_{k=1}^{N} u_j^k(x-y) h_k(y)\right] dy,$$

$$p(x) = (2\pi)^{-N} \int_{T_N} \left[\sum_{k=1}^{N} q_k(x-y) h_k(y)\right] dy,$$

and $\mathbf{v} = (v_1, ..., v_N)$. Then $\mathbf{v} \in \left[L^1(T_N)\right]^N$, $p \in L^1(T_N)$, and the pair $(\mathbf{v}, p)$ is a distribution solution of the Stokes equations given in (1.24).

Proof of Lemma 1.5. To prove the lemma, we set for $t > 0$,

$$u_j^k(x, t) = \sum_{m \in \Lambda_N} \widehat{u}_j^k(m) e^{im \cdot x - |m|t}$$

and

$$v_j(x, t) = (2\pi)^{-N} \int_{T_N} \left[\sum_{k=1}^{N} u_j^k(x-y, t) h_k(y)\right] dy.$$

Likewise, using (1.22) and (1.27), we set

$$p_j(x, t) = (2\pi)^{-N} \int_{T_N} \left[\sum_{k=1}^{N} q_k(x-y, t) h_k(y)\right] dy$$

and

$$h_j(x, t) = \sum_{m \in \Lambda_N} \widehat{h}_j(m) e^{im \cdot x - |m|t}.$$

It is clear that $v_j(x, t)$, $p_j(x, t)$, and $h_j(x, t)$ are in $C^\infty(\mathbf{R}^N)$ and periodic of period 2π in each variable. A computation shows that

$$-\nu \Delta v_j(x, t) + \frac{\partial p_j(x, t)}{\partial x_j} =$$

$$\sum_{m \in \Lambda_N} \sum_{k=1}^{N} [\nu |m|^2 \widehat{u}_j^k(m) + im_j \widehat{q}_k(m)] \widehat{h}_k(m) e^{im \cdot x - |m|t}.$$

A further computation shows that the item in brackets above has the value δ_j^k. So

$$-\nu \Delta v_j(x, t) + \frac{\partial p_j(x, t)}{\partial x_j} = \sum_{m \in \Lambda_N} \sum_{k=1}^{N} \delta_j^k \widehat{h}_k(m) e^{im \cdot x - |m|t}$$

$$= \sum_{m \in \Lambda_N} \widehat{h}_j(m) e^{im \cdot x - |m|t}$$

$$= h_j(x, t).$$

Likewise, we see that

$$\sum_{j=1}^{N} \frac{\partial v_j(x,t)}{\partial x_j} = \sum_{m \in \Lambda_N} \sum_{k=1}^{N} [\sum_{j=1}^{N} im_j \widehat{u}_j^k(m)] \widehat{h}_k(m) e^{im \cdot x - |m|t}.$$

But the item in brackets in the above equation is zero. So, we have that

$$\sum_{j=1}^{N} \frac{\partial v_j(x,t)}{\partial x_j} = 0.$$

We conclude from the above computations that $\mathbf{v}(x,t)$ and $p(x,t)$ are classical solutions of the Stokes equations when in (1.24), $\mathbf{h}(x)$ is replaced by $\mathbf{h}(x,t)$. Because they are classical solutions, it follows that they are distribution solutions. So in particular, we have that

$$-\int_{T_N} [\nu \mathbf{v}(x,t) \cdot \Delta \boldsymbol{\phi}(x) + p(x,t) \nabla \cdot \boldsymbol{\phi}(x)] dx$$

(1.29)
$$= \int_{T_N} [\mathbf{h}(x,t) \cdot \boldsymbol{\phi}(x)] dx,$$

$$\int_{T_N} \mathbf{v}(x,t) \cdot \nabla \xi(x) \, dx = 0,$$

for $t > 0$, $\boldsymbol{\phi} \in [C^\infty(T_N)]^N$, and $\xi \in C^\infty(T_N)$.

Now from the hypothesis of the lemma, we see that $v_j(x) \in L^1(T_N)$ and $p(x) \in L^1(T_N)$. So it follows from (1.19) and (1.20) above that

$$\lim_{t \to 0} \int_{T_N} |v_j(x,t) - v_j(x)| \, dx = 0 \quad \text{for} \quad j = 1, ..., N,$$

$$\lim_{t \to 0} \int_{T_N} |p(x,t) - p(x)| \, dx = 0.$$

Likewise,

$$\lim_{t \to 0} \int_{T_N} |h_j(x,t) - h_j(x)| \, dx = 0 \quad \text{for} \quad j = 1, ..., N.$$

Consequently, on taking the limit as $t \to 0$ on both sides of the two equations in (1.29), we obtain that

$$-\int_{T_N} [\nu \mathbf{v}(x) \cdot \Delta \boldsymbol{\phi}(x) + p(x) \nabla \cdot \boldsymbol{\phi}(x)] dx$$

$$= \int_{T_N} [\mathbf{h}(x) \cdot \boldsymbol{\phi}(x)] dx,$$

$$\int_{T_N} \mathbf{v}(x) \cdot \nabla \xi(x) \, dx = 0,$$

for $\boldsymbol{\phi} \in [C^\infty(T_N)]^N$ and $\xi \in C^\infty(T_N)$, which concludes the proof of the lemma. ∎

Next, we use a clever Galerkin technique to establish a basic lemma for the stationary Navier-Stokes equations. But first we need some more facts from the theory of multiple Fourier series.

We say a real-valued, vector-valued function $\boldsymbol{\psi} \in J(T_N)$ if

$$\boldsymbol{\psi} \in [C^\infty(T_N)]^N \text{ and } \nabla \cdot \boldsymbol{\psi}(x) = 0 \quad \forall x \in T_N.$$

We claim that there is a real-valued sequence $\{\boldsymbol{\psi}^n\}_{n=1}^\infty$ with $\boldsymbol{\psi}^n \in J(T_N)$ such that

$$(1.30) \qquad \int_{T_N} \boldsymbol{\psi}^n \cdot \boldsymbol{\psi}^l \, dx = \delta_l^n$$

$n, l = 1, 2, \dots$. Furthermore, this sequence can be chosen so that it has the following additional property: given $\boldsymbol{\psi} \in J(T_N)$ and $\varepsilon > 0$, there are constants $c_1, \dots, c_n$ such that

$$(1.31) \qquad \begin{aligned} &\left| \boldsymbol{\psi}(x) - \sum_{l=1}^n c_l \boldsymbol{\psi}^l(x) \right| < \varepsilon \\[2mm] &\left| \frac{\partial \boldsymbol{\psi}(x)}{\partial x_k} - \sum_{l=1}^n c_l \frac{\partial \boldsymbol{\psi}^l(x)}{\partial x_k} \right| < \varepsilon \end{aligned}$$

for $x \in T_N$ and $k = 1, \dots, N$. Also,

$$(1.31\,') \quad \boldsymbol{\psi}^1 = (\alpha_N, 0, \dots, 0), \; \boldsymbol{\psi}^2 = (0, \alpha_N, 0, \dots, 0), \dots, \boldsymbol{\psi}^N = (0, 0, \dots, 0, \alpha_N)$$

where

$$\alpha_N = (2\pi)^{-N/2}.$$

$\cdot$

To establish this claim, we first note from (1.19) and (1.20) above that

$$\left\{ e^{im \cdot x} \right\}_{m \in \Lambda_N} \text{ is a complete orthogonal system for } L^1(T_N).$$

By complete, we mean the following: given $f_1 \in L^1(T_N)$, then

$$\widehat{f_1}(m) = 0 \; \forall m \in \Lambda_N \Rightarrow f_1 = 0 \text{ for a.e. } x \in T_N.$$

Also, we note that if $g_1 \in C^\infty(T_N)$, then given $\varepsilon > 0$, $\exists R_0 > 0$ such that

$$\left| g_1(x) - \sum_{|m| \leq R} \widehat{g_1}(m) e^{im \cdot x} \right| \leq \varepsilon,$$

$$\left| \frac{\partial g_1(x)}{\partial x_k} - \sum_{|m| \leq R} im_k \widehat{g_1}(m) e^{im \cdot x} \right| \leq \varepsilon$$

for $R \geq R_0$, for $x \in T_N$, and $k = 1, \dots, N$. This follows from the fact that

$$\widehat{g_1}(m) = O\left(|m|^{-4N} \right) \text{ as } |m| \to \infty.$$

We can replace the orthogonal system $\left\{ e^{im \cdot x} \right\}_{m \in \Lambda_N}$ in the above by the real orthogonal system

$$\{\cos m \cdot x\}_{m \in \Lambda_N^\diamond \cup \{0\}} \bigcup \{\sin m \cdot x\}_{m \in \Lambda_N^\diamond},$$

where

$$\Lambda_N^\Diamond = \{m \in \Lambda_N^+ : m_1 > 0, \ m_1 = 0 \text{ and } m_2 > 0,$$
$$m_1 = m_2 = 0 \text{ and } m_3 > 0, \ \ldots,$$
$$m_1 = m_2 = \cdots = m_{N-1} = 0 \text{ and } m_N > 0\}$$

and

$$\Lambda_N^+ = \{m \in \Lambda_N : m_j \geq 0, \ j = 1, ..., N\}.$$

In other words, if we define for $f_1 \in L^1(T_N)$,

$$\widehat{f_1^c}(m) = \frac{2}{(2\pi)^N} \int_{T_N} f_1(x) \cos m \cdot x \, dx,$$

$$\widehat{f_1^s}(m) = \frac{2}{(2\pi)^N} \int_{T_N} f_1(x) \sin m \cdot x \, dx,$$

then

$$\widehat{f_1^c}(m) = 0 \text{ for } m \in \Lambda_N^\Diamond \cup \{0\} \text{ and } \widehat{f_1^s}(m) = 0 \text{ for } m \in \Lambda_N^\Diamond$$

implies that $f_1(x) = 0$ for a.e. $x \in T_N$.

Likewise, if $g_1 \in C^\infty(T_N)$ with $\widehat{g_1^c}(0) = 0$, then given $\varepsilon > 0$, $\exists R_0 > 0$ such that

$$\left| g_1(x) - \sum_{|m| \leq R, m \in \Lambda_N^\Diamond} [\widehat{g_1^c}(m) \cos m \cdot x + \widehat{g_1^s}(m) \sin m \cdot x] \right| \leq \varepsilon,$$

$$\left| \frac{\partial g_1(x)}{\partial x_k} - \sum_{|m| \leq R, m \in \Lambda_N^\Diamond} m_k [\widehat{g_1^s}(m) \cos m \cdot x - \widehat{g_1^c}(m) \sin m \cdot x] \right| \leq \varepsilon$$

for $R \geq R_0$, for $x \in T_N$, and $k = 1, ..., N$.

To get the sequence $\{\psi^n\}_{n=1}^\infty$ for $N = 2$, we proceed as follows. We first set

$$J^o(T_N) = \left\{ \mathbf{f} \in J(T_N) : \int_{T_N} f_j(x) \, dx = 0 \ \ j = 1, ..., N \right\}$$

and observe that

(1.32) $(f_1, f_2) \in J^o(T_2) \Rightarrow \widehat{f_1}(m) = 0 \ \ \forall m_2 = 0.$

This follows from the fact that

$$\frac{\partial f_1(x)}{\partial x_1} + \frac{\partial f_2(x)}{\partial x_2} = 0 \ \ \forall x \in T_N$$

implies that

$$im_1 \widehat{f_1}(m) + im_2 \widehat{f_2}(m) = 0 \ \ \forall m \in \Lambda_N.$$

Next, we observe that using (1.32),

$$\left\{ \left(\frac{im_2 e^{im \cdot x}}{|m|}, \frac{-im_1 e^{im \cdot x}}{|m|} \right) \right\}_{m \in \Lambda_2 \setminus \{0\}}$$

is a complete orthogonal system for $J^o(T_2)$.

Likewise, as above, it follows that

$$\left\{\left(\frac{m_2 \cos m \cdot x}{|m|}, \frac{-m_1 \cos m \cdot x}{|m|}\right)\right\}_{m \in \Lambda_2^\Diamond}$$
$$\bigcup \left\{\left(\frac{m_2 \sin m \cdot x}{|m|}, \frac{-m_1 \sin m \cdot x}{|m|}\right)\right\}_{m \in \Lambda_2^\Diamond}$$

is a real complete orthogonal system for $J^o(T_2)$.

From this last observation, it is easy to see that the claims asserted in (1.30) and (1.31) are true for $J(T_2)$.

To validate these claims for $J(T_3)$, we use (see Exercise 5 below) the following real complete orthogonal system for $J^o(T_3)$:

$$\left\{\left(\frac{m_2 \cos m \cdot x}{|m|}, \frac{-m_1 \cos m \cdot x}{|m|}, 0\right)\right\}_{m \in \Lambda_3^\Diamond}$$
$$\bigcup \left\{\left(\frac{-m_3 \cos m \cdot x}{|m|}, 0, \frac{m_1 \cos m \cdot x}{|m|}\right)\right\}_{m \in \Lambda_3^\Diamond}$$
$$\bigcup \left\{\left(0, \frac{m_3 \cos m \cdot x}{|m|}, \frac{-m_2 \cos m \cdot x}{|m|}\right)\right\}_{m \in \Lambda_3^\Diamond}$$
$$\bigcup \left\{\left(\frac{m_2 \sin m \cdot x}{|m|}, \frac{-m_1 \sin m \cdot x}{|m|}, 0\right)\right\}_{m \in \Lambda_3^\Diamond}$$
$$\bigcup \left\{\left(\frac{-m_3 \sin m \cdot x}{|m|}, 0, \frac{m_1 \sin m \cdot x}{|m|}\right)\right\}_{m \in \Lambda_3^\Diamond}$$
$$\bigcup \left\{\left(0, \frac{m_3 \sin m \cdot x}{|m|}, \frac{-m_2 \sin m \cdot x}{|m|}\right)\right\}_{m \in \Lambda_3^\Diamond}.$$

A similar situation prevails for $J^o(T_N)$, $N \geq 4$. So, the claims asserted in (1.30) and (1.31) are true for $J(T_N)$.

Using (1.30) and (1.31), we will prove a basic lemma for the stationary Navier-Stokes equations. But first we introduce more notation.

For $\mathbf{v}, \mathbf{w} \in [H^1(T_N)]^N$ and $\phi \in [C^\infty(T_N)]^N$, we set

$$[\mathbf{v}, \mathbf{w}] = \int_{T_N} \left[\sum_{k=1}^{N} \frac{\partial \mathbf{v}(x)}{\partial x_k} \cdot \frac{\partial \mathbf{w}(x)}{\partial x_k}\right] dx$$

(1.33) $$(\mathbf{v}, \mathbf{w}) = \int_{T_N} \mathbf{v} \cdot \mathbf{w} \, dx$$

$$\{\phi, \mathbf{v}, \mathbf{w}\} = \int_{T_N} \phi \cdot (\mathbf{v} \cdot \nabla) \mathbf{w} \, dx,$$

and observe that if $\nabla \cdot \mathbf{v} = 0$, then

(1.34) $$\{\phi, \mathbf{v}, \mathbf{w}\} = -\{\mathbf{w}, \mathbf{v}, \phi\}.$$

We will need the following lemma in the sequel:

Lemma 1.6. *Let* $\mathbf{v} \in \left[L^1\left(T_N\right)\right]^N$ *with the property that*

$$\int_{T_N} v_j\left(x\right)dx = 0 \quad for \quad j = 1,...N.$$

Suppose futhermore that

$$\int_{T_N} \mathbf{v}\left(x\right)\cdot\nabla\xi\left(x\right)dx = 0 \quad \forall\xi \in C^\infty\left(T_N\right).$$

Also, suppose that

$$\int_{T_N} \mathbf{v}\left(x\right)\cdot\Delta\boldsymbol{\psi}\left(x\right)dx = 0 \quad \forall\boldsymbol{\psi} \in J\left(T_N\right).$$

Then

$$v_j\left(x\right) = 0 \quad a.e. \; in \; T_N,$$

for $j=1,...N.$

Proof of Lemma 1.6. We designate the Fourier series of v_j by

$$\sum_{m\in\Lambda_N} \widehat{v}_j(m)e^{im\cdot x}.$$

Writing $e^{im\cdot x} = \cos\left(m\cdot x\right) + i\sin\left(m\cdot x\right)$, we see from the second hypothesis in the lemma that

$$\int_{T_N} \mathbf{v}\left(x\right)\cdot\nabla e^{im\cdot x}dx = 0 \quad \forall m \in \Lambda_N.$$

Consequently,

$$\sum_{j=1}^N im_j\widehat{v}_j(m) = 0 \quad \forall m \in \Lambda_N.$$

So, if we set

$$v_j(x,t) = \sum_{m\in\Lambda_N} \widehat{v}_j(m)e^{im\cdot x-|m|t}$$

for $t > 0$ and

$$\mathbf{v}\left(x,t\right) = \left(v_1(x,t),...,v_N(x,t)\right),$$

we see that $v_j(x,t) \in C^\infty\left(T_N\right)$ and that

$$\mathbf{v}\left(x,t\right) \in J\left(T_N\right) \quad \text{for } t > 0.$$

Hence, it follows from the third hypothesis in the lemma that

$$\int_{T_N} \mathbf{v}\left(x\right)\cdot\Delta\mathbf{v}\left(x,t\right)dx = 0 \quad \forall\mathbf{t} > \mathbf{0}.$$

But then we obtain that

$$\sum_{m\in\Lambda_N} |m|^2 \left(\sum_{j=1}^N |\widehat{v}_j(m)|^2\right)e^{-|m|t} = 0 \quad \forall\mathbf{t} > \mathbf{0}.$$

This last fact coupled with the first hypothesis in the lemma shows that

$$\widehat{v}_j(m) = 0 \quad \forall m \in \Lambda_N \quad \text{and } j = 1, ..., N.$$

This fact in turn implies that

$$v_j(x) = 0 \quad a.e. \text{ in } T_N \quad \text{for } j = 1, ..., N,$$

completing the proof to the lemma.   ∎

With $\{\boldsymbol{\psi}^n\}_{n=1}^{\infty}$, the sequence in (1.30) and (1.31) above, we next prove the following

Lemma 1.7. *Let $F(x) \in L^1(T_N)$ be a nonnegative function and let $\mathbf{f}(x,s)$ satisfy $(f-1)$ and $(f-2)$. Suppose that*

$$|f_j(x,\mathbf{s})| \leq F(x) \quad \forall x \in T_N \quad \text{and } \forall s \in \mathbf{R}^N,$$

$j = 1, ..., N$. *Then if n is a given positive integer, there is a vector-valued function $\mathbf{v} = \gamma_1 \boldsymbol{\psi}^1 + \cdots + \gamma_n \boldsymbol{\psi}^n$ such that*

$$(1.35) \quad \nu\left[\boldsymbol{\psi}^l, \mathbf{v}\right] + \left\{\boldsymbol{\psi}^l, \mathbf{v}, \mathbf{v}\right\} + \left(\boldsymbol{\psi}^l, \mathbf{v}\right) n^{-1} = \int_{T_N} \boldsymbol{\psi}^l(x) \cdot \mathbf{f}(x, \mathbf{v}(x)) \, dx$$

for $l = 1, .., n$.

Proof of Lemma 1.7.   For each $\alpha = (\alpha_1, ..., \alpha_n) \in \mathbf{R}^n$, we introduce the components of an $n \times n$ matrix

$$(1.36) \qquad A_{il}(\alpha) = \nu\left[\boldsymbol{\psi}^i, \boldsymbol{\psi}^l\right] + \left\{\boldsymbol{\psi}^i, \sum_{k=1}^{n} \alpha_k \boldsymbol{\psi}^k, \boldsymbol{\psi}^l\right\} + \left(\boldsymbol{\psi}^i, \boldsymbol{\psi}^l\right) n^{-1},$$

$i, l = 1, ..., n$.
We see that $A(\alpha)$ gives rise to a linear transformation on $\mathbf{R}^n$ sending $\beta = (\beta_1, ..., \beta_n)$ into

$$A(\alpha)\beta = \left(\sum_{l=1}^{n} A_{1l}(\alpha)\beta_l, ..., \sum_{l=1}^{n} A_{nl}(\alpha)\beta_l\right)$$

where

$$\sum_{l=1}^{n} A_{il}(\alpha)\beta_l = \nu\left[\boldsymbol{\psi}^i, \sum_{l=1}^{n} \beta_l \boldsymbol{\psi}^l\right] + \left\{\boldsymbol{\psi}^i, \sum_{k=1}^{n} \alpha_k \boldsymbol{\psi}^k, \sum_{l=1}^{n} \beta_l \boldsymbol{\psi}^l\right\}$$

$$+ \left(\boldsymbol{\psi}^i, \sum_{l=1}^{n} \beta_l \boldsymbol{\psi}^l\right) n^{-1}.$$

Observing from (1.34) that $\{\boldsymbol{\psi}, \mathbf{w}, \boldsymbol{\psi}\} = 0$ for $\boldsymbol{\psi}, \mathbf{w} \in J(T_N)$, we see from this last equality that

$$
\begin{aligned}
\beta \cdot A(\alpha)\beta &= \sum_{l=1}^{n} \beta_i A_{il}(\alpha)\beta_l \\
&= \nu \left[\sum_{i=1}^{n} \beta_i \boldsymbol{\psi}^i, \sum_{l=1}^{n} \beta_l \boldsymbol{\psi}^l\right] + \left(\sum_{i=1}^{n} \beta_i \boldsymbol{\psi}^i, \sum_{l=1}^{n} \beta_l \boldsymbol{\psi}^l\right) n^{-1}.
\end{aligned}
$$

From (1.30), we see that $(\boldsymbol{\psi}^i, \boldsymbol{\psi}^l) = \delta_l^i$. Consequently, we obtain from this last computation that

$$
n^{-1} |\beta|^2 \leq |\beta| \, |A(\alpha)\beta| \quad \forall \beta \in \mathbf{R}^n.
$$

Therefore, $A(\alpha)^{-1}$ exists for $\alpha \in \mathbf{R}^n$ and

(1.37) $$\left\|[A(\alpha)]^{-1}\right\|_{\mathcal{M}} \leq n \quad \forall \alpha \in \mathbf{R}^n.$$

Next, for $\alpha \in \mathbf{R}^n$, set $S(\alpha) = (S_1(\alpha), ..., S_n(\alpha))$ where

(1.38) $$S_i(\alpha) = \int_{T_N} \boldsymbol{\psi}^i(x) \cdot \mathbf{f}\left(x, \sum_{l=1}^{n} \alpha_l \boldsymbol{\psi}^l(x)\right) dx$$

for $i = 1, ..., n$. Since $\mathbf{f}$ meets $(f-1)$ and $(f-2)$, it follows that S is a continuous mapping of $\mathbf{R}^n$ into $\mathbf{R}^n$. Also, since

$$
|f_j(x, s)| \leq F(x) \in L^1(T_N)
$$

and $\boldsymbol{\psi}^i \in C^\infty(T_N)$, it is clear from (1.38) that S maps $\mathbf{R}^n$ into a compact subset of $\mathbf{R}^n$, i.e., there is a positive integer M such that

$$
|S(\alpha)| \leq M \quad \forall \alpha \in \mathbf{R}^n.
$$

Consequently, it follows from (1.37) that

$$
\left|A(\alpha)^{-1} S(\alpha)\right| \leq nM \quad \forall \alpha \in \mathbf{R}^n.
$$

Therefore,

$$
|\alpha| \leq nM \Rightarrow \left|A(\alpha)^{-1} S(\alpha)\right| \leq nM.
$$

Since $A(\alpha)^{-1} S(\alpha)$ is clearly continuous as a function of α, it follows from the Brower fixed point theorem that there is a $\gamma \in \mathbf{R}^n$ such that

$$
A(\gamma)^{-1} S(\gamma) = \gamma.
$$

Consequently, $S(\gamma) = A(\gamma)\gamma$. But then

(1.39) $$S_i(\gamma) = \sum_{l=1}^{n} A_{il}(\gamma)\gamma_l$$

for $i = 1, ..., n$. We set

$$
\mathbf{v} = \sum_{l=1}^{n} \gamma_l \boldsymbol{\psi}^l
$$

and see from (1.36) and (1.39) that

$$S_i(\gamma) = \nu\left[\boldsymbol{\psi}^i, \mathbf{v}\right] + \left\{\boldsymbol{\psi}^i, \mathbf{v}, \mathbf{v}\right\} + \left(\boldsymbol{\psi}^i, \mathbf{v}\right) n^{-1}.$$

We conclude from (1.38) that

$$\nu\left[\boldsymbol{\psi}^i, \mathbf{v}\right] + \left\{\boldsymbol{\psi}^i, \mathbf{v}, \mathbf{v}\right\} + \left(\boldsymbol{\psi}^i, \mathbf{v}\right) n^{-1} = \int_{T_N} \boldsymbol{\psi}^i(x) \cdot \mathbf{f}(x, \mathbf{v})\, dx$$

for $i = 1, ..., n$. This last expression is exactly (1.35), and the proof of the lemma is complete. ∎

We are now ready to prove Theorem 1.1.

Proof of Theorem 1.1.   We will establish the necessary condition first. Suppose then the pair $(\mathbf{v}, p)$ where $\mathbf{v} \in \left[L^2(T_N)\right]^N$ and $p \in L^1(T_N)$ satisfies (1.2) where $\mathbf{f}(x, \mathbf{v}(x))$ is replaced by $\mathbf{f}(x)$ and

$$\mathbf{f}(x) = (f_1(x), ..., f_N(x))$$

with $\mathbf{f} \in \left[L^2(T_N)\right]^N$.

   In (1.2), take

$$\phi(x) = \boldsymbol{\psi}^j(x) \quad \text{for } j = 1, ..., N$$

where $\boldsymbol{\psi}^j(x)$ is defined in (1.31′). Then it follows that the left-hand side of of (1.2) is zero. So we obtain that

$$\int_{T_N} \mathbf{f}(x) \cdot \boldsymbol{\psi}^j(x)\, dx = 0 \quad \text{for } j = 1, ..., N.$$

But we see from (1.31′) that this last fact is the same as the ★-condition in the theorem, which establishes the necessary condition of the theorem.

   To prove the sufficiency part of the theorem, we invoke Lemma 1.7 with $\mathbf{f}(x, s)$ replaced by $\mathbf{f}(x)$ where $\mathbf{f}(x)$ meets the ★-condition in the theorem. Then for each positive integer n, we obtain a vector-valued function

$$\mathbf{v}^n = \gamma_1^n \boldsymbol{\psi}^1 + \cdots + \gamma_n^n \boldsymbol{\psi}^n$$

such that

(1.40) $$\nu\left[\boldsymbol{\psi}^l, \mathbf{v}^n\right] + \left\{\boldsymbol{\psi}^l, \mathbf{v^n}, \mathbf{v^n}\right\} + \left(\boldsymbol{\psi}^l, \mathbf{v}^n\right) n^{-1} = \int_{T_N} \boldsymbol{\psi}^l(x) \cdot \mathbf{f}(x)\, dx$$

for $l = 1, ..., n$.

   Next, recalling (1.31′) and (1.34) and the ★-condition in the theorem, we see from (1.40) that

(1.41) $$\left(\boldsymbol{\psi}^l, \mathbf{v}^n\right) n^{-1} = 0 \text{ for } l = 1, ..., N \text{ and } \forall n.$$

   We proceed with the proof by multiplying each side of the equation in (1.40) by γ_j^n and sum on j from 1 to n to obtain

$$\nu\left[\mathbf{v}^n, \mathbf{v}^n\right] + \left\{\mathbf{v}^n, \mathbf{v^n}, \mathbf{v}^n\right\} + \left(\mathbf{v}^n, \mathbf{v}^n\right) n^{-1} = \int_{T_N} \mathbf{v}^n(x) \cdot \mathbf{f}(x)\, dx.$$

Since $\{\mathbf{v}^n, \mathbf{v^n}, \mathbf{v}^n\} = 0$, we obtain from this last equation that

$$(1.42) \qquad \nu\,[\mathbf{v}^n, \mathbf{v}^n] \leq \sum_{j=1}^{N} \left\| v_j^n \right\|_{L^2} \left\| f_j \right\|_{L^2}.$$

From (1.41), we see that $\widehat{v}_j^n\,(0) = 0$. So

$$\left\| v_j^n \right\|_{L^2}^2 \leq \sum_{k=1}^{N} \left\| \partial v_j^n / \partial x_k \right\|_{L^2}^2.$$

We conclude from (1.33) and (1.42) that there is a constant K such that

$$\sum_{j=1}^{N} \left\| v_j^n \right\|_{H^1}^2 \leq K \quad \forall n.$$

It follows from Lemma 2.6 in Chapter 5 that there exists $v_j \in H^1\,(T_N)$ such that

$$(i) \;\; v_j^n \rightharpoonup v_j \text{ in } H^1\,(T_N),$$

$$(ii) \;\; v_j^n \to v_j \text{ in } L^2\,(T_N),$$

(1.43)

$$(iii) \;\; v_j^n \to v_j \quad \text{a.e. in } T_N,$$

$$(iv) \;\; \left| v_j^n\,(x) \right| \leq G\,(x) \quad \text{a.e. in } T_N \;\; \forall n,$$

for $j = 1, ..., N$ where $G \in L^2\,(T_N)$ and where we have used the full sequence for ease of notation:

$\mathbf{v}^n \in J\,(T_N)$. Therefore, from (1.34) and (1.43), we see that for fixed k,

$$\lim_{n \to \infty} \left\{ \boldsymbol{\psi}^k, \mathbf{v^n}, \mathbf{v}^n \right\} = -\lim_{n \to \infty} \left\{ \mathbf{v}^n, \mathbf{v^n}, \boldsymbol{\psi}^k \right\} = -\left\{ \mathbf{v}, \mathbf{v}, \boldsymbol{\psi}^k \right\} = \left\{ \boldsymbol{\psi}^k, \mathbf{v}, \mathbf{v} \right\}.$$

So it follows from (1.40) and (1.43) that

$$(1.44) \qquad \nu\,[\boldsymbol{\psi}^n, \mathbf{v}] + \{\boldsymbol{\psi}^n, \mathbf{v}, \mathbf{v}\} = \int_{T_N} \boldsymbol{\psi}^n\,(x) \cdot \mathbf{f}\,(x) \; dx$$

for $n = 1, 2, ...,$ where $\mathbf{v} \in \left[H^1\,(T_N) \right]^N$.

Also, we see from (1.41) and (1.43) that

$$(1.45) \qquad \int_{T_N} v_j\,(x)\,dx = 0 \text{ for } j = 1, ..., N.$$

Next, we set

$$(1.46) \qquad h_j\,(x) = f_j\,(x) - \sum_{k=1}^{N} v_k\,(x)\,\partial v_j\,(x)\,/\partial x_k$$

for $j = 1, ...N.$

Since $\mathbf{v}^n \in J(T_N)$, it follows that

$$(1.47) \qquad \int_{T_N} \mathbf{v}^n \cdot \nabla \xi dx = 0 \quad \forall \xi \in C^\infty(T_N)$$

and consequently that

$$\int_{T_N} \sum_{k=1}^{N} v_k^{n_1}(x) \, \partial v_j^{n_2}(x) / \partial x_k \, dx = 0 \quad \text{for } n_1 \text{ and } n_2 \text{ positive integers,}$$

and $j = 1, ..., N$.

But then it follows from (1.43) that

$$\int_{T_N} \sum_{k=1}^{N} v_k(x) \, \partial v_j(x) / \partial x_k \, dx = 0,$$

and we conclude from (1.46) and the ★-condition in the theorem that

$$\int_{T_N} h_j(x) \, dx = 0,$$

for $j = 1, ...N$.

Using h_j, we invoke Lemma 1.5 and obtain a $\mathbf{w} = (w_1, ..., w_N)$ with $\mathbf{w} \in \left[L^1(T_N) \right]^N$ and a $p \in L^1(T_N)$ such that the pair $(\mathbf{w}, p)$ is a distribution solution of the Stokes equations given in (1.24) with $\mathbf{w}$ replacing $\mathbf{v}$, i.e., the equations in (1.25) hold with $\mathbf{w}$ replacing $\mathbf{v}$. In particular, from Lemma 1.5, we have that

$$(1.47\,') \qquad \begin{aligned} &(i) \ \ w_j(x) = (2\pi)^{-N} \int_{T_N} \left[\sum_{k=1}^{N} u_j^k(x-y) h_k(y) \right] dy, \\[2mm] &(ii) \ \ p(x) = (2\pi)^{-N} \int_{T_N} \left[\sum_{k=1}^{N} q_k(x-y) h_k(y) \right] dy, \end{aligned}$$

for $j = 1, ..., N$.

We replace ϕ by $\psi \in J(T_N)$ in the first equation in (1.25) and obtain

$$-\int_{T_N} \nu \mathbf{w} \cdot \Delta \psi \, dx = \int_{T_N} \mathbf{h}(x) \cdot \psi(x) \, dx.$$

Also, from (1.44) and (1.46), we see that

$$-\int_{T_N} \nu \mathbf{v} \cdot \Delta \psi \, dx = \int_{T_N} \mathbf{h}(x) \cdot \psi(x) \, dx \quad \forall \psi \in J(T_N).$$

In addition, from (1.43), (1.47), and the second equation in (1.25) for $\mathbf{w}$, we obtain that

$$\int_{T_N} (\mathbf{v} - \mathbf{w}) \cdot \nabla \xi dx = 0 \quad \forall \xi \in C^\infty(T_N).$$

It is also clear from the definition of $u_j^k(x)$, (1.47$\,'$), and (1.45) that

$$\int_{T_N} [v_j(x) - w_j(x)] \, dx = 0 \quad \text{for } j = 1, ..., N.$$

We conclude from these last four equations and Lemma 1.6 that

$$v_j(x) = w_j(x) \quad \text{a.e. for } x \in T_N$$

for $j = 1, ..., N$.

Since $\mathbf{w}$ satisfies the equations in (1.25), it follows from this last established fact that

(1.48)
$$-\int_{T_N}[\nu\mathbf{v} \cdot \Delta\phi + p\nabla \cdot \phi]dx$$
$$= \int_{T_N}[\mathbf{h}(x) \cdot \phi(x)]dx$$

$$\int_{T_N}\mathbf{v} \cdot \nabla\xi dx = 0$$

for all $\phi \in [C^\infty(T_N)]^N$ and $\xi \in C^\infty(T_N)$.

Since $\mathbf{v} \in \mathbf{H}^1(T_N)$, we obtain from the second equation in (1.48) that

$$\sum_{k=1}^N \partial v_k/\partial x_k = 0 \text{ in } L^2(T_N).$$

Consequently,

$$\{\phi, \mathbf{v}, \mathbf{v}\} = -\{\mathbf{v}, \mathbf{v}, \phi\} \quad \forall\phi \in [C^\infty(T_N)]^N.$$

So, we conclude from (1.46) and the first equation in (1.48) that

$$-\int_{T_N}[\nu\mathbf{v} \cdot \Delta\phi + \mathbf{v} \cdot (\mathbf{v} \cdot \nabla\phi) + p\nabla \cdot \phi]dx = \int_{T_N}[\mathbf{f}(x) \cdot \phi(x)]dx$$

$\forall\phi \in [C^\infty(T_N)]^N$.

This fact coupled with the second equation in (1.48) shows that the pair $(\mathbf{v}, p)$ is a distribution solution of the stationary Navier-Stokes equations (1.1) and completes the proof of the theorem. ■

Next, we establish the following lemma:

Lemma 1.8. *Let n be a given positive integer and let $\mathbf{f}(x, \mathbf{s})$ satisfy (f-1) and (f-2). Suppose there is a nonnegative function $F(x) \in L^2(T_N)$ such that*

(1.49)
$$f_j(x, \mathbf{s}) \le s_j/2n + F(x) \quad \text{for } 0 \le s,$$

$$f_j(x, \mathbf{s}) \ge s_j/2n - F(x) \quad \text{for } s \le 0$$

for $\mathbf{s} \in \mathbf{R}^N$, $x \in T_N$, and $j = 1, ..., N$. Then there is a vector-valued function $\mathbf{v}(x) = \gamma_1\psi_1 + \cdots + \gamma_n\psi_n$ such that (1.35) holds.

Proof of Lemma 1.8. Let M be a positive integer. Set

$$\mathbf{t}(\mathbf{s}, j, \pm M) = (t_1(\mathbf{s}, j, \pm M), ..., t_N(\mathbf{s}, j, \pm M))$$

for $j = 1, ..., N$ where

$$t_k (\mathbf{s}, j, \pm M) = s_k, \quad k \neq j$$
$$= \pm M, \quad k = j,$$

for $k = 1, ..., N$.

Next, we set

$$f_j^M (x, \mathbf{s}) = f_j (x, \mathbf{t} (\mathbf{s}, j, M)) \qquad M \leq s_j$$

(1.50)
$$= f_j (x, \mathbf{s}) \qquad |s_j| \leq M$$

$$= f_j x, \mathbf{t} (\mathbf{s}, j, -M) \qquad s_j \leq -M.$$

Since $\mathbf{f}$ meets condition $(f - 2)$, it follows from (1.50) that there is an $F^M \in L^2 (T_N)$ such that

$$\left| f_j^M (x, \mathbf{s}) \right| \leq F^M (x)$$

for $x \in T_N$, $\mathbf{s} \in \mathbf{R}^N$, and $j = 1, ..., N$.

Since $\mathbf{f}^M (x, s)$ also meets condition $(f - 1)$ and $(f - 2)$, it follows from Lemma 1.7 that if n is a given positive integer, there is a sequence of vector-valued functions $\{\mathbf{v}^M\}_{M=1}^{\infty}$ such that

(1.51)
$$\nu \left[\boldsymbol{\psi}^l, \mathbf{v}^M \right] + \left\{ \boldsymbol{\psi}^l, \mathbf{v^M}, \mathbf{v}^M \right\} + \left(\boldsymbol{\psi}^l, \mathbf{v}^M \right) n^{-1} = \int_{T_N} \boldsymbol{\psi}^l (x) \cdot \mathbf{f}^M (x, \mathbf{v}^M (x)) \, dx$$

for $l = 1, ..., n$ where

(1.52)
$$\mathbf{v^M} (x) = \sum_{l=1}^{n} \gamma_l^M \boldsymbol{\psi}^l (x).$$

Next, we observe from (1.50) that the inequalities in (1.49) still hold if we replace the left-side with $f_j^M (x, \mathbf{s})$. But then it follows that

(1.53)
$$s_j f_j^M (x, \mathbf{s}) \leq s_j^2 / 2n + |s_j| F (x)$$

for $x \in T_N$, $s \in \mathbf{R^N}$, and $j = 1, ..., N$, and $M = 1, 2, ...$ Since

$$\left\{ \mathbf{v}^M, \mathbf{v}^M, \mathbf{v}^M \right\} = 0,$$

we consequently infer from (1.51) and (1.52) that

$$\nu \left[\mathbf{v}^M, \mathbf{v}^M \right] + \left(\mathbf{v}^M, \mathbf{v}^M \right) n^{-1} = \int_{T_N} \mathbf{v}^M \cdot \mathbf{f}^M (x, \mathbf{v}^M (x)) \, dx.$$

But then it follows from (1.53) that

$$\nu \left[\mathbf{v}^M, \mathbf{v}^M \right] + \left(\mathbf{v}^M, \mathbf{v}^M \right) / n \leq \left(\mathbf{v}^M, \mathbf{v}^M \right) / 2n + N \int_{T_N} \left| \mathbf{v}^M (x) \right| F (x) \, dx.$$

This in turn gives that

(1.54)
$$\nu \left[\mathbf{v}^M, \mathbf{v}^M \right] + \left(\mathbf{v}^M, \mathbf{v}^M \right) / 2n \leq N \left(\mathbf{v}^M, \mathbf{v}^M \right)^{1/2} \| F \|_{L^2}$$

for $M = 1, 2, ...$

From this last fact, we obtain that

$$\left(\mathbf{v}^M, \mathbf{v}^M\right)^{1/2} \leq 2nN \left\|F\right\|_{L^2}$$

and next from (1.52) that

$$\left\{\gamma_l^M\right\}_{M=1}^{\infty} \quad \text{is a uniformly bounded sequence}$$

for $l = 1, ..., n$.

This in turn implies that there is a positive constant K such that

$$\left|v_j^M(x)\right| \leq K \quad \forall x \in T_N$$

for $j = 1, ..., N$ and $M = 1, 2, ...$ It then follows from (1.50) that for $M > K$,

$$f_j^M\left(x, \mathbf{v}^M(x)\right) = f_j\left(x, \mathbf{v}^M(x)\right) \quad \forall x \in T_N$$

for $j = 1, ..., N$.

But (1.51) and this last fact imply that $\mathbf{v}^M$ is indeed a solution of (1.35) for every $M > K$, and the proof of the lemma is complete. ∎

Next, for $v \in \left[H^1(T_N)\right]^N$, we set

(1.55) $\left\|\mathbf{v}\right\|_1^2 = [\mathbf{v} \cdot \mathbf{v}] + (\mathbf{v}, \mathbf{v})$

and establish the following lemma:

Lemma 1.9. *Suppose that the components of* $\mathbf{f}(x, \mathbf{s})$ *satisfy (f-1) and (f-2) and there is a nonnegative function* $F(x) \in L^2(T_N)$ *such that*

$$f_j(x, \mathbf{s}) \leq s_j + F(x) \quad \text{for} \ \ 0 \leq s,$$

(1.56)

$$f_j(x, \mathbf{s}) \geq s_j - F(x) \quad \text{for} \ \ s \leq 0$$

for $\mathbf{s} \in \mathbf{R}^N$, $x \in T_N$, *and* $j = 1, ..., N$. *Suppose also that for every positive integer* n, *there is a*

$$\mathbf{v}^n = \gamma_1^n \boldsymbol{\psi}^1 + \cdots + \gamma_n^n \boldsymbol{\psi}^n$$

which satisfies

(1.57)

$$\nu\left[\boldsymbol{\psi}^l, \mathbf{v}^n\right] + \left\{\boldsymbol{\psi}^l, \mathbf{v^n}, \mathbf{v}^n\right\} + \left(\boldsymbol{\psi}^l, \mathbf{v}^n\right) n^{-1} = \int_{T_N} \boldsymbol{\psi}^l(x) \cdot \mathbf{f}(x, \mathbf{v}^n(x)) \, dx$$

for $l = 1, .., n$. *Suppose, furthermore, there is a constant* K *such that*

(1.58) $\left\|\mathbf{v}^n\right\|_1 \leq K \quad \forall n.$

Then there is a constant K^* *such that*

(1.59) $\int_{T_N} \left|f_j(x, \mathbf{v}^n(x))\right| \left|v_j^n(x)\right| \leq K^*$

for $j = 1, ..., N$ *and* $n = 1, 2, ...$

Proof of Lemma 1.9. Multiplying both sides of (1.57) by γ_l^n and summing on l from 1 to n, we obtain that

$$(1.60) \qquad \nu\left[\mathbf{v}^n, \mathbf{v}^n\right] + \left(\mathbf{v}^n, \mathbf{v}^n\right) n^{-1} = \int_{T_N} \mathbf{v}^n\left(x\right) \cdot \mathbf{f}\left(x, \mathbf{v}^n\left(x\right)\right) dx.$$

In particular, we obtain from this last equality that

$$(1.61) \qquad 0 \leq \int_{T_N} \mathbf{v}^n\left(x\right) \cdot \mathbf{f}\left(x, \mathbf{v}^n\left(x\right)\right) dx.$$

Next, we note from (1.56) that

$$(1.62) \qquad \mathbf{s} \cdot \mathbf{f}\left(x, \mathbf{s}\right) \leq \mathbf{s} \cdot \mathbf{s} + N\left|\mathbf{s}\right| F\left(x\right)$$

for $x \in T_N$ and $\mathbf{s} \in \mathbf{R}^N$. Also, we introduce the sets A_n and B_n as follows:

$$A_n = \{x \in T_N : \mathbf{v}^n\left(x\right) \cdot \mathbf{f}\left(x, \mathbf{v}^n\left(x\right)\right) \leq 0\}$$

and

$$B_n = \{x \in T_N : \mathbf{v}^n\left(x\right) \cdot \mathbf{f}\left(x, \mathbf{v}^n\left(x\right)\right) > 0\}.$$

From (1.55), (1.58), and (1.62), we see that there is a constant K_1 such that

$$\int_{B_n} \mathbf{v}^n\left(x\right) \cdot \mathbf{f}\left(x, \mathbf{v}^n\left(x\right)\right) dx \leq K_1.$$

Also, from (1.61), we see that

$$-\int_{A_n} \mathbf{v}^n\left(x\right) \cdot \mathbf{f}\left(x, \mathbf{v}^n\left(x\right)\right) dx \leq K_1.$$

We conclude from these last two inequalities that

$$(1.63) \qquad \int_{T_N} \left|\mathbf{v}^n\left(x\right) \cdot \mathbf{f}\left(x, \mathbf{v}^n\left(x\right)\right)\right| dx \leq 2K_1$$

for $n = 1, 2, \ldots$.

Next, we set

$$(1.64) \qquad F_j^n\left(x\right) = \mathbf{v}^n\left(x\right) \cdot \mathbf{f}\left(x, \mathbf{v}^n\left(x\right)\right) - v_j^n\left(x\right) f_j\left(x, \mathbf{v}^n\left(x\right)\right)$$

and

$$(1.65) \qquad \begin{aligned} C_j^n &= \left\{x \in T_N : v_j^n\left(x\right) f_j\left(x, \mathbf{v}^n\left(x\right)\right) > 0\right\} \\ D_j^n &= \left\{x \in T_N : F_j^n\left(x\right) > 0\right\}. \end{aligned}$$

From (1.56) and (1.64), we see that

$$v_j^n\left(x\right) f_j^n\left(x, \mathbf{v}^n\left(x\right)\right) \leq \left|v_j^n\left(x\right)\right|^2 + \left|v_j^n\left(x\right)\right| \left|F\left(x\right)\right|$$

and

$$F_j^n\left(x\right) \leq \left|\mathbf{v}^n\left(x\right)\right|^2 - \left|v_j^n\left(x\right)\right|^2 + \left(N-1\right) \left|\mathbf{v}^n\left(x\right)\right| \left|F\left(x\right)\right|.$$

Consequently, we conclude from (1.58) and (1.65) that there is a constant K_2 such that

$$\int_{C_j^n} \left| v_j^n(x) f_j(x, \mathbf{v}^n(x)) \right| dx \leq K_2,$$

(1.66)

$$\int_{D_j^n} \left| F_j^n(x) \right| dx \leq K_2.$$

Next, from (1.63), (1.64), and (1.66), we see that

$$\int_{C_j^n} \left| v_j^n(x) f_j(x, \mathbf{v}^n(x)) - F_j^n(x) \right| dx \leq 2K_1 + 2K_2,$$

(1.67)

$$\int_{D_j^n} \left| v_j^n(x) f_j^n(x, \mathbf{v}^n(x)) - F_j^n(x) \right| dx \leq 2K_1 + 2K_2,$$

for $j = 1, ..., N$ and $n = 1, 2, ...$.

Also, we have for $x \in T_N \backslash C_j^n \cap T_N \backslash D_j^n$,

$$
\begin{aligned}
\left| v_j^n(x) f_j(x, \mathbf{v}^n(x)) - F_j^n(x) \right| &\leq -v_j^n(x) f_j(x, \mathbf{v}^n(x)) - F_j^n(x) \\
&\leq -\mathbf{v}^n(x) \cdot \mathbf{f}(x, \mathbf{v}^n(x)) \\
&\leq \left| \mathbf{v}^n(x) \cdot \mathbf{f}(x, \mathbf{v}^n(x)) \right|.
\end{aligned}
$$

So, we obtain from (1.63) that

$$\int_{T_N \backslash C_j^n \cap T_N \backslash D_j^n} \left| v_j^n(x) f_j^n(x, \mathbf{v}^n(x)) - F_j^n(x) \right| dx \leq 2K_1.$$

Since

$$T_N = C_j^n \cup \left(T_N \backslash C_j^n \cap D_j^n \right) \cup \left(T_N \backslash C_j^n \cap T_N \backslash D_j^n \right),$$

we conclude from (1.67) and this last inequality that

$$\int_{T_N} \left| v_j^n(x) f_j^n(x, \mathbf{v}^n(x)) - F_j^n(x) \right| dx \leq 6K_1 + 4K_2.$$

Next, utilizing the fact that

$$2|a| \leq |a + b| + |a - b|,$$

we conclude from (1.63) and this last inequality

$$2 \int_{T_N} \left| v_j^n(x) f_j(x, \mathbf{v}^n(x)) \right| dx \leq 8K_1 + 4K_2$$

for $j = 1, ..., N$ and $n = 1, 2, ...$.

This establishes (1.59) with $K^* = 4K_1 + 2K_2$, and the proof of the lemma is complete. ∎

Next, we establish the following lemma, which will be needed in the proof of Theorem 1.2:

Lemma 1.10. *Suppose the conditions in the hypothesis of Lemma 1.9 hold. Then the sequence $\{f_j(x, \mathbf{v}^n(x))\}_{n=1}^{\infty}$ is absolutely equi-integrable for $j = 1, ..., N$.*

Proof of Lemma 1.10.   For the definition of absolutely equi-integrable, we refer the reader to below (1.70) in Chapter 5.

Proceeding with the proof of the lemma, given $\varepsilon > 0$, we first choose $r > 0$ so that

$$(1.68) \qquad\qquad\qquad K^*/r < \varepsilon/2,$$

where K^* is the constant given in Lemma 1.9.

Next, using $(f - 2)$, we choose $\zeta_r(x) \in L^2(T_N)$ such that

$$(1.69) \qquad\qquad |f_j(x, \mathbf{s})| \leq \zeta_r(x) \quad \text{for } |s_j| \leq r$$

for $x \in T_N$, $s_k \in \mathbf{R}$, $k \neq j$, $k = 1, ..., N$, $j = 1, ..., N$. Also, we set

$$A(n, j) = \left\{x \in T_N : \left|v_j^n(x)\right| \leq r\right\}$$

and

$$B(n, j) = \left\{x \in T_N : \left|v_j^n(x)\right| > r\right\}.$$

In addition, we choose $\delta > 0$ so that

$$(1.70) \qquad E \subset T_N \text{ and } |E| \leq \delta \Rightarrow \int_E |\zeta_r(x)|\, dx \leq \varepsilon/2.$$

Now suppose $E \subset T_N$ and $|E| \leq \delta$. Then it follows from Lemma 1.9, (1.69), and (1.70) that

$$
\begin{aligned}
\int_E |f_j(x, \mathbf{v}^n(x))|\, dx \;\leq\; & \int_{E \cap A(n,j)} |\zeta_r(x)|\, dx \\
& + r^{-1} \int_{E \cap B(n,j)} \left|v_j^n(x) f_j(x, \mathbf{v}^n(x))\right| dx \\
\leq\; & \varepsilon/2 + K^*/r,
\end{aligned}
$$

for $j = 1, ..., N$ and $n = 1, 2, ...$

From (1.68), we see that the right-hand side of this last inequality is less than ε. Consequently, the sequence $\{f_j(x, \mathbf{v}^n(x))\}_{n=1}^{\infty}$ is absolutely equi-integrable, and the proof of the lemma is complete. ∎

Proof of Theorem 1.2. Since $\mathbf{f}$ satisfies $(f - 2)$ and $(f - 3)$, it is easy to see from (1.4) that for every $\varepsilon > 0$, there exists $F^\varepsilon \in L^2(T_N)$ with $F^\varepsilon \geq 0$ such that

$$
\begin{aligned}
f_j(x, \mathbf{s}) &\leq \varepsilon s_j + F^\varepsilon(x) && \text{for } 0 \leq s_j \\[-2pt]
(1.71)\\[-10pt]
&\geq \varepsilon s_j - F^\varepsilon(x) && \text{for } 0 \geq s_j
\end{aligned}
$$

for $x \in T_N$, $\mathbf{s} \in \mathbf{R}^N$, $j = 1, ..., N$. Consequently, it follows from Lemma 1.8 that there is a sequence $\{\mathbf{v}^n\}_{n=1}^{\infty}$ with the following properties:

$$(1.72) \qquad \mathbf{v}^n = \gamma_1^n \boldsymbol{\psi}^1 + \cdots + \gamma_n^n \boldsymbol{\psi}^n;$$

(1.73)
$$\nu \left[\boldsymbol{\psi}^l, \mathbf{v}^n \right] + \{ \boldsymbol{\psi}^l, \mathbf{v}^n, \mathbf{v}^n \} + \left(\boldsymbol{\psi}^l, \mathbf{v}^n \right) n^{-1}$$

$$= \int_{T_N} \boldsymbol{\psi}^l (x) \cdot \mathbf{f} (x, \mathbf{v}^n (x)) \, dx$$

for $l = 1, ..., n$ and $n = 1, 2, ...$, where $\{\boldsymbol{\psi}^l\}_{l=1}^{\infty}$ is the sequence of functions in $J(T_N)$ that satisfy (1.30) and (1.31).

We claim there is a constant K_1 such that

$$(1.74) \qquad \|\mathbf{v}^n\|_1^2 \le K_1 \quad \forall n,$$

where $\|\mathbf{v}^n\|_1^2 = [\mathbf{v}^n . \mathbf{v}^n] + (\mathbf{v}^n, \mathbf{v}^n)$.

Suppose (1.74) is false. Then there is a subsequence of $\{\|\mathbf{v}^n\|_1\}_{n=1}^{\infty}$ which tends to ∞. We consequently see from Lemma 2.6 in Chapter 5 and well-known facts from Hilbert space theory that the following prevails:

$$(1.75) \quad \exists \left\{ \mathbf{w}^M \right\}_{M=1}^{\infty} \text{ such that } \mathbf{w}^M = \mathbf{v}^{n_M} \text{ with } lim_{M \to \infty} \left\| \mathbf{w}^M \right\|_1 = \infty;$$

with $\mathbf{W}^M = \frac{\mathbf{w}^M}{\|\mathbf{w}^M\|_1}$, $\exists \mathbf{V} \in \left[H^1 (T_N) \right]^N$ such that

(1.76)
$$lim_{M \to \infty} \left(\mathbf{W}^M - \mathbf{V}, \mathbf{W}^M - \mathbf{V} \right) = 0;$$

$$(1.77) \qquad lim_{M \to \infty} \mathbf{W}^M (x) = \mathbf{V} (x) \quad \text{ for a.e. } x \in T_N;$$

$$(1.78) \qquad \lim_{M \to \infty} \int_{T_N} \frac{\partial W_j^M}{\partial x_k} \xi dx = \int_{T_N} \frac{\partial V_j}{\partial x_k} \xi dx \quad \forall \xi \in C^{\infty} (T_N),$$

for $j, k = 1, ..., N$.

It follows from (1.72) and (1.73) that

$$\nu \left[\mathbf{W}^M, \mathbf{W}^M \right] + \left(\mathbf{W}^M, \mathbf{W}^M \right) n_M^{-1}$$

(1.79)
$$= \left\| \mathbf{w}^M \right\|_1^{-2} \int_{T_N} \mathbf{w}^M \cdot \mathbf{f} \left(x, \mathbf{w}^M (x) \right) dx.$$

From (1.71), we see that

$$(1.80) \qquad \mathbf{w}^M (x) \cdot \mathbf{f} \left(x, \mathbf{w}^M (x) \right) \le \varepsilon \left| \mathbf{w}^M (x) \right|^2 + N \left| \mathbf{w}^M (x) \right| F^{\varepsilon} (x)$$

for $x \in T_N$ and $\forall M$. It follows therefore from (1.75) and (1.79) that

$$\lim_{M \to \infty} \left[\mathbf{W}^M, \mathbf{W}^M \right] \le \varepsilon/\nu,$$

and since $\varepsilon > 0$ is arbitrary that

$$(1.81) \qquad \lim_{M \to \infty} \left[\mathbf{W}^M, \mathbf{W}^M \right] = 0.$$

This fact in conjunction with (1.78) gives us that

$$(1.82) \qquad \frac{\partial V_j}{\partial x_k}(x) = 0 \quad \text{for a.e. } x \in T_N \quad j, k = 1, ..., N.$$

From (1.76), we see that

$$\left[\mathbf{W}^M, \mathbf{W}^M\right] + \left(\mathbf{W}^M, \mathbf{W}^M\right) = 1.$$

So we obtain from (1.81) that

$$(1.83) \qquad \lim_{M \to \infty} \left(\mathbf{W}^M, \mathbf{W}^M\right) = 1,$$

and consequently from (1.76) that $(\mathbf{V}, \mathbf{V}) = 1$. This fact in conjunction with (1.82) tells us that

$$V_j(x) = c_j \quad \text{for a.e. } x \in T_N \quad j = 1, ..., N,$$

$$(1.84)$$

$$c_1^2 + \cdots + c_N^2 = (2\pi)^{-N}.$$

Next, since

$$\lim_{M \to \infty} n_M = \infty,$$

it follows from (1.79), (1.81), and (1.83) that

$$(1.85) \qquad \lim_{M \to \infty} -\left\|\mathbf{w}^M\right\|_1^{-2} \int_{T_N} \mathbf{w}^M \cdot \mathbf{f}\left(x, \mathbf{w}^M(x)\right) dx = 0.$$

We see from (1.80) with $\varepsilon = 1$ that

$$0 \leq -\mathbf{w}^M(x) \cdot \mathbf{f}\left(x, \mathbf{w}^M(x)\right) + \left|\mathbf{w}^M(x)\right|^2 + N\left|\mathbf{w}^M(x)\right| F^1(x).$$

Therefore, we obtain from Fatou's lemma, (1.76), (1.77), and (1.85) that

$$(i) \quad \liminf_{M \to \infty} \left[-\mathbf{w}^M \cdot \mathbf{f}\left(x, \mathbf{w}^M(x)\right) \left\|\mathbf{w}^M\right\|_1^{-2}\right] \in L^1(T_N),$$

$$(1.87)$$

$$(ii) \quad \int_{T_N} \liminf_{M \to \infty} \left[-\mathbf{w}^M \cdot \mathbf{f}\left(x, \mathbf{w}^M(x)\right) \left\|\mathbf{w}^M\right\|_1^{-2}\right] dx \leq 0.$$

We next set

$$(1.88) \qquad a_j^M(x) = -w_j^M(x) f_j\left(x, \mathbf{w}^M(x)\right) \left\|\mathbf{w}^M\right\|_1^{-2}$$

for $j = 1, ..., N$ and observe from (1.71) that

$$s_j f_j(x, \mathbf{s}) \leq \varepsilon s^2 + |s| F^\varepsilon(x)$$

for $x \in T_N$ and $\mathbf{s} \in \mathbf{R}^N$. Consequently, it follows from (1.77) and (1.88) that

$$(1.89) \qquad \liminf_{M \to \infty} a_j^M(x) \geq 0 \quad \text{for a.e. } \mathbf{x} \in \mathrm{T}_N.$$

Since

$$\liminf_{M \to \infty} \left[-\mathbf{w}^M \cdot \mathbf{f}\left(x, \mathbf{w}^M(x)\right) \left\|\mathbf{w}^M\right\|_1^{-2}\right] \geq \sum_{j=1}^{N} \liminf_{M \to \infty} a_j^M(x),$$

we conclude first from (1.87)(i) and (1.89) that

$$\liminf_{M\to\infty} a_j^M(x) \in L^1(T_N) \quad \text{for} \quad j = 1, ..., N,$$

and next from (1.87)(ii) that

$$(1.90) \qquad \liminf_{M\to\infty} a_j^M(x) = 0 \quad \text{for a.e. } \mathbf{x} \in T_N$$

for $j = 1, ..., N$.

Continuing, we see from (1.84) that at least one of the $c_j \neq 0$. For ease of notation, we will suppose

$$(1.91) \qquad c_1 \neq 0$$

and will arrive at a contradiction of (1.90) when $j = 1$. A similar line of reasoning prevails in case we were dealing with other values of j.

From condition (1.8) in the hypothesis of the theorem and (1.7), we see that there are positive integers K and l such that

$$(1.92) \qquad |E_1(\mathbf{f}, K, l)| = \eta \neq 0.$$

Also, from (1.77) and (1.84), we see that

$$\lim_{M\to\infty} w_1^M(x) / \|\mathbf{w}^M\|_1 = c_1 \quad \text{for a.e. } \mathbf{x} \in E_1(\mathbf{f}, K, l).$$

Consequently, we obtain from Egoroff's theorem and (1.92) that there is a subset

$$E_1'(\mathbf{f}, K, l) \subset E_1(\mathbf{f}, K, l)$$

and an $M_0 > 0$ such that

$$(i) \quad |E_1'(\mathbf{f}, K, l)| \geq \eta/2,$$

$$(1.93)$$

$$(ii) \quad \frac{|w_1^M(x)|}{\|\mathbf{w}^M\|_1} \geq \left|\frac{c_1}{2}\right| \text{ for } x \in E_1'(\mathbf{f}, K, l)$$

for $M \geq M_0$.

Since $\|\mathbf{w}^M\|_1 \to \infty$, it follows from (1.91) and (1.93)(ii) that there is an $M_1 > M_0$ such that

$$(1.94) \qquad |w_1^M(x)| > l \quad \text{for } x \in E_1'(\mathbf{f}, K, l) \text{ and } M > M_1.$$

Now we know from (1.6) that that for $x \in E_1'(\mathbf{f}, K, l)$ and $|s_1| > l$,

$$(1.95) \qquad -f(x, \mathbf{s})/s_1 > K^{-1}$$

for $s_k \in \mathbf{R}$, $k = 2, ..., N$. Also, it follows from (1.88) that for $x \in E_1'(\mathbf{f}, K, l)$ and $M > M_1$,

$$a_1^M(x) = \frac{|w_1^M(x)|^2}{\|\mathbf{w}^M\|_1^2} \left[-f_j(x, \mathbf{w}^M(x))/w_1^M(x)\right].$$

So we obtain from (1.93)(ii), (1.94), and (1.95) that

$$a_1^M(x) \geq \left|\frac{c_1}{2}\right|^2 K^{-1}$$

for $x \in E'_1 (\mathbf{f}, K, l)$ and $M > M_1$. Therefore,

$$\liminf_{M \to \infty} a_j^M (x) \geq \left| \frac{c_1}{2} \right|^2 K^{-1} \quad \text{for } x \in E'_1 (\mathbf{f}, K, l).$$

But by (1.93)(i), $|E'_1 (\mathbf{f}, K, l)| > 0$. Likewise by (1.91), $\left| \frac{c_1}{2} \right|^2 K^{-1} > 0$. So this last inequality is a direct contradiction to (1.90) for $j = 1$. We conclude that there is no subsequence of $\{ \|\mathbf{v}^n\|_1 \}_{n=1}^{\infty}$, which tends to ∞, and that the inequality in (1.74) is indeed true.

Proceeding with the proof of the theorem, we see from (1.74) and Lemma 2.6 in Chapter 5 that there exists a subsequence $\{\mathbf{v}^{n_M}\}_{M=1}^{\infty}$ and a vector-valued function $\mathbf{v} \in \left[H^1 (T_N) \right]^N$ such that the following holds:

(1.96)

$$(i) \quad \mathbf{w}^M = \mathbf{v}^{n_M} \quad M = 1, 2, ...;$$

$$(ii) \quad \lim_{M \to \infty} \mathbf{w}^M (x) = \mathbf{v} (x) \quad \text{for a.e. } x \in T_N;$$

$$(iii) \quad \lim_{M \to \infty} \left(\mathbf{w}^M - \mathbf{v}, \mathbf{w}^M - \mathbf{v} \right) = 0;$$

$$(iv) \quad \lim_{M \to \infty} \int_{T_N} \frac{\partial w_j^M}{\partial x_k} \xi dx = \int_{T_N} \frac{\partial v_j}{\partial x_k} \xi$$

for $j, k = 1, ..., N$ and $\xi \in C^{\infty} (T_N)$.

We next use $(f - 1)$ and (1.96)(ii) to obtain

(1.97) $$\lim_{M \to \infty} \mathbf{f} \left(x, \mathbf{w}^M (x) \right) = \mathbf{f} (x, \mathbf{v} (x)) \quad \text{for a.e. } x \in T_N,$$

and (1.74) in conjunction with Lemma 1.10   to obtain

(1.98) $$\left\{ \mathbf{f} \left(x, \mathbf{w}^M (x) \right) \right\}_{M=1}^{\infty} \text{ is absolutely equi-integrable.}$$

From this last fact, we see there is a constant K_2 such that

$$\int_{T_N} \left| \mathbf{f} \left(x, \mathbf{w}^M (x) \right) \right| dx \leq K_2 \quad \forall M.$$

Hence, it follows from (1.97) and Fatou's lemma that

(1.99) $$\mathbf{f} (x, \mathbf{v} (x)) \in \left[L^1 (T_N) \right]^N.$$

Consequently, we obtain from Egoroff's theorem in conjunction with (1.97)-(1.99) that

(1.100) $$\lim_{M \to \infty} \int_{T_N} \mathbf{f} \left(x, \mathbf{w}^M (x) \right) \cdot \boldsymbol{\phi} dx = \int_{T_N} \mathbf{f} (x, \mathbf{v} (x)) \cdot \boldsymbol{\phi} dx$$

$\forall \boldsymbol{\phi} \in \left[C^{\infty} (T_N) \right]^N$.

Next, we recall that $\mathbf{v} \in \left[H^1 (T_N) \right]^N$ and that $\mathbf{w}^M = \mathbf{v}^{n_M} \in J (T_N)$. Therefore, we obtain from (1.96)(iv) that

(1.101) $$\nabla \cdot \mathbf{v} \in L^2 (T_N) \quad \text{and} \quad \nabla \cdot \mathbf{v} = 0.$$

Using this fact in conjunction with (1.33) and (1.34) and (1.96)(iii) gives us

$$(1.102) \qquad \lim_{M \to \infty} \left\{ \boldsymbol{\psi}^l, \mathbf{w}^M, \mathbf{w}^M \right\} = \left\{ \boldsymbol{\psi}^l, \mathbf{v}, \mathbf{v} \right\} \quad \forall l.$$

Next, we see from (1.73) and (1.96)(i) that

$$\nu \left[\boldsymbol{\psi}^l, \mathbf{w}^M \right] + \left\{ \boldsymbol{\psi}^l, \mathbf{w}^M, \mathbf{w}^M \right\} + \left(\boldsymbol{\psi}^l, \mathbf{w}^M \right) n^{-1}$$

$$= \int_{T_N} \boldsymbol{\psi}^l (x) \cdot \mathbf{f} \left(x, \mathbf{w}^M (x) \right) dx.$$

Hence, it follows from the above that

$$\nu \left[\boldsymbol{\psi}^l, \mathbf{v} \right] + \left\{ \boldsymbol{\psi}^l, \mathbf{v}, \mathbf{v} \right\} = \int_{T_N} \boldsymbol{\psi}^l (x) \cdot \mathbf{f} \left(x, \mathbf{v} (x) \right) dx$$

for $l = 1, 2, \dots$. But then from (1.31) and (1.99), we have that

$$(1.103) \qquad \nu \left[\boldsymbol{\psi}, \mathbf{v} \right] + \left\{ \boldsymbol{\psi}, \mathbf{v}, \mathbf{v} \right\} = \int_{T_N} \boldsymbol{\psi} (x) \cdot \mathbf{f} \left(x, \mathbf{v} (x) \right) dx$$

for $\boldsymbol{\psi} \in J (T_N)$.

From (1.101), we have that

$$\left\{ \boldsymbol{\psi}, \mathbf{v}, \mathbf{v} \right\} = - \left\{ \mathbf{v}, \mathbf{v}, \boldsymbol{\psi} \right\}$$

for $\boldsymbol{\psi} \in J (T_N)$. Using this fact in conjunction with (1.31′) and (1.103) gives us that

$$\int_{T_N} f_j (x, \mathbf{v} (x)) \, dx = 0 \quad \text{for} \quad j = 1, \dots N.$$

Likewise, we see from the fact that $\mathbf{v} \in \left[H^1 (T_N) \right]^N$ and $\nabla \cdot \mathbf{v} = 0$ that

$$\int_{T_N} \sum_{k=1}^{N} v_k (x) \, \partial v_j (x) / \partial x_k dx = 0 \quad \text{for} \quad j = 1, \dots N.$$

Next, we set

$$h_j (x) = f_j (x, \mathbf{v} (x)) - \sum_{k=1}^{N} v_k (x) \, \partial v_j (x) / \partial x_k$$

for $j = 1, \dots N$, and observe from the above that

$$h_j \in L^1 (T_N)$$

and that

$$\int_{T_N} h_j (x) \, dx = 0 \quad \text{for} \quad j = 1, \dots N.$$

Using h_j, we invoke Lemma 1.5 and obtain a $\mathbf{w} = (w_1, \dots, w_N)$ with $\mathbf{w} \in \left[L^1 (T_N) \right]^N$ and a $p \in L^1 (T_N)$ such that the pair $(\mathbf{w}, p)$ is a distribution solution of the Stokes equations given in (1.24) with $\mathbf{w}$ replacing $\mathbf{v}$, i.e., the equations in (1.25) hold with $\mathbf{w}$ replacing $\mathbf{v}$. Also, $\mathbf{h} = (h_1, ...h_N)$. In particular, we have that

(1.104)
$$-\int_{T_N}[\nu\mathbf{w} \cdot \Delta\boldsymbol{\phi} + p\nabla \cdot \boldsymbol{\phi}]dx$$
$$= \int_{T_N}[\mathbf{h}(x) \cdot \boldsymbol{\phi}(x)]dx$$

$$\int_{T_N} \mathbf{w} \cdot \nabla\xi dx = 0$$

for all $\boldsymbol{\phi} \in [C^\infty(T_N)]^N$ and $\xi \in C^\infty(T_N)$.

We replace $\boldsymbol{\phi}$ by $\boldsymbol{\psi} \in J(T_N)$ in the first equation in (1.104) and obtain

$$-\int_{T_N} \nu\mathbf{w} \cdot \Delta\boldsymbol{\psi}\, dx = \int_{T_N} \mathbf{h}(x) \cdot \boldsymbol{\psi}(x)\, dx.$$

From (1.103) and the definition of $\mathbf{h}$, we furthermore see that

$$-\int_{T_N} \nu\mathbf{v} \cdot \Delta\boldsymbol{\psi}\, dx = \int_{T_N} \mathbf{h}(x) \cdot \boldsymbol{\psi}(x)\, dx \quad \forall\boldsymbol{\psi} \in J(T_N).$$

In addition, from (1.101) and the second equation in (1104) for $\mathbf{w}$, we obtain that

$$\int_{T_N} (\mathbf{v} - \mathbf{w}) \cdot \nabla\xi dx = 0 \quad \forall\xi \in C^\infty(T_N).$$

Setting

$$\gamma_j = (2\pi)^{-N} \int_{T_N} v_j(x)\, dx \quad \text{for } j = 1, ..., N$$

and $\boldsymbol{\gamma} = (\gamma_1, ..., \gamma_N)$, and observing that

(1.105)
$$\int_{T_N} \boldsymbol{\gamma} \cdot \Delta\boldsymbol{\phi}\, dx = 0 \quad \forall\boldsymbol{\phi} \in [C^\infty(T_N)]^N,$$

$$\int_{T_N} \boldsymbol{\gamma} \cdot \nabla\xi dx = 0 \quad \forall\xi \in C^\infty(T_N),$$

it follows from these last six equations and Lemma 1.6 that

$$\mathbf{v}(x) - \boldsymbol{\gamma} - \mathbf{w}(x) \quad \text{for a.c. } x \subset T_N.$$

Next, using this last equation in conjunction with (1.104) and (1.105) gives us that

(1.106)
$$-\int_{T_N}[\nu\mathbf{v} \cdot \Delta\boldsymbol{\phi} + p\nabla \cdot \boldsymbol{\phi}]dx$$
$$= \int_{T_N}[\mathbf{h}(x) \cdot \boldsymbol{\phi}(x)]dx$$

$$\int_{T_N} \mathbf{v} \cdot \nabla\xi dx = 0$$

for all $\boldsymbol{\phi} \in [C^\infty(T_N)]^N$ and $\xi \in C^\infty(T_N)$. Recalling from the above that

$$\mathbf{h}(x) = \mathbf{f}(x, \mathbf{v}(x)) - (\mathbf{v}(x) \cdot \nabla)\mathbf{v}(x),$$

we see that the first equation in (1.106) is actually

$$-\int_{T_N}[\nu\mathbf{v} \cdot \Delta\boldsymbol{\phi} - (\mathbf{v} \cdot \nabla)\mathbf{v} \cdot \boldsymbol{\phi} + p\nabla \cdot \boldsymbol{\phi}]dx = \int_{T_N}[\mathbf{f}(x, \mathbf{v}) \cdot \boldsymbol{\phi}(x)]dx.$$

But using (1.34) and the second equation in (1.106), we obtain that

$$\int_{T_N} (\mathbf{v} \cdot \nabla) \mathbf{v} \cdot \boldsymbol{\phi} \, dx = - \int_{T_N} \mathbf{v} \cdot (\mathbf{v} \cdot \nabla) \boldsymbol{\phi} \, dx.$$

So the first equation in (1.106) is really

$$-\int_{T_N} [\nu \mathbf{v} \cdot \Delta \boldsymbol{\phi} + \mathbf{v} \cdot (\mathbf{v} \cdot \nabla) \boldsymbol{\phi} + p \nabla \cdot \boldsymbol{\phi}] dx = \int_{T_N} [\mathbf{f}(x, \mathbf{v}) \cdot \boldsymbol{\phi}(x)] dx.$$

This last equation is the same as the first equation in (1.2), and we conclude that $(\mathbf{v}, p)$ is indeed a distribution solution of the stationary Navier-Stokes equations (1.1).

All that remains to complete the proof of the theorem is to show that

(1.107)
$$\int_{T_N} |v_j \, f_j(x, \mathbf{v})| \, dx < \infty$$

for $j = 1, ..., N$.

To establish (1.107), we use (1.74) and Lemma 1.9 to obtain the existence of a positive constant K^* such that

$$\int_{T_N} |f_j(x, \mathbf{w}^M(x))| \, |w_j^M(x)| \, dx \leq K^* \quad \forall M.$$

But it follows from (1.96)(ii), Fatou's lemma, and this last inequality that

$$\int_{T_N} |f_j(x, \mathbf{v}(x))| \, |v_j(x)| \, dx \leq K^*.$$

This gives inequality (1.107) and completes the proof of the theorem. ∎

Proof of Theorem 1.3. We are given that $\mathbf{f}(x, \mathbf{s}) = \mathbf{g}(x, \mathbf{s}) - \mathbf{h}(x)$ where the components of $\mathbf{g}$ satisfy $(f-4)$ and $\mathbf{h} \in [L^2(T_N)]^N$. Consequently, we see that

(1.108)
$$f_j(x, s) \leq \zeta(x) + |h_j(x)| \quad \text{for } s_j \geq 0$$
$$\geq -\zeta(x) - |h_j(x)| \quad \text{for } s_j \leq 0,$$

for $x \in T_N$, $\mathbf{s} \in \mathbf{R}^N$, $j = 1, ..., N$, where $\zeta \in L^2(T_N)$.

In particular,

$$f_j(x, s) \leq n^{-1} s_j + \zeta(x) + |h_j(x)| \quad \text{for } s_j \geq 0$$
$$\geq n^{-1} s_j - \zeta(x) - |h_j(x)| \quad \text{for } s_j \leq 0,$$

for $n = 1, 2, ...$ Hence $\mathbf{f}(x, \mathbf{s})$ meets the condition (1.49) in the hypothesis of Lemma 1.8, and we can invoke this lemma to obtain a sequence $\{\mathbf{v}^n\}_{n=1}^\infty$ with the following properties:

(1.109)
$$\mathbf{v}^n = \gamma_1^n \boldsymbol{\psi}^1 + \cdots + \gamma_n^n \boldsymbol{\psi}^n;$$

$$\nu \left[\boldsymbol{\psi}^l, \mathbf{v}^n\right] + \left\{\boldsymbol{\psi}^l, \mathbf{v^n}, \mathbf{v^n}\right\} + \left(\boldsymbol{\psi}^l, \mathbf{v}^n\right) n^{-1}$$

(1.110)

$$= \int_{T_N} \boldsymbol{\psi}^l \left(x\right) \cdot \mathbf{f} \left(x, \mathbf{v}^n \left(x\right)\right) dx$$

for $l = 1, ..., n$ and $n = 1, 2, ...$, where $\left\{\boldsymbol{\psi}^l\right\}_{l=1}^{\infty}$ is the sequence of functions in $J\left(T_N\right)$ that satisfy (1.30) and (1.31).

As in the proof of Theorem 1.2, we claim that there is a constant K_1 such that

(1.111)
$$\left\|\mathbf{v}^n\right\|_1^2 \leq K_1 \quad \forall n,$$

where $\left\|\mathbf{v}^n\right\|_1^2 = \left[\mathbf{v}^n, \mathbf{v}^n\right] + \left(\mathbf{v}^n, \mathbf{v}^n\right)$.

Suppose (1.111) is false. Then there is a subsequence of $\left\{\left\|\mathbf{v}^n\right\|_1\right\}_{n=1}^{\infty}$ which tends to ∞. We consequently see from Lemma 2.6 in Chapter 5 that the following prevails:

(1.112)
$$\exists \left\{\mathbf{w}^M\right\}_{M=1}^{\infty} \text{ such that } \mathbf{w}^M = \mathbf{v}^{n_M} \text{ with } lim_{M \to \infty} \left\|\mathbf{w}^M\right\|_1 = \infty;$$

with $\mathbf{W}^M = \dfrac{\mathbf{w}^M}{\left\|\mathbf{w}^M\right\|_1}, \exists \mathbf{V} \in \left[H^1\left(T_N\right)\right]^N$ such that

(1.113)

$$lim_{M \to \infty} \left(\mathbf{W}^M - \mathbf{V}, \mathbf{W}^M - \mathbf{V}\right) = 0;$$

(1.114)
$$lim_{M \to \infty} \mathbf{W}^M \left(x\right) = \mathbf{V} \left(x\right) \quad \text{for a.e. } x \in T_N;$$

(1.115)
$$\lim_{M \to \infty} \int_{T_N} \frac{\partial W_j^M}{\partial x_k} \xi dx = \int_{T_N} \frac{\partial V_j}{\partial x_k} \xi dx \quad \forall \xi \in C^{\infty}\left(T_N\right)$$

for $j, k = 1, ..., N$.

Also, we see from (1.108) that

(1.116)
$$\mathbf{w}^M \cdot \mathbf{f}\left(x, \mathbf{w}^M\left(x\right)\right) \leq N \left|\mathbf{w}^M\left(x\right)\right| \left[\zeta\left(x\right) + \mathbf{h}\left(x\right)\right]$$

for $x \in T_N$ and $M = 1, 2, ...$. We conclude from (1.110), (1.112), and (1.116) as in the proof of Theorem 1.2 that

(1.117)
$$(i) \ \lim_{M \to \infty} \left[\mathbf{W}^M, \mathbf{W}^M\right] = 0,$$

$$(ii) \ \lim_{M \to \infty} \left(\mathbf{W}^M, \mathbf{W}^M\right) = 1.$$

From (1.115) and (1.117)(i), we see that

$$\frac{\partial V_j \left(x\right)}{\partial x_k} = 0 \quad \text{for a.e. } x \in T_N$$

for $j, k = 1, ..., N$. Hence, with c_j being constants, we obtain from (1.113) and (1.117)(ii) that

(1.118)
$$(i) \ V_j \left(x\right) = c_j \ \text{for a.e. } x \in T_N,$$

$$(ii) \ c_1^2 + \cdots + c_N^2 = \left(2\pi\right)^{-N}.$$

Next, we return to (1.110) and (1.113) and observe that

$$\nu \left[\mathbf{W}^M, \mathbf{W}^M \right] + \left(\mathbf{W}^M, \mathbf{W}^M \right) n^{-1}$$

$$= \left\| \mathbf{w}^M \right\|_1^{-1} \int_{T_N} \mathbf{W}^M (x) \cdot \mathbf{f} \left(x, \mathbf{w}^M (x) \right) dx.$$

Consequently,

$$(1.119) \qquad \int_{T_N} \mathbf{W}^M (x) \cdot \left[\mathbf{g} \left(x, \mathbf{w}^M (x) \right) - \mathbf{h} (x) \right] dx \geq 0 \quad \forall M.$$

Also, since $\mathbf{g}$ satisfies $(f - 4)$, we see that

$$(1.120) \qquad s_j g_j (x, \mathbf{s}) \leq |s_j| \zeta (x)$$

for $x \in T_N$, $\mathbf{s} \in \mathbf{R}^N$, and $J = 1, ..., N$. Furthermore, from (1.108), we obtain that

$$0 \leq - \mathbf{W}^M (x) \cdot \left[\mathbf{g} \left(x, \mathbf{w}^M (x) \right) - \mathbf{h} (x) \right] + N \left| \mathbf{W}^M (x) \right| \left[\zeta (x) + |\mathbf{h} (x)| \right]$$

for $x \in T_N$ and $\forall M$.

So from Fatou's lemma and (1.119), we obtain that

$$(1.121) \quad \begin{array}{l} (i) \ \ \limsup_{M \to \infty} \mathbf{W}^M (x) \cdot \left[\mathbf{g} \left(x, \mathbf{w}^M (x) \right) - \mathbf{h} (x) \right] \in L^1 (T_N), \\[2mm] (ii) \ \int_{T_N} \limsup_{M \to \infty} \mathbf{W}^M (x) \cdot \left[\mathbf{g} \left(x, \mathbf{w}^M (x) \right) - \mathbf{h} (x) \right] dx \geq 0. \end{array}$$

Next, we set

$$(1.122) \qquad b_j^M (x) = \frac{w_j^M (x)}{\left\| \mathbf{w}^M \right\|_1} \left[g_j \left(x, \mathbf{w}^M (x) \right) - h_j (x) \right]$$

and recall that $\left\| \mathbf{w}^M \right\|_1 \to \infty$ and that

$$\lim_{M \to \infty} \frac{w_j^M (x)}{\left\| \mathbf{w}^M \right\|_1} = c_j \quad \text{for a.e. } x \in T_N.$$

Now $c_j = 0$ or $c_j > 0$ or $c_j < 0$.

If $c_j = 0$, then it follows from (1.120) that

$$(1.123) \qquad \limsup_{M \to \infty} b_j^M (x) \leq 0 \quad \text{for a.e. } x \in T_N.$$

If $c_j > 0$ or if $c_j < 0$, then it follows from the fact that $(\mathbf{g}, \mathbf{h})$ satisfies $(f - 5)$ that (1.123) holds. So we conclude that the inequality in (1.123) is valid for $j = 1, ..., N$.

Next, we observe from (1.122), (1.123), and the fact that

$$\limsup_{M \to \infty} \sum_{j=1}^{N} b_j^M (x) \leq \sum_{j=1}^{N} \limsup_{M \to \infty} b_j^M (x)$$

that

$$(1.124) \quad \limsup_{M \to \infty} \mathbf{W}^M (x) \cdot \left[\mathbf{g} \left(x, \mathbf{w}^M (x) \right) - \mathbf{h} (x) \right] \leq 0 \quad \text{for a.e. } x \in T_N.$$

But then we obtain from (1.121)(ii) that

$$\int_{T_N} \limsup_{M \to \infty} \mathbf{W}^M (x) \cdot \left[\mathbf{g} \left(x, \mathbf{w}^M (x) \right) - \mathbf{h} (x) \right] dx = 0,$$

and consequently, from (1.124) that

$$\limsup_{M \to \infty} \mathbf{W}^M (x) \cdot \left[\mathbf{g} \left(x, \mathbf{w}^M (x) \right) - \mathbf{h} (x) \right] = 0 \quad \text{for a.e. } x \in T_N.$$

We conclude from (1.123) that

(1.125) $$\limsup_{M \to \infty} b_j^M (x) = 0 \ \text{ for a.e. } x \in T_N$$

for $j = 1, ..., N$.

From (1.118)(ii), we see that at least one $c_j \neq 0$. For ease of notation, we will suppose that it is c_1. Also, we will suppose that c_1 is positive, with a similar line of reasoning prevailing incase c_1 is negative. So we suppose

(1.126) $$c_1 > 0.$$

We will now use these last two facts to arrive at a contradiction.

By assumption, (1.14) in the hypothesis of the theorem is

$$\left| E_1^+ (\mathbf{g}, \mathbf{h}) \right| > 0.$$

So it follows from (1.11) and (1.13) that there are positive integers K and l such that the set $E_1^+ (\mathbf{g}, \mathbf{h}, K, l)$ has a positive Lebesgue measure, i.e.,

$$E_1^+ (\mathbf{g}, \mathbf{h}, K, l) = \{ x \in T_N : g_1 (x, \mathbf{s}) - h_1 (x) < -K^{-1}$$

$$\text{for } s_1 > l \text{ and for } s_k \in \mathbf{R}, \ k = 2, ..., N \}$$

has a positive Lebesgue measure $\eta > 0$.

Since $b_1^M (x) = \frac{w_1^M (x)}{\|\mathbf{w}^M\|_1} \left[g_1 \left(x, \mathbf{w}^M (x) \right) - h_1 (x) \right]$ and

$$\lim_{M \to \infty} \frac{w_1^M (x)}{\|\mathbf{w}^M\|_1} = c_1 \quad \text{for a.e. } x \in T_N,$$

it follows from Egoroff's theorem there is an $M_1 > 0$ and a set

$$E_1^{\prime +} (\mathbf{g}, \mathbf{h}, K, l) \subset E_1^+ (\mathbf{g}, \mathbf{h}, K, l)$$

such that

(i) $\left| E_1^{\prime +} (\mathbf{g}, \mathbf{h}, K, l) \right| \geq \eta/2,$

(ii) $w_1^M (x) > l$ for $x \in E_1^{\prime +} (\mathbf{g}, \mathbf{h}, K, l),$

(1.127)

(iii) $\frac{w_1^M (x)}{\|\mathbf{w}^M\|_1} > c_1/2$ for $x \in E_1^{\prime +} (\mathbf{g}, \mathbf{h}, K, l),$

(iv) $- b_1^M (x) > c_1/2K$ for $x \in E_1^{\prime +} (\mathbf{g}, \mathbf{h}, K, l),$

for $M \geq M_1$.

As a consequence of (1.127)(iv), we see that

$$\liminf_{M\to\infty} \left[-b_1^M(x) \right] \geq c_1/2K \quad \text{for } x \in E_1'^+(\mathbf{g},\mathbf{h},K,l).$$

This in turn implies that

$$\limsup_{M\to\infty} b_1^M(x) \leq -c_1/2K \quad \text{for } x \in E_1'^+(\mathbf{g},\mathbf{h},K,l).$$

But from (1.127)(i), we see that $E_1'^+(\mathbf{g},\mathbf{h},K,l)$ has a positive Lebesgue measure. Since c_1 and K are both positive constants, this last inequality is a direct contradiction to the statement in (1.125). So we conclude that the inequality in (1.111) is indeed true.

Hence, it follows that

$$\|\mathbf{v}^n\|_1^2 \leq K_1 \quad \forall n,$$

where $\|\mathbf{v}^n\|_1^2 = [\mathbf{v}^n,\mathbf{v}^n] + (\mathbf{v}^n,\mathbf{v}^n)$, and from (1.110) that

$$\nu\left[\boldsymbol{\psi}^l,\mathbf{v}^n\right] + \left\{\boldsymbol{\psi}^l,\mathbf{v^n},\mathbf{v}^n\right\} + \left(\boldsymbol{\psi}^l,\mathbf{v}^n\right)n^{-1}$$

$$= \int_{T_N} \boldsymbol{\psi}^l(x) \cdot \mathbf{f}(x,\mathbf{v}^n(x))\,dx$$

for $l = 1,...,n$ and $n = 1,2,...$, where $\{\boldsymbol{\psi}^l\}_{l=1}^{\infty}$ is the sequence of functions in $J(T_N)$ that satisfy (1.30) and (1.31).

Using these last two facts, the rest of the proof of this theorem proceeds exactly as it does for the proof of Theorem 1.2 from (1.96) onward. No changes have to be made. We leave the details to the reader, and consider the proof of this theorem complete. ■

Proof of Theorem 1.4.   We prove the sufficiency part of the theorem first and set $\mathbf{f}(x,\mathbf{s}) = \mathbf{g}(\mathbf{s}) - \mathbf{h}(x)$. Also, we set

$$(1.128) \qquad \zeta(x) = \sum_{j=1}^{N} |g_j(\infty)| + |g_j(-\infty)| \quad \text{for } x \in T_N.$$

From (1.17) in the hypothesis of the theorem, we see that

$$-\zeta(x) < g_j(s_j) < \zeta(x)$$

for $x \in T_N$ and $s_j \in \mathbf{R}$, $j = 1,...,N$.

Since $f_j(x,\mathbf{s}) = g_j(s_j) - h_j(x)$, we obtain from this last inquality that

$$f_j(x,s) \leq \zeta(x) + |h_j(x)| \quad \text{for } s_j \geq 0$$

$$\geq -\zeta(x) - |h_j(x)| \quad \text{for } s_j \leq 0,$$

for $x \in T_N$ and $s_j \in \mathbf{R}$, $j = 1,...,N$. Consequently, the analogue of (1.108) in the proof of Theorem 1.3 holds and the proof here proceeds along the lines of Theorem 1.3. In particular, all the material from (1.108)-(1.121) remains valid. The proof of the sufficiency will be complete if we can show that (1.121)(ii) leads to a contradiction, i.e., we have to show that

$$(1.129) \qquad \int_{T_N} \limsup_{M \to \infty} \mathbf{W}^M(x) \cdot \left[\mathbf{g}\left(x, \mathbf{w}^M(x)\right) - \mathbf{h}(x) \right] dx \geq 0$$

is false using the condition in (1.18) in the hypothesis of the theorem.

Recall from (1.112), (1.113), (1.114), and (1.118), we have that

$$(i) \ \ lim_{M \to \infty} \left\| \mathbf{w}^M \right\|_1 = \infty,$$

$$(ii) \ \ \mathbf{W}^M(x) = \mathbf{w}^M(x) / \left\| \mathbf{w}^M \right\|_1,$$

$$(1.130) \qquad (iii) \ lim_{M \to \infty} \frac{w_j^M(x)}{\left\| \mathbf{w}^M \right\|_1} = c_j \quad \text{for a.e. } x \in T_N,$$

$$(iv) \ \ c_1^2 + \cdots + c_N^2 = (2\pi)^{-N},$$

$$(v) \left| \mathbf{W}^M(x) \right| \leq G(x) \quad \text{for a.e. } x \in T_N \text{ and } \forall M$$

where $G \in L^2(T_N)$.

Now

$$(1.131) \ \ \mathbf{W}^M(x) \cdot \left[\mathbf{g}\left(x, \mathbf{w}^M(x)\right) - \mathbf{h}(x) \right] = \sum_{j=1}^{N} \frac{w_j^M(x)}{\left\| \mathbf{w}^M \right\|_1} \left[g\left(w_j^M\right) - h_j(x) \right].$$

Also, from (1.17) and (1.130), we see that the limit as $M \to \infty$ of the expression inside the summation sign in (1.131) is

$$c_j \left[g(\infty) - h_j(x) \right] \qquad \text{if } c_j > 0$$

$$(1.132) \qquad\qquad 0 \qquad\qquad\qquad \text{if } c_j = 0$$

$$c_j \left[g(-\infty) - h_j(x) \right] \qquad \text{if } c_j < 0$$

a.e. in T_N. Furthermore, from (1.18), it follows that

$$\int_{T_N} c_j \left[g(\infty) - h_j(x) \right] dx < 0 \qquad \text{if } c_j > 0$$

$$(1.133) \qquad \int_{T_N} c_j \left[g(\infty) - h_j(x) \right] dx = 0 \qquad \text{if } c_j = 0$$

$$\int_{T_N} c_j \left[g(-\infty) - h_j(x) \right] dx < 0 \quad \text{if } c_j < 0.$$

We consequently obtain from (1.131)-(1.133) that

$$\int_{T_N} \limsup_{M \to \infty} \mathbf{W}^M(x) \cdot \left[\mathbf{g}\left(x, \mathbf{w}^M(x)\right) - \mathbf{h}(x) \right] dx$$

$$\leq \sum_{j \in A_N} \int_{T_N} c_j \left[g(\infty) - h_j(x) \right] dx$$

$$+ \sum_{j \in B_N} \int_{T_N} c_j \left[g(-\infty) - h_j(x) \right] dx,$$

where

$$
\begin{aligned}
A_N &= \{j : c_j > 0 \text{ for } j = 1, ..., N\}, \\
B_N &= \{j : c_j < 0 \text{ for } j = 1, ..., N\}.
\end{aligned}
$$

From (1.130)(iv), at least one of A_N or B_N is nonempty. Hence, it follows from (1.133) that

$$
\int_{T_N} \limsup_{M \to \infty} \mathbf{W}^M (x) \cdot [\mathbf{g}(x, \mathbf{w}^M(x)) - \mathbf{h}(x)] \, dx \ < 0.
$$

This is a direct contradiction of the inequality in (1.129), and the proof of the sufficiency condition of the theorem is complete.

To establish the necessary part of the theorem, we suppose that the pair (v, p) satisfies (1.2) where

$$
f_j (x, \mathbf{v}(x)) = g_j (v_j(x)) - h_j(x)
$$

for $j = 1, ..., N$. Also, we have that $v_j \in H^1(T_N)$ and $p \in L^1(T_N)$ for $j = 1, ..., N$. Fixing j and taking $\phi_k = 0$ for $k \neq j$ and $\phi_j = 1$, we obtain in particular from (1.2) that

$$
(1.134) \qquad \int_{T_N} g_j (v_j(x)) \, dx = \int_{T_N} h_j(x) \, dx.
$$

But then from (1.17), we see that

$$
g_j (\infty) < g_j (v_j(x)) < g_j (-\infty)
$$

a.e. in T_N.

Applying this set of inequalities to the integral on the left-hand side of the equal sign in (1.134), we obtain that

$$
(2\pi)^N g_j (\infty) < \int_{T_N} h_j(x) \, dx < (2\pi)^N g_j (-\infty).
$$

But this is precisely the statement in (1.18). The proof of the necessary part of the theorem is established, and the proof of the theorem is complete.

Exercises.

1. Prove that if $\mathbf{v} \in [H^1(T_3)]^3$ with $\widehat{v}_j(0) = 0$ for $j = 1, 2, 3$, and

$$
-\int_{T_N} [\nu \mathbf{v} \cdot \Delta\phi + \mathbf{v} \cdot (\mathbf{v} \cdot \nabla)\phi + p\nabla \cdot \phi] dx = 0
$$

$$
\int_{T_N} \mathbf{v} \cdot \nabla \xi \, dx = 0
$$

for all $\phi \in [C^\infty(T_N)]^N$ and $\xi \in C^\infty(T_N)$, then

$$
v_j(x) = 0 \quad \text{for a.e. } x \in T_3, \ j = 1, 2, 3.
$$

2. Prove that with $N = 4$, $\mathbf{f} = (f_1, ..., f_N)$ where

$$
f_j (x, s) = -s_j \eta_j(x) + \eta_j(x) \quad \text{for} \ \ j = 1, ..., N
$$

and η_j is a nonnegative function, which is in $L^2(T_N)$ with

$$\int_{T_N} \eta_j(x)\,dx > 0 \qquad \text{for } j = 1, ..., N,$$

meets the conditions in the hypothesis of Theorem 1.2.

3. Prove that $\mathbf{f} = (f_1, ..., f_N)$ where

$$f_j(x, \mathbf{s}) = -\eta_j(x)\, s_j / \left(1 + s_j^2\right)^{1/2} - \eta_j(x)/2$$

and η_j is a nonnegative function, which is in $L^2(T_N)$ with

$$\int_{T_N} \eta_j(x)\,dx > 0 \qquad \text{for } j = 1, ..., N,$$

meets the conditions in the hypothesis of Theorem 1.3 but not the conditions in the hypothesis of Theorem 1.2.

4. Prove that

$$[\nu\,|m|^2\,\widehat{u}_j^k(m) + im_j\widehat{q}_k(m)] = \delta_k^j \qquad \text{for } m \in \Lambda_N \backslash \{0\}$$

where $\widehat{u}_j^k(m)$ is defined below (1.26) and $\widehat{q}_k(m)$ is defined below (1.28).

5. Let $\{\phi_n\}_{n=1}^{\infty}$ be an enumeration of the following system:

$$\left\{ \left(\frac{im_2 e^{im \cdot x}}{|m|}, \frac{-im_1 e^{im \cdot x}}{|m|}, 0 \right) \right\}_{m \in \Lambda_3 \backslash \{0\}}$$
$$\cup \left\{ \left(\frac{im_3 e^{im \cdot x}}{|m|}, 0, \frac{-im_1 e^{im \cdot x}}{|m|} \right) \right\}_{m \in \Lambda_3 \backslash \{0\}}$$
$$\cup \left\{ \left(0, \frac{im_3 e^{im \cdot x}}{|m|}, \frac{-im_2 e^{im \cdot x}}{|m|} \right) \right\}_{m \in \Lambda_3 \backslash \{0\}}.$$

Prove that if $\mathbf{f} = (f_1, f_2, f_3) \in J^o(T_3)$ and also

$$\int_{T_3} \mathbf{f} \cdot \phi_n\,dx = 0 \quad \forall n,$$

then $f_j(x). = 0$ for $x \in T_3$ and $j = 1, 2, 3$.

2. Classical Solutions

In this section, we deal with classical solutions of the following systems of equations:

(2.1)
$$-\nu \Delta \mathbf{v} + (\mathbf{v} \cdot \nabla)\,\mathbf{v} + \nabla p = \mathbf{f}$$

$$(\nabla \cdot \mathbf{v}) = 0$$

where ν is a positive constant, $\mathbf{v}$ and $\mathbf{f}$ are vector-valued functions, and $N = 2$ or $N = 3$.

We say $f_1 \in C(T_N)$, provided f_1 is a real-valued function in $C(\mathbf{R}^N)$, which is periodic of period 2π in each variable.

Given $\mathbf{f} \in C[(T_N)]^N$, we will say the pair $(\mathbf{v},p)$ is a periodic classical solution of the Navier-Stokes system (2.1) provided:

$$\mathbf{v} \in \left[C^2(T_N)\right]^N \quad \text{and} \quad p \in C^1(T_N)$$

and

(2.2)
$$-\nu \Delta \mathbf{v}(x) + (\mathbf{v}(x) \cdot \nabla)\mathbf{v}(x) + \nabla p(x) = \mathbf{f}(x) \qquad \forall x \in T_N$$

$$(\nabla \cdot \mathbf{v})(x) = 0 \qquad\qquad \forall x \in T_N.$$

To obtain classical solutions of the Navier-Stokes system, we will require slightly more for the driving force $\mathbf{f}$ than periodic continuity. In particular, we say $f_1 \in C^\alpha(T_N)$, $0 < \alpha < 1$, provided the following holds:

(i) $f_1 \in C(T_N)$;
(ii) $\exists\ c_1 > 0$ s. t. $|f_1(x) - f_1(y)| \le c_1 |x - y|^\alpha \qquad \forall x, y \in \mathbf{R}^N$.

$g_1 \in C^{2+\alpha}(T_N)$ means $g_1 \in C^2(T_N)$ and each of its second partial derivatives $\partial^2 g_1/\partial x_j \partial x_k \in C^{2+\alpha}(T_N)$ for $j, k = 1, ..., N$.

We will say $\mathbf{f} = (f_1, ... f_N) \in [C^\alpha(T_N)]^N$ provided each of its components $f_j \in C^\alpha(T_N)$. In order to obtain classical solutions of the Navier-Stokes system here in this section, we will assume that $\mathbf{f} \in [C^\alpha(T_N)]^N$.

Likewise, we will say $\mathbf{v} = (v_1, ... v_N) \in \left[C^{2+\alpha}(T_N)\right]^N$ provided each of its components $v_j \in C^{2+\alpha}(T_N)$.

We will prove the following theorem:

Theorem 2.1. *Suppose there exists an α such that $\mathbf{f} \in [C^\alpha(T_N)]^N$ where $0 < \alpha < 1$, where $\mathbf{f} = (f_1, ... f_N)$, and $N = 2$ or $N = 3$. Suppose also that*

(2.3)
$$\int_{T_N} f_j(x)\, dx = 0 \quad \text{for} \quad j = 1, ..., N.$$

Then there exists a pair $(\mathbf{v},p)$ with $\mathbf{v} \in \left[C^{2+\alpha}(T_N)\right]^N$ and $p \in C^{1+\alpha}(T_N)$ such that the pair $(\mathbf{v},p)$ is a periodic classical solution of the Navier-Stokes equations (2.1).

To prove the theorem, we will first need the following four lemmas that are true for $N \ge 2$:

Lemma 2.2. *Suppose $h_j \in L^r(T_N)$, $1 < r < \infty$, and $\widehat{h}_j(0) = 0$ for $j = 1, ..., N$, $N \ge 2$. Set*

$$v_j(x) = (2\pi)^{-N} \int_{T_N} \left[\sum_{k=1}^N u_j^k(x - y) h_k(y)\right] dy,$$

for $j = 1, ..., N$ where u_j^k is defined in (1.26). Then $v_j \in W^{2,r}(T_N)$ for $j = 1, ..., N$.

Lemma 2.3. *Suppose $h_j \in C(T_N)$ and $\widehat{h}_j(0) = 0$ for $j = 1, ..., N$, $N \geq 2$. Set*

$$v_j(x) = (2\pi)^{-N} \int_{T_N} \left[\sum_{k=1}^{N} u_j^k(x-y) h_k(y) \right] dy,$$

for $j = 1, ..., N$ where u_j^k is defined in (1.26). Then $v_j \in C^{1+\alpha}(T_N)$, $0 < \alpha < 1$, for $j = 1, ..., N$.

Lemma 2.4. *Suppose there exists an α such that $\mathbf{h} \in [C^\alpha(T_N)]^N$ where $0 < \alpha < 1$, where $\mathbf{h} = (h_1, ... h_N)$, and where $\widehat{h}_j(0) = 0$ for $j = 1, ..., N$, $N \geq 2$. Set*

$$v_j(x) = (2\pi)^{-N} \int_{T_N} \left[\sum_{k=1}^{N} u_j^k(x-y) h_k(y) \right] dy,$$

for $j = 1, ..., N$ where u_j^k is defined in (1.26). Then $v_j \in C^{2+\alpha}(T_N)$, for $j = 1, ..., N$.

Lemma 2.5. *Suppose there exists an α such that $\mathbf{h} \in [C^\alpha(T_N)]^N$ where $0 < \alpha < 1$, where $\mathbf{h} = (h_1, ... h_N)$, and where $\widehat{h}_j(0) = 0$ for $j = 1, ..., N$, $N \geq 2$. Set*

$$p(x) = (2\pi)^{-N} \int_{T_N} \left[\sum_{k=1}^{N} H_k(x-y) h_k(y) \right] dy,$$

where H_k is defined in Lemma A in §1. Then $p \in C^{1+\alpha}(T_N)$.

Proof of Lemma 2.2. For ease of notation, we will prove the lemma for $v_1(x)$. A similar proof will prevail for the other values of j.

We first of all notice that

(2.4) $$\widehat{v}_1(m) = \sum_{k=1}^{N} \widehat{u}_1^k(m) \widehat{h}_k(m) \text{ for } m \in \Lambda_N$$

where

$$\begin{aligned} \widehat{u}_1^k(m) &= \left[\delta_1^k - m_1 m_k |m|^{-2} \right] |m|^{-2} \nu^{-1} \quad \text{for } m \neq 0 \\ &= 0 \text{ for } m = 0. \end{aligned}$$

We see from Lemma B in §1 that for each k,

$$\sum_{|m|>0} |m|^2 \widehat{u}_1^k(m) \widehat{h}_k(m) e^{im \cdot x}$$

is the Fourier series of a function in $L^r(T_N)$. Hence, it follows that there is

(2.5) $$w_1 \in L^r(T_N)$$

such that

$$S[w_1] = \sum_{|m|>0} |m|^2 [\sum_{k=1}^{N} \widehat{u}_1^k(m) \widehat{h}_k(m)]e^{im\cdot x},$$

where $S[w_1]$ stands for the Fourier series of w_1. From (2.4), we also have

$$S[v_1] = \sum_{|m|>0} [\sum_{k=1}^{N} \widehat{u}_1^k(m) \widehat{h}_k(m)]e^{im\cdot x}.$$

Furthermore, from (1.21) and Lemma A, we have that $H_0 \in L^1(T_N)$ and

$$S[H_0] = \sum_{|m|>0} |m|^{-2} e^{im\cdot x}.$$

Consequently, it follows from these last three equalities that

$$v_1(x) = (2\pi)^{-N} \int_{T_N} H_0(x-y) w_1(y) \, dy$$

for a.e. $x \in T_N$.

We conclude from (2.5) and Theorem 6.1 in Chapter 2 that

$$v_1 \in W^{2,r}(T_N).$$

This completes the proof of the lemma. ■

Proof of Lemma 2.3.   For ease of notation, we will prove the lemma for $v_1(x)$. A similar proof will prevail for the other values of j.

We recall that $v_1 \in C^{1+\alpha}(T_N)$ means that $v_1 \in C^1(T_N)$ and

$$\frac{\partial v_1}{\partial x_j} \in C^{\alpha}(T_N) \quad \text{for} \quad j = 1, ... N.$$

We start the proof by invoking Lemma 2.2 and obtaining that $v_1 \in W^{2,r}(T_N)$ for $1 < r < \infty$. Hence, from the very definition of $W^{2,r}(T_N)$, we see there exists $g_j \in W^{1,r}(T_N)$ such that

(2.6) $$\int_{T_N} v_1(x) \frac{\partial \xi(x)}{\partial x_j} dx = -\int_{T_N} g_j(x) \xi(x) \, dx \quad \forall \xi \in C^{\infty}(T_N)$$

for $j = 1, ..., N$.

As in the proof of Lemma 2.2, (see (2.4)), we have

(2.7)
$$\widehat{v}_1(m) = \sum_{k=1}^{N} \widehat{u}_1^k(m) \widehat{h}_k(m) \text{ for } m \in \Lambda_N,$$

$$\widehat{g}_j(m) = im_j[\sum_{k=1}^{N} \widehat{u}_1^k(m) \widehat{h}_k(m)] \text{ for } m \in \Lambda_N.$$

Next, we set

(2.8) $$w_{jk}(x) = (2\pi)^{-N} \int_{T_N} H_j(x-y) h_k(y) \, dy$$

for $j, k = 1, ..., N$ where $H_j(x)$ is defined in Lemma A. It follows from the hypothesis of the lemma and Theorem 4.1 in Appendix B that

$$(2.9) \qquad w_{jk}(x) \in C^\alpha(T_N), 0 < \alpha < 1,$$

for $j, k = 1, ..., N$.

From (2.8), we see that

$$(2.10) \qquad \widehat{w}_{jk}(m) = im_j\widehat{h}_k(m) / |m|^2 \quad \text{for } m \in \Lambda_N \backslash \{0\}.$$

Also, it follows from the definition of $\widehat{u}_1^k(m)$ given below (1.26) and Lemma C that

$$\widehat{w}_{jk}(m)\widehat{u}_1^k(m)|m|^2 \quad \text{for } m \in \Lambda_N \backslash \{0\}$$

is the Fourier coefficient of a function in $C^\alpha(T_N)$. We consequently obtain from (2.9) and (2.10)

$$\sum_{|m|>0} im_j[\sum_{k=1}^N \widehat{h}_k(m)\widehat{u}_1^k(m)]e^{im\cdot x}$$

is the Fourier series of a function in $C^\alpha(T_N)$.

We conclude from (2.7) that

$$(2.11) \qquad g_j \in C^\alpha(T_N) \quad \text{for } j = 1, ...N.$$

Next, an easy computation using the Fourier coefficients in (2.7) shows that

$$v_1(x) = -(2\pi)^{-N} \sum_{j=1}^N \int_{T_N} g_j(x-y) H_j(y)\, dy$$

where $H_j(x)$ is defined in Lemma A for $j = 1, ..., N$.

Since, in particular, $H_j \in L^1(T_N)$, we see from (2.11) that

$$v_1 \in C^\alpha(T_N).$$

It remains to show that $v_1 \in C^1(T_N)$ and that

$$(2.12) \qquad \frac{\partial v_1(x)}{\partial x_j} = g_j(x) \quad \text{for } x \in T_N$$

for $j = 1, ...N$, where g_j is the function defined by (2.6).

For ease of notation, we will establish (2.12) for $j = 1$. A similar proof will prevail for other values of j.

We let $\sigma_n^\Diamond(v_1, x)$ designate the n-th iterated Fejer sum of v_1 as defined in (2.6) of Chapter 1. Also, let x^* be a fixed but arbitrary point in T_N. We know from (2.7) that $\widehat{g}_1(m) = im_1\widehat{v}_1(m) \ \forall m \in \Lambda_N$. Since $\sigma_n^\Diamond(v_1, x)$ is a trigonometric polynomial, it follows that

$$\partial\sigma_n^\Diamond(v_1, x)/\partial x_1 = \sigma_n^\Diamond(g_1, x) \quad \text{for } x \in \mathbf{R}^N,$$

and therefore that

$$\sigma_n^{\Diamond}(v_1, x_1^* + t, x_2^*, ..., x_N^*) - \sigma_n^{\Diamond}(v_1, x_1^*, x_2^*, ..., x_N^*)$$

$$= \int_{x_1^*}^{x_1^*+t} \sigma_n^{\Diamond}(g_1, s, x_2^*, ..., x_N^*) ds$$

for $t \in \mathbf{R}$.

Now both v_1 and g_1 are in $C^{\alpha}(T_N)$. So by Theorem 2.1 in Chapter 1, $\lim_{n \to \infty} \sigma_n^{\Diamond}(g_1, x) = g_1(x)$ uniformly for $x \in \mathbf{R}^N$. Using this uniformity, we obtain from this last equality that

$$v_1(x_1^* + t, x_2^*, ..., x_N^*) - v_1(x_1^*, x_2^*, ..., x_N^*) = \int_{x_1^*}^{x_1^*+t} g_1(s, x_2^*, ..., x_N^*) ds$$

for $t \in \mathbf{R}$.

Next, we divide both sides of this last equality by $t \neq 0$, pass to the limit as $t \to 0$, and conclude that

$$\frac{\partial v_1(x^*)}{\partial x_1} = g_1(x^*).$$

This establishes (2.12) and shows that indeed $v_1 \in C^1(T_N)$. Since in (2.11), we have shown that $g_j \in C^{\alpha}(T_N)$, it follows from (2.12) that

$$v_1 \in C^{1+\alpha}(T_N).$$

This fact concludes the proof of the lemma. ■

Proof of Lemma 2.4.   For ease of notation, we will prove the lemma for $v_1(x)$, i.e., we will show that

$$v_1 \in C^{2+\alpha}(T_N).$$

A similar proof will prevail for the other values of j.

From Lemma 2.3, we know that

$$v_1 \in C^{1+\alpha}(T_N).$$

Using the notation employed in the proof of Lemma 2.3, we show in (2.12) that

$$(2.12) \qquad \frac{\partial v_1(x)}{\partial x_j} = g_j(x) \quad \text{for } x \in T_N$$

for $j = 1, ...N$, where the Fourier coefficients of g_j are given by (see (2.7))

$$(2.13) \quad \widehat{g_j}(m) = im_j |m|^{-2} \left[\sum_{k=1}^{N} |m|^2 \widehat{u}_1^k(m) \widehat{h}_k(m) \right] \quad \text{for } m \in \Lambda_N \backslash \{0\}$$

with $\widehat{g_j}(0) = 0$.

So the proof of this lemma will be complete if we show that

$$g_j \in C^{1+\alpha}(T_N)$$

for $j = 1, ..., N$. In other words, we have to show that

$$\frac{\partial g_j(x)}{\partial x_k} \text{ exists for } x \in T_N$$

$$\text{and } \frac{\partial g_j(x)}{\partial x_k} \in C^\alpha(T_N)$$

for $k = 1, ..., N$.

Once again, for ease of notation, we will establish these last two facts for $k = 1$. A similar proof will prevail for the other values of k.

So the proof of this lemma will be complete when we demonstrate that

(2.14)

$$(i) \quad \frac{\partial g_j(x)}{\partial x_1} \text{ exists for } x \in T_N,$$

$$(ii) \quad \frac{\partial g_j(x)}{\partial x_1} \in C^\alpha(T_N)$$

for $j = 1, ..., N$.

Now we know from the fact that $h_k \in C^\alpha(T_N)$ in conjunction with Lemma C that

$$\sum_{k=1}^N |m|^2 \widehat{u}_1^k(m) \widehat{h}_k(m) \quad \text{for } m \in \Lambda_N$$

is the Fourier coefficient of a function in $C^\alpha(T_N)$ where $\widehat{u}_1^k(m)$ is defined below (1.26). Hence, using Lemma C once again, we obtain that

$$-\frac{m_1 m_j}{|m|^2} \sum_{k=1}^N |m|^2 \widehat{u}_1^k(m) \widehat{h}_k(m) \quad \text{for } m \in \Lambda_N \backslash \{0\}$$

is the Fourier coefficient of a function in $C^\alpha(T_N)$ for $j = 1, ..., N$.

We set

(2.15)
$$S[w_j] = -\sum_{|m|>0} \frac{m_1 m_j}{|m|^2} [\sum_{k=1}^N |m|^2 \widehat{u}_1^k(m) \widehat{h}_k(m)] e^{im \cdot x}$$

and have that

(2.16)
$$w_j \in C^\alpha(T_N)$$

for $j = 1, ..., N$.

Also, we have from (2.13) that

(2.17)
$$S[g_j] = \sum_{|m|>0} \frac{im_j}{|m|^2} [\sum_{k=1}^N |m|^2 \widehat{u}_1^k(m) \widehat{h}_k(m)] e^{im \cdot x},$$

and from (2.12) that

(2.18)
$$g_j \in C^\alpha(T_N).$$

Next, we let $\sigma_n^\Diamond(g_j, x)$ designate the n-th iterated Fejer sum of g_j as defined in (2.6) of Chapter 1. Likewise, let $\sigma_n^\Diamond(w_j, x)$ designate the n-th iterated Fejer sum of w_j. We note from (2.15) and (2.17) that

$$\widehat{g_j}(m) = im_1 \widehat{w}_j(m) \quad \text{for } m \in \Lambda \backslash \{0\}$$

for $j = 1, ..., N$. Consequently, since $\sigma_n^\Diamond(g_j, x)$ is a trigonometric polynomial, we have

(2.19)
$$\frac{\partial \sigma_n^\Diamond(g_j, x)}{\partial x_1} = \sigma_n^\Diamond(w_j, x) \quad \forall x \in T_N.$$

Also, we have from (2.16) and (2.18) and Theorem 2.1 in Chapter 1 that

(2.20)
$$(i) \ lim_{n \to \infty} \sigma_n^\Diamond(g_j, x) = g_j(x) \quad \text{uniformly for } x \in \mathbf{R}^N,$$

$$(ii) \ lim_{n \to \infty} \sigma_n^\Diamond(w_j, x) = w_j(x) \quad \text{uniformly for } x \in \mathbf{R}^N.$$

Let $x^* = (x_1^*, ..., x_N^*)$ be a fixed but arbitrary point in T_N. It follows from (2.19) that

$$\sigma_n^\Diamond(g_j, x_1^* + t, x_2^*, ..., x_N^*) - \sigma_n^\Diamond(g_j, x_1^*, x_2^*, ..., x_N^*)$$

$$= \int_{x_1^*}^{x_1^* + t} \sigma_n^\Diamond(w_j, s, x_2^*, ..., x_N^*) ds$$

for $t \in \mathbf{R}$.

Using (2.20) in conjunction with this last equality, we obtain that

$$g_j(x_1^* + t, x_2^*, ..., x_N^*) - g_j(x_1^*, x_2^*, ..., x_N^*)$$

$$= \int_{x_1^*}^{x_1^* + t} w_j(s, x_2^*, ..., x_N^*) ds.$$

Dividing both sides of this last equality by t and passing to the limit as $t \to 0$, we obtain

$$\frac{\partial g_j(x^*)}{\partial x_1} = w_j(x^*).$$

Since x^* is a fixed but arbitrary point in T_N, we conclude that

$$\frac{\partial g_j(x)}{\partial x_1} = w_j(x) \quad \forall x \in T_N$$

for $j = 1, ..., N$.

Since by (2.16), $w_j \in C^\alpha(T_N)$, this last equality establishes both (i) and (ii) of (2.14), and the proof of the lemma is complete. ■

Proof of Lemma 2.5.   To prove the lemma, we have to show that

(2.20 ')
$$(i) \ \frac{\partial p_j(x)}{\partial x_k} \text{ exists for } x \in T_N,$$

$$(ii) \ \frac{\partial p_j(x)}{\partial x_k} \in C^\alpha(T_N)$$

for $k = 1, ..., N$.

For ease of notation, we will do this for the special case $k = 1$. A similar proof prevails for other values of k.

Since $H_k \in L^1(T_N)$, it follows from the fact that $h_k \in C^\alpha(T_N)$ and the definition of $p(x)$ that

$$(2.21) \qquad\qquad\qquad p \in C^\alpha(T_N).$$

Also, it follows from Lemma A that

$$(2.22) \qquad\qquad \widehat{p}(m) = \sum_{k=1}^{N} \frac{im_k}{|m|^2} \widehat{h}_k(m) \quad \text{for } m \in \Lambda_N \setminus \{0\}$$

with $\widehat{p}(0) = 0$.

Since $h_k \in C^\alpha(T_N)$, it follows from Lemma C that

$$- \sum_{k=1}^{N} \frac{m_1 m_k}{|m|^2} \widehat{h}_k(m) \quad \text{for } m \in \Lambda_N \setminus \{0\}$$

are the Fourier coefficients of a function in $C^\alpha(T_N)$. We designate this function by $f_1(x)$ and have that

$$(2.23) \qquad\qquad\qquad f_1 \in C^\alpha(T_N),$$

and that

$$(2.24) \qquad\qquad S[f_1] = - \sum_{|m|>0} [\sum_{k=1}^{N} \frac{m_1 m_k}{|m|^2} \widehat{h}_k(m)] e^{im \cdot x}.$$

Also, from (2.22), we have that

$$(2.25) \qquad\qquad S[p] = \sum_{|m|>0} [\sum_{k=1}^{N} \frac{im_k}{|m|^2} \widehat{h}_k(m)] e^{im \cdot x}.$$

We let $\sigma_n^\Diamond(p, x)$ designate the n-th iterated Fejer sum of p as defined in (2.6) of Chapter 1. Likewise, we let $\sigma_n^\Diamond(f_1, x)$ designate the n-th iterated Fejer sum of f_1. It follows from (2.24) and (2.25) that

$$\frac{\partial \sigma_n^\Diamond(p, x)}{\partial x_1} = \sigma_n^\Diamond(f_1, x) \quad \forall x \in T_N.$$

Because we have (2.21) and (2.23), we proceed exactly as we did at this point in the proof of Lemma 2.4 and obtain that

$$\frac{\partial p(x)}{\partial x_1} = f_1(x) \quad \forall x \in T_N.$$

Since $f_1 \in C^\alpha(T_N)$, this last fact establishes $(2.20')(i)$ and (ii) for the special case $k = 1$ and completes the proof of the lemma. ∎

Proof of Theorem 2.1. In particular, $f_j \in C(\mathbf{R}^N)$ and is periodic of period 2π in each variable for $j = 1, ..., N$. Therefore, $f_j \in L^2(T_N)$, and we can invoke Theorem 1.1 to obtain a pair $(\mathbf{v}, p)$ with $v_j \in H^1(T_N)$ for

$j = 1, ...N$ and $p \in L^2(T_N)$ such that $(\mathbf{v}, p)$ is a distribution solution of the stationary Navier-Stokes equations (2.1), i.e.,

$$- \int_{T_N} [\nu \mathbf{v} \cdot \Delta \phi + \mathbf{v} \cdot (\mathbf{v} \cdot \nabla) \phi + p \nabla \cdot \phi] dx$$
$$= \int_{T_N} [\mathbf{f}(x) \cdot \phi(x)] dx$$

(2.26)

$$\int_{T_N} \mathbf{v} \cdot \nabla \xi dx = 0$$

for all $\phi \in [C^\infty(T_N)]^N$ and $\xi \in C^\infty(T_N)$.

Next, as in the proof of Theorem 1.1, we set

(2.27)
$$h_j(x) = f_j(x) - \sum_{k=1}^{N} v_k(x) \, \partial v_j(x) / \partial x_k,$$

and obtain from the proof Theorem 1.1 that

(2.28)
$$\int_{T_N} h_j(x) \, dx = 0$$

for $j = 1, ..., N$.

Also, we obtain from the proof of Theorem 1.1 (see (1.47$'$) and (1.48)) that

$$- \int_{T_N} [\nu \mathbf{v} \cdot \Delta \phi + p \nabla \cdot \phi] dx$$
$$= \int_{T_N} [\mathbf{h}(x) \cdot \phi(x)] dx$$

(2.29)

$$\int_{T_N} \mathbf{v} \cdot \nabla \xi dx = 0$$

for all $\phi \in [C^\infty(T_N)]^N$ and $\xi \in C^\infty(T_N)$ and that

$$v_j(x) = (2\pi)^{-N} \int_{T_N} \left[\sum_{k=1}^{N} u_j^k(x-y) h_k(y) \right] dy,$$

(2.30)

$$p(x) = (2\pi)^{-N} \int_{T_N} \left[\sum_{k=1}^{N} q_k(x-y) h_k(y) \right] dy$$

for $j = 1, ..., N$.

The trick in the proof of this theorem is to show that somehow by a bootstrap argument that

(2.30 $'$) $v_j \in C^{2+\alpha}(T_N)$ for $j = 1, ..., N$ and $p \in C^{1+\alpha}(T_N)$.

This is the case because once (2.30$'$) is established, it follows from (2.26) that

$$- \int_{T_N} [\nu \Delta \mathbf{v} \cdot \phi - (\mathbf{v} \cdot \nabla) \mathbf{v} \cdot \phi - \nabla p \cdot \phi] dx$$
$$= \int_{T_N} [\mathbf{f}(x) \cdot \phi(x)] dx$$

$$\int_{T_N} \left(\sum_{k=1}^{N} \partial v_k / \partial x_k \right) \xi dx = 0$$

for all $\phi \in [C^\infty(T_N)]^N$ and $\xi \in C^\infty(T_N)$.

But then it is easy to see that the statement in (2.2) holds. So, the pair $(\mathbf{v}, p)$ is indeed a periodic classical solution of the Navier-Stokes equations (2.1).

Assuming that

$$N = 2 \text{ or } N = 3,$$

we now show by a bootstrap argument that $(2.30')$ is indeed valid.

To do this, we first observe from the fact that $v_j \in H^1(T_N)$ for $j = 1, ..., N$ that

$$v_j \in W^{1,2}(T_N).$$

Therefore, from the standard Sobolev inequalities in dimension 2 or dimension 3, we obtain that

$$v_j \in L^6(T_N),$$

(2.31)
$$\frac{\partial v_j}{\partial x_k} \in L^2(T_N),$$

$$v_k \frac{\partial v_j}{\partial x_k} \in L^{3/2}(T_N)$$

for $j, k = 1, ..., N$. Consequently, we have from (2.27) that

$$h_j \in L^{3/2}(T_N).$$

But then we obtain from (2.30) and Lemma 2.2 that

$$v_j \in W^{2,3/2}(T_N),$$

(2.32)
$$\frac{\partial v_j}{\partial x_k} \in W^{1,3/2}(T_N).$$

We apply the standard Sobolev inequalities in dimension 2 or dimension 3 to (2.32) to obtain that

$$v_j \in L^r(T_N) \quad \forall r > 1,$$

(2.33)
$$\frac{\partial v_j}{\partial x_k} \in L^3(T_N).$$

Using (2.27) once again, we see from (2.33) that

$$h_j \in L^{3-\varepsilon}(T_N) \quad \forall \varepsilon > 0.$$

We use this last established fact in conjunction with Lemma 2.2 to obtain that

$$v_j \in W^{2,3-\varepsilon}(T_N) \quad \forall \varepsilon > 0,$$

(2.34)
$$\frac{\partial v_j}{\partial x_k} \in W^{1,3-\varepsilon}(T_N) \quad \forall \varepsilon > 0.$$

We apply the standard Sobolev inequalities in dimension 2 or dimension 3 to (2.34) to obtain that

$$v_j \in C\left(T_N\right),$$

(2.35)

$$\frac{\partial v_j}{\partial x_k} \in L^r\left(T_N\right) \quad \forall r > 1.$$

Using (2.27) once again, we see from (2.35) that

$$h_j \in L^r\left(T_N\right) \quad \forall r > 1.$$

Using this last fact in conjunction with Lemma 2.2 enables us to obtain that

$$v_j \in W^{2,r}\left(T_N\right) \quad \forall r > 1,$$

$$\frac{\partial v_j}{\partial x_k} \in W^{1,r}\left(T_N\right) \quad \forall r > 1$$

and consequently, from the standard Sobolev inequalities in dimension 2 or 3 that

$$v_j \in C\left(T_N\right),$$

$$\frac{\partial v_j}{\partial x_k} \in C\left(T_N\right).$$

Using (2.27) once again, we see that

$$h_j \in C\left(T_N\right).$$

We now apply Lemma 2.3 in conjunction with this last fact and (2.25) to obtain that

$$v_j \in C^{1+\alpha}\left(T_N\right),$$

$$\frac{\partial v_j}{\partial x_k} \in C^{\alpha}\left(T_N\right).$$

Using (2.27) for the last time, we obtain from this last established fact that

(2.36) $$h_j \in C^{\alpha}\left(T_N\right).$$

We now apply (2.30) in conjunction with (2.36) and Lemma 2.4 to obtain that

$$v_j \in C^{2+\alpha}\left(T_N\right),$$

(2.37)

$$\frac{\partial v_j}{\partial x_k} \in C^{1+\alpha}\left(T_N\right).$$

Next, from (1.22), (1.28), and from (1.47$'$)(ii) in the proof of Theorem 1.1, we see that

$$p\left(x\right) = -\left(2\pi\right)^{-N}\int_{T_N} \left[\sum_{k=1}^{3} H_k\left(x-y\right) h_k\left(y\right)\right] dy.$$

Consequently, we have from (2.36) and Lemma 2.5 that

$$p \in C^{1+\alpha}\left(T_N\right).$$

This last fact, along with (2.37), shows that the assertion in (2.30′) is indeed valid, and the proof of the theorem is complete. ∎

Exercises.

1. With $W^{1,2}(T_3)$ defined in §6 of Chapter 2, prove, using the Sobolev inequalities for $W_0^{1,p}(\Omega)$ where $\Omega \subset \mathbf{R}^3$ is a bounded open set (see [Ev, Theorem 3, p.265]) that

$$u \in W^{1,2}(T_3) \Rightarrow u \in L^6\left(T_3\right).$$

2. Suppose $w_1 \in L^r\left(T_3\right), 1 < r < \infty$, with $\widehat{w}_1\left(0\right) = 0.$ Set

$$v_1\left(x\right) = \int_{T_3} w_1\left(x - y\right) H_1\left(y\right) dy,$$

$$u_1\left(x\right) = \int_{T_3} v_1\left(x - y\right) H_2\left(y\right) dy$$

for $x \in T_3$ where H_1 and H_2 are defined Lemma A of §1. Prove that

$$u_1 \in W^{2,r}(T_3).$$

3. Suppose $u_1, v_1 \in C\left(T_3\right)$, and

$$\widehat{v}_1\left(m\right) = im_1\widehat{u}_1\left(m\right) \quad \forall m \in \Lambda_3.$$

Prove that $\frac{\partial u_1}{\partial x_1}\left(x\right)$ exists for $x \in T_3$ and

$$\frac{\partial u_1}{\partial x_1}\left(x\right) = v_1\left(x\right).$$

3. Further Results and Comments

1. There are a number of results which deal with removable singularities for the stationary Navier-Stokes equations, and some of the results are even unexpected. Consider the system

$$-\nu\Delta\mathbf{v}\left(x\right) + \left(\mathbf{v}\left(x\right) \cdot \nabla\right)\mathbf{v}\left(x\right) + \nabla p\left(x\right) = 0$$

(3.1)

$$\left(\nabla \cdot \mathbf{v}\right)\left(x\right) = 0.$$

With $B(x_0, r)$, the open N-ball with center x_0 and radius r, the pair $(\mathbf{v}, p)$ is said to be a classical solution of (3.1) in $B(x_0, r)$, provided $v_j \in C^2\left[B(x_0, r)\right]$ for $j = 1, ..., N$, $p \in C^1\left[B(x_0, r)\right]$, and the system of equations (3.1) is satisfied for all $x \in B(x_0, r)$.

The following result (with a best possible corollary in dimension $N = 2$) prevails for removable singularities in dimensions $N \geq 2$ [Sh21]:

Theorem 1. *Let the pair* $(\mathbf{v}, p)$ *be a classical solution of (3.1) in* $B(0,1) \setminus \{0\}$ *for* $N \geq 2$. *Suppose that*

(i) $\exists \beta > N$ *such that* $|\mathbf{v}| \in L^\beta (B(0,1))$,

(ii) *for* $N = 2$, $\lim_{r \to 0} \left| r^2 \log r \right|^{-1} \int_{B(0,r)} |\mathbf{v}| \, dx = 0$.

Then $(\mathbf{v}, p)$ *can be defined at 0 so that* $(\mathbf{v}, p)$ *is a classical solution of (3.1) in* $B(0,1)$.

It is to be observed that in dimension $N = 2$, the following best possible result is an immediate corollary to Theorem 1.

Corollary 2. *Let the pair* $(\mathbf{v}, p)$ *be a classical solution of (3.1) in* $B(0,1) \setminus \{0\}$ *for* $N=2$. *Suppose that*

(3.2) $\qquad v_j (x) = o (|\log |x||) \qquad$ *as* $|x| \to 0$ *for* $j = 1, 2$.

Then $(\mathbf{v}, p)$ *can be defined at 0 so that* $(\mathbf{v}, p)$ *is a classical solution of (3.1) in* $B(0,1)$.

In the paper "A Counter-example in the Theory of Planar Viscous Incompressible Flow" published in the J. Diff. Eqns. [Sh22], it is shown that if assumption (3.2) is replaced with

$$v_1 (x) = O (|\log |x||) \text{ and } v_2 (x) = o (|\log |x||) \qquad \text{as } |x| \to 0,$$

then the conclusion to Corollary 2 is false.

It is to be observed that both in Theorem 1 and in Corollary 2, no growth condition has been put on the pressure p. If a growth condition is put on p, then the unexpected result mentioned earlier occurs. In particular, the following theorem prevails in dimension $N = 2$ [Sh23];

Theorem 3. *In dimension* $N=2$, *let the pair* $(\mathbf{v}, p)$ *be a classical solution of (3.1) in* $B(0,1) \setminus \{0\}$. *Suppose that*

(i) $|v(x)| = o \left(|x|^{-\frac{1}{2}} \right) \qquad$ *as* $|x| \to 0$,

(ii) $|p(x)| = o \left(|x|^{-1} \right) \qquad$ *as* $|x| \to 0$.

Then $(\mathbf{v}, p)$ *can be defined at 0 so that* $(\mathbf{v}, p)$ *is a classical solution of (3.1) in* $B(0,1)$.

This result is unexpected when compared with a similar situation involving removable singularities for harmonic functions in punctured disks.

Theorem 3 is also best possible because in *(ii)* the little "*o*" cannot be replaced by big "*O*." All this is shown in [Sh22], described above, which contains the proof of the following theorem:

Theorem 4. *In dimension N=2, there exists a pair* $(\mathbf{v}, p)$ *which is a classical solution of (3.1) in* $B(0,1) \setminus \{0\}$ *such that* $v_2(x)$ *and* $|x| \, p(x)$ *are uniformly bounded in* $B(0,1) \setminus \{0\}$ *and such that*

$$\lim_{|x| \to 0} \frac{v_1(x)}{\log |x|} = \gamma \ \text{ where } \gamma \text{ is a finite-valued nonzero constant.}$$

All of the above theorems make strong use of multiple Fourier series.

Theorem 3 was motivated by an earlier paper on removable singularities of the stationary Navier-Stokes equations by Dyer and Edmunds [DE].

2. There are some interesting results involving removable sets of capacity zero and the stationary Navier-Stokes equations. For simplicity in discussing these matters, we will restrict our attention to dimension $N = 3$.

In the sequel, $\Omega \subset \mathbf{R}^3$ will be a bounded open connected set.

A closed set $Z \subset \Omega$ will be said to be of Newtonian capacity zero provided the following holds:

$$\int_\Omega \int_\Omega |x - y|^{-1} \, d\mu(x) \, d\mu(y) = \infty$$

for every μ that is a nonnegative finite Borel measure on Ω with $\mu(\Omega) = 1$ and $\mu(\Omega \setminus Z) = 0$. (This is essentially the same definition that we have given previously in §6 of Chapter 1 for ordinary capacity zero on T_3.)

Designating the stationary Navier-Stokes equations with a driving force in dimension 3 by

(3.3)
$$-\nu \Delta w_j + \sum_{k=1}^N w_k \frac{\partial w_j}{\partial x_k} + \frac{\partial p}{\partial x_j} = f_j \quad j = 1, 2, 3,$$

$$\sum_{k=1}^N \frac{\partial w_j}{\partial x_k} = 0,$$

it turns out that the following theorem involving capacity holds for this set of equations in Ω:

Theorem 5. *Let $\boldsymbol{f}$, p, $\boldsymbol{v}$, and $\boldsymbol{u}$ be respectively in* $[L^1(\Omega)]^3, L^2(\Omega)$, $[L^\infty(\Omega)]^3$, *and* $[W^{1,2}(\Omega)]^3$. *Also, let $\boldsymbol{w}=\boldsymbol{v}+u$ and let $Z \subset \Omega$ be a closed set of Newtonian capacity zero. Suppose that the pair $(\mathbf{w}, p)$ is a distribution solution of (3.3) in $\Omega \setminus Z$. Then the pair $(\mathbf{w}, p)$ is a distribution solution of (3.3) in Ω.*

This theorem is established in [Sh24].

Also, this last reference, in dimension $N = 3$, has a statement classical solutions of (3.1) and capacity zero.

Let $Z \subset \Omega$ be a closed set. Call Z an (L^∞, L^2)-removable set for the stationary Navier-Stokes equations (3.1) if the following obtains:

Let $\mathbf{v}$ and p be in $[L^\infty(\Omega)]^3$ and $L^2(\Omega)$ respectively. Suppose that the pair $(\mathbf{v}, p)$ is a classical solution of (3.1) in $\Omega \backslash Z$. Then $\mathbf{v}$ and p can be defined at the points of Z so that the pair $(\mathbf{v}, p)$ is a classical solution of (3.1) in Ω.

Theorem 6. *Let $N = 3$, and let $Z \subset \Omega$ be a closed set. Then a necessary and sufficient condition that Z be an (L^∞, L^2)-removable set for the staionary Navier-Stokes equations (3.1) is that Z be of Newtonian capacity zero.*

For the proof of the necessary condition in Theorem 6, some ideas from [Sh25] are needed.

For a paper treating more general type capacities and the stationary Navier-Stokes equations, see [ShWe].

3. Theorem 2.1, involving classical solutions of the stationary Navier-Stokes equations that, we prove in dimensions $N = 2, 3$, actually is true in dimension $N = 4$. The first part of the proof, however, that we have provided for $N = 2, 3$ will not work for $N = 4$.

We do know in all three cases that the pair $(\mathbf{v}, p)$ is a distribution solution of the equations in (2.2) with $\mathbf{v} \in [W^{1,2}(T_N)]^N$ and $p \in L^2(T_N)$. In dimensions $N = 2, 3$, we then obtain from the Sobolev inequalities that $v_k \partial v_j / \partial x_k \in L^{3/2}(T_N)$ for $j, k = 1, ..., N$ and consequently from (2.27) and (2.30) that $\mathbf{v} \in [W^{2,3/2}(T_N)]^N$. But this procedure will not work in dimension $N = 4$.

The best we can get in dimension $N = 4$ using the Sobolev inequalities is that $v_k \partial v_j / \partial x_k \in L^{4/3}(T_N)$ and consequently from (2.27) and (2.30) that $\mathbf{v} \in [W^{2,4/3}(T_4)]^4$. But this is not good enough to carry out the rest of the boot-strap argument. However, if we adopt the ideas in the clever paper of Gerhardt [Ge], it can be shown directly that $\mathbf{v} \in [W^{2,2}(T_4)]^4$, and this is sufficient to carry out the rest of the boot-strap argument in the proof of Theorem 2.1.

The main lemma in [Ge] adapted for the situation in hand is the following:

Lemma 7. *Let $\varepsilon > 0$ and let $g \in L^2(T_4)$. Then there exists a positive constant C_ε depending only on ε and g such that*

$$\int_{T_4} |g| \, |u|^2 \, dx \leq \varepsilon \|u\|^2_{W^{1,2}(T_4)} + C_\varepsilon \|u\|^2_{L^2(T_4)} \quad \forall u \in W^{1,2}(T_4).$$

Using an extension of this lemma and the ideas developed in Gerhardt's paper, we can obtain with no difficulty the following result:

Theorem 8. *Let the pair $(\boldsymbol{v}, p)$ be a distribution solution of the stationary Navier-Stokes equations (2.2) where $N=4$, $\mathbf{f} \in [C^\alpha (T_4)]^4$, $\boldsymbol{v} \in [W^{1,2} (T_4)]^4$, and $p \in L^2 (T_4)$. Then $\boldsymbol{v} \in [W^{2,2} (T_4)]^4$ and $p \in W^{1,2} (T_4)$.*

$\boldsymbol{v} \in [W^{2,2} (T_4)]^4$ is sufficient to get the boot-strap argument to work. So indeed Theorem 2.1 is true for $N = 4$.

Integrals and Identities

1. Integral Identities

In this section, we establish various integral identities concerning Bessel functions that we need. The Bessel function of the first kind of order ν is defined as follows:

$$(1.1) \qquad J_\nu(t) = \sum_{n=0}^{\infty} \frac{(-1)^n (\frac{t}{2})^{\nu+2n}}{n!\Gamma(\nu+n+1)} \quad \text{for } t > 0 \quad \text{where } \nu > -1.$$

The first integral identity that we need is the following:

$$(1.2) \quad J_\nu(t) = \frac{t^\nu}{2^{\nu-1}\Gamma(\nu+\frac{1}{2})\Gamma(\frac{1}{2})} \int_0^{\pi/2} \cos(t\cos\theta)(\sin\theta)^{2\nu} d\theta \quad \text{for } t > 0$$

where $\nu > -\frac{1}{2}$. To establish this integral identity, we use two well-known identities for the Gamma function, namely,

$$(1.3) \qquad (i) \ \frac{\Gamma(\alpha)\Gamma(\beta)}{\Gamma(\alpha+\beta)} = 2\int_0^{\pi/2}(\cos\theta)^{2\alpha-1}(\sin\theta)^{2\beta-1}d\theta,$$

$$(ii) \ \Gamma(2t)\Gamma(\tfrac{1}{2}) = 2^{2t-1}\Gamma(t)\Gamma(t+\tfrac{1}{2}),$$

which can be found in [Ti1, pp. 56-57] and many other places. The second of the above formulas is referred to as "the duplication formula for the Gamma function."

To show that (1.2) holds, we observe from (1.3)(i) that

$$\frac{\Gamma(\nu+\frac{1}{2})\Gamma(n+\frac{1}{2})}{2\Gamma(n+\nu+1)} = \int_0^{\pi/2}(\cos\theta)^{2n}(\sin\theta)^{2\nu}d\theta,$$

and hence from (1.1) and (1.3)(ii) that

$$\begin{aligned}
J_\nu(t) &= \frac{t^\nu}{2^{\nu-1}\Gamma(\nu+\frac{1}{2})} \sum_{n=0}^{\infty} \int_0^{\pi/2} \frac{(-1)^n(\frac{t}{2})^{2n}}{\Gamma(n+1)\Gamma(n+\frac{1}{2})}(\cos\theta)^{2n}(\sin\theta)^{2\nu}d\theta \\
&= \frac{t^\nu}{2^{\nu-1}\Gamma(\nu+\frac{1}{2})} \sum_{n=0}^{\infty} \int_0^{\pi/2} \frac{(-1)^n(t)^{2n}}{\Gamma(2n+1)\Gamma(\frac{1}{2})}(\cos\theta)^{2n}(\sin\theta)^{2\nu}d\theta.
\end{aligned}$$

Using the series expansion for $\cos(t\cos\theta)$ in this last sum on the right validates the integral identity (1.2).

The second integral identity that we need is the following:
(1.4)

$$J_{\mu+\nu+1}(t) = \frac{t^{\nu+1}}{2^\nu \Gamma(\nu+1)} \int_0^{\pi/2} J_\mu(t\sin\theta)(\sin\theta)^{\mu+1}(\cos\theta)^{2\nu+1}d\theta \quad \text{for } t > 0,$$

where $\mu > -1$ and $\nu > -1$.

To show that (1.4) holds, we refer to the right-hand side of the equality in (1.4) as $I(t)$ and see from (1.1) and (1.3)(i) that

$$
\begin{aligned}
I(t) &= \frac{t^{\nu+1}}{2^\nu \Gamma(\nu+1)} \sum_{n=0}^\infty \int_0^{\pi/2} \frac{(-1)^n(\frac{t\sin\theta}{2})^{\mu+2n}}{n!\Gamma(\mu+n+1)}(\sin\theta)^{\mu+1}(\cos\theta)^{2\nu+1}d\theta \\
&= \frac{t^{\mu+\nu+1}}{2^{\mu+\nu}\Gamma(\nu+1)} \sum_{n=0}^\infty \frac{(-1)^n(\frac{t}{2})^{2n}}{n!\Gamma(\mu+n+1)} \int_0^{\pi/2} (\sin\theta)^{2(\mu+n)+1}(\cos\theta)^{2\nu+1}d\theta \\
&= \frac{t^{\mu+\nu+1}}{2^{\mu+\nu+1}} \sum_{n=0}^\infty \frac{(-1)^n(\frac{t}{2})^{2n}}{n!\Gamma(\mu+\nu+n+2)} \\
&= \sum_{n=0}^\infty \frac{(-1)^n(\frac{t}{2})^{2n+\mu+\nu+1}}{n!\Gamma(\mu+\nu+n+2)}.
\end{aligned}
$$

The last sum is $J_{\mu+\nu+1}(t)$, and the identity (1.4) is established.

We will use both of these integral identities in a slightly different form than presented. In particular, the one in (1.2) will be used in the form

$$(1.5) \qquad J_\nu(t) = \frac{t^\nu}{2^\nu \Gamma(\nu+\frac{1}{2})\Gamma(\frac{1}{2})} \int_0^\pi e^{it\cos\theta}(\sin\theta)^{2\nu}d\theta \quad \text{for } t > 0,$$

which follows from (1.2) because of the respective evenness and oddness of $\cos(t\cos\theta)\,(\sin\theta)^{2\nu}$ and $\sin(t\cos\theta)\,(\sin\theta)^{2\nu}$ around $\pi/2$.

Likewise, the one in (1.4) will be used in the form

$$(1.6) \qquad J_{\mu+\nu+1}(t) = \frac{t^{\nu-\mu-1}}{2^\nu \Gamma(\nu+1)} \int_0^t J_\mu(s)s^{\mu+1}(1-\frac{s^2}{t^2})^\nu ds \quad \text{for } t > 0,$$

which follows from (1.4) when the variable θ in the integral is replaced by $s = t\,\sin\theta$ with $ds = t\,\cos\theta\,d\theta$.

Next, we establish the following integral identity involving Bessel functions:

$$(1.7) \qquad \int_0^\infty e^{-ar} J_\nu(br)r^{\nu+1}dr = \frac{2a(2b)^\nu \Gamma(\nu+3/2)}{(a^2+b^2)^{\nu+3/2}\Gamma(1/2)} \quad \text{for } \nu > -1,$$

where a and b are positive real numbers.

To do this, we start out by observing that for $n \geq 1$,

$$\Gamma(\nu+n+\frac{3}{2}) = (\nu+n-1+\frac{3}{2})\cdots(\nu+\frac{3}{2})\Gamma(\nu+\frac{3}{2}),$$

and hence, from the familiar formula for the binomial series [Ru1, p. 201] that

$$(1.8) \qquad \sum_{n=0}^{\infty} (-1)^n \Gamma(\nu + n + \frac{3}{2}) \frac{t^n}{n!} = \Gamma(\nu + \frac{3}{2})(1+t)^{-(\nu + \frac{3}{2})}$$

for $0 < t < 1$. Next, using the definition of the Gamma function, we observe that

$$\int_0^{\infty} e^{-ar}(br)^{\nu + 2n} r^{\nu+1} dr = \frac{1}{b^{\nu+2}} (\frac{b}{a})^{2(\nu+n+1)} \Gamma(2\nu + 2n + 2).$$

Applying this equality in conjunction with the fact obtained from (1.3)(ii) that

$$\Gamma(2\nu + 2n + 2) = 2^{2\nu + 2n + 1} \Gamma(\nu + n + 1) \Gamma(\nu + n + \frac{3}{2}) / \Gamma(\frac{1}{2})$$

and the definition of $J_\nu(t)$ in (1.1) gives us that

$$\int_0^{\infty} e^{-ar} J_\nu(br) r^{\nu+1} dr = \sum_{n=0}^{\infty} \frac{(-1)^n}{\Gamma(\frac{1}{2})n!} \Gamma(\nu + n + \frac{3}{2}) \frac{2^{\nu+1}}{b^{\nu+2}} (\frac{b}{a})^{2(\nu+n+1)}.$$

It is easy to see using (1.8) that the right-hand side of this last equality is the same as the right-hand side of the equality in (1.7), and the integral identity in (1.7) is established for $0 < b < a$. By analytic continuation, it extends to the other positive values of b.

We shall also need the following integral identity, which can be obtained from the one in (1.7), namely

$$(1.9) \qquad \int_0^{\infty} e^{-ar} J_{\nu+1}(r) r^{\nu+1} dr = \frac{2^{\nu+1} \Gamma(\nu + 3/2)}{(a^2+1)^{\nu+3/2} \Gamma(1/2)}$$

for $\nu > -1$ and $0 < a < 1$.

To establish this identity, we set b=1 in (1.7) and integrate by parts using the familiar fact that $\int J_\nu(r) r^{\nu+1} dr = J_{\nu+1}(r) r^{\nu+1}$. The integral identity in (1.9) easily follows from this.

Next, define

$$(1.10) \qquad H_R^\alpha(x) = (2\pi)^{-N} \int_{B(0,R)} e^{iy \cdot x} (1 - |y|^2 / R^2)^\alpha \, dy$$

for $\alpha > (N-1)/2$. We shall need the following identity for $H_R^\alpha(x)$:

$$(1.11) \qquad H_R^\alpha(x) = c(N, \alpha) J_{\frac{N}{2} + \alpha}(R|x|) R^{\frac{N}{2} - \alpha} / |x|^{\frac{N}{2} + \alpha},$$

where $c(N, \alpha) = (2\pi)^{-N} \omega_{N-2} 2^\alpha \Gamma(\alpha + 1) = 2^\alpha \Gamma(\alpha + 1)/(2\pi)^{N/2}$ and ω_{N-2} is defined below.

To establish this identity, we see from the spherical coordinate notation introduced in Chapter 1 that

$$
\begin{aligned}
(2\pi)^N H_R^\alpha(x) &= \int_{B(0,R)} e^{iy\cdot x}(1 - |y|^2/R^2)^\alpha \, dy \\
&= |S_{N-2}| \int_0^R (1 - \frac{r^2}{R^2})^\alpha r^{N-1} \int_0^\pi e^{i|x|r\cos\theta}(\sin\theta)^{N-2} d\theta \\
&= \omega_{N-2} \int_0^R (1 - \frac{r^2}{R^2})^\alpha r^{N-1} \frac{J_{(N-2)/2}(|x|\,r)}{(|x|\,r)^{(N-2)/2}} dr \\
&= \frac{\omega_{N-2}}{|x|^N} \int_0^{R|x|} (1 - \frac{s^2}{(R\,|x|)^2})^\alpha s^{N/2} J_{(N-2)/2}(s) ds,
\end{aligned}
$$

where we have made use of (1.5) and where

$$
\omega_{N-2} = |S_{N-2}|\, 2^{(N-2)/2}\Gamma((N-1)/2)\Gamma(\frac{1}{2}) = (2\pi)^{N/2},
$$

and $|S_{N-2}| = 2(\pi)^{(N-1)/2}/\Gamma\left(\frac{N-1}{2}\right)$ is defined two lines above (3.4) in Chapter 1.

We have tacitly assumed $N \geq 2$ in this last computation. For $N = 1$, we can easily see directly that the equality just established for $H_R^\alpha(x)$ is still valid with $\omega_{-1} = (2\pi)^{1/2}$ because, as is well-known,

$$
\cos t = (\pi/2)^{1/2}t^{1/2}J_{-\frac{1}{2}}(t) \quad \text{for } t > 0.
$$

Consequently, we see from this last computation and (1.6) with $\mu + 1 = N/2$ and $\nu = \alpha$ that

$$
H_R^\alpha(x) = c(N,\alpha)J_{\frac{N}{2}+\alpha}(R\,|x|)R^{\frac{N}{2}-\alpha}/\,|x|^{\frac{N}{2}+\alpha},
$$

where $c(N,\alpha) = (2\pi)^{-N}\,\omega_{N-2}2^\alpha\Gamma(\alpha + 1) = 2^\alpha\Gamma(\alpha + 1)/(2\pi)^{N/2}$, and the identity in (1.11) is established.

We also need the following integral identity:

$$
(1.12) \qquad \int_0^\infty e^{-s^2}\cos 2ts \, ds = \frac{\pi^{\frac{1}{2}}}{2}e^{-t^2}
$$

for $t \in \mathbf{R}$.

To establish this identity, we start by observing that

$$
\int_0^\infty e^{-s^2} ds = \frac{\pi^{\frac{1}{2}}}{2}
$$

and then set

$$
\int_0^\infty e^{-s^2}\cos 2ts \, ds = I(t).
$$

An integration by parts shows that

$$
\frac{dI(t)}{dt} = -\int_0^\infty e^{-s^2} 2s \sin 2ts \, ds = -2t \int_0^\infty e^{-s^2}\cos 2ts \, ds.
$$

Therefore, we see that $\frac{dI(t)}{dt} = -2tI(t)$ for $t \in \mathbf{R}$ and obtain from the uniqueness theory of ODE that

$$\int_0^\infty e^{-s^2} \cos 2ts \; ds = \frac{\pi^{\frac{1}{2}}}{2} e^{-t^2},$$

and (1.12) is established.

2. Estimates for Bessel Functions

We will need estimates for the way Bessel functions act in a neighborhood of the origin and in the neighborhood of infinity. In particular, we will establish the following two estimates: $\exists K_\nu > 0$ such that

$$(2.1) \qquad |J_\nu(t)| \le K_\nu t^\nu \text{ for } 0 < t \le 1 \text{ and } \nu > -\frac{1}{2};$$

$$(2.2) \qquad |J_\nu(t)| \le K_\nu t^{-\frac{1}{2}} \text{ for } 1 \le t < \infty \text{ and } \nu > -1.$$

The estimate in (2.1) follows immediately from the integral identity in (1.2). To establish the estimate in (2.2), we observe, after an easy calculation using (1.1) that for $\nu > -1$, $J_\nu(t)$ satisfies the differential equation

$$t^2 \frac{d^2y}{dt^2} + t\frac{dy}{dt} + (t^2 - \nu^2)y = 0 \text{ for } t > 0.$$

As a consequence of this last fact, it is easy to see that

$$(2.3) \qquad t^2 \frac{d^2[t^{\frac{1}{2}} J_\nu(t)]}{dt^2} - (\nu^2 - \frac{1}{4})[t^{\frac{1}{2}} J_\nu(t)] + t^2[t^{\frac{1}{2}} J_\nu(t)] = 0 \text{ for } t > 0$$

and $\nu > -1$. But then in [CH, pp. 331-2], it is shown that a function $y(t)$, which satisfies an equation of the form

$$\frac{d^2y}{dt^2} + y + \rho(t)y = 0 \quad \text{for } t > 0,$$

is uniformly bounded on the interval $[1, \infty)$ provided $\rho \in C^1((0, \infty))$ with $|\rho(t)| < \beta t^{-1}$ and $|d\rho(t)/dt| < \beta t^{-2}$ on $[1, \infty)$ where β is a positive constant. The equation in (2.3) is such an equation with $\rho(t) = -(\nu^2 - \frac{1}{4})/t^2$. We conclude that $t^{\frac{1}{2}} J_\nu(t)$ is uniformly bounded for $1 \le t < \infty$, and the estimate in (2.2) is established.

Next, with $\nu = (N - 2)/2$, we set

$$(2.4) \qquad A_n^\nu(t) = \frac{\Gamma(n/2)}{2^{N/2}\Gamma[(N+n)/2]} \int_0^\infty e^{-st} J_{\nu+n}(s)s^{\nu+1}ds$$

and obtain the following estimate for $A_n^\nu(1/r)$, which we state as a theorem:

Theorem 2.1.   *Let $N \geq 2$ and let n be any positive integer. Also, set $\nu = (N-2)/2$. Then there exists a constant $C(N,n)$ depending on N and n such that*

$$|A_n^\nu(1/r) - 1| \leq C(N,n)\, r^{-\frac{1}{2}}   \text{ for } 2 \leq r < \infty.$$

Proof.   To prove the theorem, we first take $n = 1$ and observe from (1.9) that

$$A_1^\nu(t) - 1 = \frac{1}{(1 + t^2)^{\nu + 3/2}} - 1.$$

Hence, using (1.8), we see that there is a constant c depending on ν such that

$$|A_1^\nu(t) - 1| \leq ct \text{ for } 0 < t \leq 1/2.$$

So, the theorem follows in this case.

Next, we take $n \geq 2$, invoke the hypergeometric representation for $A_n^\nu(t)$ given in [Wa, p. 385] and then use both the integral representation for hypergeometric functions (see [Ra, p. 47] or [AAR, p. 65]) and the integral definition of the beta function (see [Ra, p.18]) to obtain that

(2.5) $\Gamma(1/2)\Gamma[(n-1)/2][A_n^\nu(1/r) - 1]/\Gamma(n/2)$

is equal to the following integral

(2.6) $\displaystyle\int_0^1 t^{\nu + (n+1)/2}(1-t)^{\frac{n-3}{2}}[(t + r^{-2})^{-(\nu + \frac{n}{2} + 1)} - t^{-(\nu + \frac{n}{2} + 1)}]dt$

for $r > 0$. The absolute value of this latter integral in turn is majorized by

$$c_1 \int_0^{1/r} t^{-1/2}dt + c_2 r^{-2} \int_{1/r}^{1/2} t^{-3/2}dt + c_3 r^{-2} \int_{1/2}^1 (1-t)^{\frac{n-3}{2}}dt$$

for $r \geq 2$ where c_1, c_2, and c_3 are constants depending on ν and n. So the theorem follows in this case also.   ∎

Besides the estimate for $A_n^\nu(t)$ in the theorem we just established, we shall need several more. In particular, with $\nu = (N-2)/2$ and $N \geq 2$, we will need the following one:

(2.7) $\exists\, c(\nu, n) > 0$ such that $|A_n^\nu(t)| \leq c(\nu, n)t^{-N}$ for $t > 0$.

To establish the estimate in (2.7), we observe from (2.1) and (2.2) that for every positive integer n, $J_n(t)$ is a bounded function on the positive real axis. In case $\nu = 0$, the estimate in (2.7) then follows immediately from the integral representation of $A_n^\nu(t)$ in (2.4). In case $\nu \geq 1/2$, we use the estimates in (2.1) and (2.2) to obtain $|J_{\nu+n}(t)|$, which is majorized by a constant multiple of t^ν for $t > 0$. Once again the estimate in (2.7) then follows immediately from the integral representation of $A_n^\nu(t)$ in (2.4).

Next, we need the following estimate for $A_n^\nu(t)$:

(2.8) $0 \le A_n^\nu(t) \le 1$ for $t > 0$.

In case $n = 1$, as we observed in the proof of Theorem 2.1, it follows from (1.9) above that

$$A_1^\nu(t) = \frac{1}{(1 + t^2)^{\nu+3/2}}$$

and the estimate in (2.8) follows in this case.

In case $n \ge 2$, we use the equality between the expressions in (2.5) and (2.6). The first inequality in (2.8) then follows from the fact that for the beta function

$$B(\frac{1}{2}, \frac{n-1}{2}) = \Gamma(\frac{1}{2})\Gamma(\frac{n-1}{2})/\Gamma(\frac{n}{2}).$$

The second inequality in (2.8) follows from the observation that

$$A_n^\nu(t) - 1 < 0.$$

Next, because of the inequalities in (2.8) joined with the conclusion of Theorem 2.1, we can make the following observation about $A_n^\nu(t)$:

(2.9) $\exists\, C^*(N, n) > 0$ such that $|A_n^\nu(t) - 1| \le C^*(N, n)t^{\frac{1}{2}}$ for $0 < t \le 1$,

for $n \ge 1$, $\nu = (N - 2)/2$, and $N \ge 2$.

Finally, we need the fact that $A_n^\nu(t)' = dA_n^\nu(t)/dt$ is uniformly bounded on the positive real-axis, i.e.,

(2.10) $\sup_{0<t<\infty} \left| A_n^\nu(t)' \right| < \infty,$

for $n \ge 1$, $\nu = (N - 2)/2$, and $N \ge 2$.

To establish the strict inequality in (2.10), we observe from (2.4) that

(2.11) $A_n^\nu(t)' = \dfrac{-\Gamma(n/2)}{2^{N/2}\Gamma[(N+n)/2]} \displaystyle\int_0^\infty e^{-st} J_{\nu+n}(s)s^{\nu+2}ds$

for $t > 0$. So for the case $n = 1$, we see from (1.7) above that $A_n^\nu(t)'$ is a constant multiple of $t/(t^2 + 1)^{\nu+5/2}$. Hence, the strict inequality in (2.10) holds in this case.

For $n \ge 2$, we see that the integral in (2.11) can be written as

$$\int_0^\infty e^{-st} J_{\nu+1+(n-1)}(s)s^{\nu+1+1}ds.$$

Consequently, from (2.4), we see that $A_n^\nu(t)'$ is a constant multiple of $A_{n-1}^{\nu+1}(t)$. So the strict inequality in (2.10) for $n \ge 2$ follows from the estimate in (2.8), and the strict inequality in (2.10) for all n is completely established.

Exercises.
   1. Given

$$|A_n^\nu(1/r) - 1|$$

$$\leq -\alpha_n \int_0^1 t^{\nu+(n+1)/2}(1-t)^{\frac{n-3}{2}}[(t+r^{-2})^{-(\nu+\frac{n}{2}+1)} - t^{-(\nu+\frac{n}{2}+1)}]dt$$

where $n \geq 2, \nu = \frac{N-2}{2}$, and α_n is a positive constant depending only on n, prove that for $r \geq 2$

$$|A_n^\nu(1/r) - 1|$$

is majorized by

$$c_1 \int_0^{1/r} t^{-1/2}dt + c_2 r^{-2} \int_{1/r}^{1/2} t^{-3/2}dt + c_3 r^{-2} \int_{1/2}^1 (1-t)^{\frac{n-3}{2}} dt$$

where c_1, c_2, and c_3 are constants depending on ν and n.

3. Surface Spherical Harmonics

In this section, we shall be interested in exploiting the connection between surface spherical harmonics and Bessel functions in order to compute the principal-valued Fourier coefficient of the periodic Calderon-Zygmund kernel of spherical harmonic type, $K^*(x)$, defined in (1.8) of Chapter 2.

With n a nonnegative integer, $Q_n(x)$ is called a spherical harmonic function of degree n if the following two facts apply:

(i) $Q_n(x)$ is a homogeneous real polynomial of degree n;
(ii) $Q_n(x)$ is a harmonic function, i.e., $\Delta Q_n(x) = 0 \ \forall \ x \in \mathbf{R}^N$, $N \geq 2$.

With $\xi = x/|x|$, $Q_n(\xi)$ is called a surface spherical harmonic of degree n. Sometimes we will write $Y_n(\xi)$ in place of $Q_n(\xi)$, so that

$$(3.1) \qquad Y_n(\xi) = Q_n(\frac{x}{|x|}) = r^{-n}Q_n(x) \quad \text{where } x = r\xi \text{ and } r = |x|.$$

A good example of a surface spherical harmonic of degree n on S_2, the unit sphere in $\mathbf{R}^3$, is $P_n(\eta^* \cdot \xi)$ where $\eta^* = (1,0,0)$, $\xi = (\xi_1, \xi_2, \xi_3)$ is any point on S_2, and $P_n(t)$ is the Legendre polynomial of degree n.

To see this using the spherical coordinate notation introduced in §3 of Chapter 1 with θ replacing θ_1, we observe that $\eta^* \cdot \xi = \cos\theta$ and our spherical harmonic function then becomes $Q_n(x) = r^n P_n(\cos\theta)$. Since $P_n(t)$ is an even or odd polynomial corresponding to whether n is an even or odd number, $Q_n(x)$ is clearly a homogeneous polynomial of degree n in x_1, x_2, x_3. To verify that $Q_n(x)$ also satisfies Laplace's equation, we need the familiar differential equation that $P_n(t)$ satisfies on the interval $(-1, 1)$, namely,

$$(3.2) \qquad (1-t^2)P_n''(t) - 2tP_n'(t) = -n(n+1)P_n(t) \quad \forall t \in (-1,1),$$

where $P_n'(t) = dP_n(t)/dt$ and $P_n''(t) = d^2P_n(t)/dt^2$. An easy calculation using (3.2) shows that $\Delta r^n P_n(\cos\theta) = 0$ for $0 < r < 1$ and $-\pi < \theta < \pi$, where in spherical coordinates, because of dependence only on r and θ, Δ has the form

$$(3.3) \qquad \Delta = \frac{1}{r^2}\frac{\partial}{\partial r}r^2\frac{\partial}{\partial r} + \frac{1}{r^2\sin\theta}\frac{\partial}{\partial \theta}\sin\theta\frac{\partial}{\partial \theta} \quad \text{for } N = 3.$$

Hence, $Q_n(x)$ is a harmonic function, and therefore $Q_n(\xi) = P_n(\eta^* \cdot \xi)$ is indeed a surface spherical harmonic of degree n.

The analogue of $P_n(\eta^* \cdot \xi)$ in dimension $N \geq 4$ is obtained via the Gegenbauer polynomials, $C_n^\nu(t)$, where $\nu = (N-2)/2$. The n-th Gegenbauer polynomial, $C_n^\nu(t)$, with $\nu \neq 0$, is defined by means of the equation (see [Ra, pp. 276-279] or [AAR, p. 302])

$$(1 - 2rt + r^2)^{-\nu} = \sum_{n=0}^{\infty} C_n^\nu(t)r^n, \qquad \text{for } 0 < r < 1 \text{ and } -1 \leq t \leq 1,$$

which is called the generating relation for $C_n^\nu(t)$. It is clear from the well-known generating relation for the Legendre polynomials [CH, p. 85] that $\nu = \frac{1}{2}$ in the above gives rise to the Legendre polynomials, i.e., $P_n(t) = C_n^{\frac{1}{2}}(t)$. Gegenbauer polynomials are also called ultraspherical polynomials.

It is shown in [Ra, p. 279] and [AAR, p.247] that the differential equation that $C_n^\nu(t)$ satisfies is

$$(3.4) \quad (1 - t^2)C_n^{\nu\prime\prime}(t) - 2t(\nu + \frac{1}{2})C_n^{\nu\prime}(t) = -n(n + 2\nu)C_n^\nu(t) \quad \forall t \in (-1,1),$$

where $C_n^{\nu\prime}(t) = dC_n^\nu(t)/dt$, which is the same as the equation in (3.2) when $\nu = \frac{1}{2}$.

For dimension $N \geq 4$, using the spherical coordinate notation introduced in §3 of Chapter 1 with θ replacing θ_1, we will now show that $C_n^\nu(\eta^* \cdot \xi)$ is a surface spherical harmonic of degree n on S_{N-1}, where $\eta^* = (1,0,...,0)$ and ξ is any point on S_{N-1}.

First, we observe as before that $\eta^* \cdot \xi = \cos\theta$. Since $C_n^\nu(\eta^* \cdot \xi)$ is a polynomial in $\cos\theta$ and since $C_n^\nu(t)$ is an even or odd polynomial in t corresponding as to whether n is an even or odd number [Ra, p. 277], it is clear that $Q_n(x) = r^n C_n^\nu(\cos\theta)$ is a homogeneous polynomial of degree n in $x_1, ..., x_N$.

Next, we see from [EMOT, p. 235], in spherical coordinate notation when Δ depends only on r and θ, Δ has the form

$$(3.5) \qquad \Delta = \frac{1}{r^{N-1}}\frac{\partial}{\partial r}r^{N-1}\frac{\partial}{\partial r} + \frac{1}{r^2(\sin\theta)^{N-2}}\frac{\partial}{\partial \theta}(\sin\theta)^{N-2}\frac{\partial}{\partial \theta}.$$

Also, we see that the differential equation in (3.4), after multiplying both sides by $(1 - t^2)^{\nu - \frac{1}{2}}$, is

$$\frac{d}{dt}[(1 - t^2)^{\nu + \frac{1}{2}}\frac{dC_n^\nu(t)}{dt}] = -n(n + 2\nu)(1 - t^2)^{\nu - \frac{1}{2}}C_n^\nu(t) \quad \forall t \in (-1,1).$$

Hence, with $t = \cos\theta$ and $2\nu = N - 2$, the equation in (3.4) becomes

$$(3.6) \qquad \frac{d}{d\theta}(\sin\theta)^{N-2}\frac{dC_n^\nu(\cos\theta)}{d\theta} = -n(n + N - 2)(\sin\theta)^{N-2}C_n^\nu(\cos\theta)$$

for $\theta \in (-\pi, \pi)$.

Using the form of Δ in (3.5) in conjunction with the differential equation in (3.6), we obtain after an easy computation that

$$\Delta r^n C_n^\nu(\cos\theta) = 0 \quad \text{for } \theta \in (-\pi, \pi).$$

Hence, $Q_n(x)$ is a harmonic function in $\mathbf{R}^N$, and therefore $C_n^\nu(\eta^* \cdot \xi)$ is a surface spherical harmonic of degree n.

Of course, what we have just shown is that for every fixed η on S_{N-1}, $C_n^\nu(\eta \cdot \xi)$ is, as a function of ξ, a surface spherical harmonic of degree n.

Before proceeding further, we discuss the notion of linear independence for surface spherical functions. Let

$$\{Y_{j,n}(\xi)\}_{j=1}^k$$

be a set of surface spherical harmonic functions of degree n. We say the set of functions is linearly independent
if $a_j \in \mathbf{R}$ for $j = 1, ..., k$ and

$$\sum_{j=1}^k a_j Y_{j,n}(\xi) = 0 \ \ \forall \xi \in S_{N-1} \Rightarrow a_j = 0 \text{ for } j = 1, ..., k.$$

We let $\mu_{n,N}$ designate the largest number of linear independent surface spherical harmonic functions of degree n and observe from [AAR, p. 450], [EMOT, p.237], and [ABR, p.78] that

$$(3.7) \qquad \mu_{n,N} = (2n + N - 2)\frac{(n + N - 3)!}{(N - 2)!n!}.$$

Next, we let $\{Y_{j,n}(\xi)\}_{j=1}^{\mu_{n,N}}$ be an orthonormal set of surface spherical harmonics of degree n, i.e.,

$$(3.8) \qquad \int_{S_{N-1}} Y_{j,n}(\xi)Y_{k,n}(\xi)dS(\xi) = \delta_{j,k} \quad \text{for } j, k = 1, ..., \mu_{n,N},$$

where $\delta_{j,k}$ is the Kronecker-δ. Then the well-known addition formula for surface spherical harmonics [EMOT, p. 243] says

$$(3.9) \qquad \frac{C_n^\nu(\xi \cdot \eta)}{C_n^\nu(1)} = \frac{|S_{N-1}|}{\mu_{n,N}}\sum_{j=1}^{\mu_{n,N}} Y_{j,n}(\xi)Y_{j,n}(\eta)$$

where $\nu = (N - 2)/2$ and $|S_{N-1}|$ designates the $(N - 1)$-dimensional volume of S_{N-1}, computed two lines above (3.4) in Chapter 1 where it is shown that

$$(3.10) \qquad |S_{N-1}| = \frac{2(\pi)^{N/2}}{\Gamma(\frac{N}{2})}.$$

Two other places to find the derivation of the addition formula given in (3.8) are [Se, p. 118] and [AAR, p. 456]. We will establish a special case of the addition formula (3.10) in Theorem 3.4 below near the end of this section. In particular, we will establish (3.9) in the special case when $N = 3$ and $\nu = 1/2$.

Because of (3.7) and (3.8), we see that if $Y_n(\xi)$ is a surface spherical harmonic of degree n, then there exists $\{a_k\}_{k=1}^{\mu_{n,N}}$ such that

$$(3.11) \qquad Y_n(\eta) = \sum_{k=1}^{\mu_{n,N}} a_k Y_{k,n}(\eta) \ \text{ for } \eta \in S_{N-1}.$$

Hence, it follows from (3.8), (3.9), and (3.11) that

$$\int_{S_{N-1}} C_n^\nu(\xi \cdot \eta) Y_n(\eta) dS(\eta) \ = \ \frac{C_n^\nu(1) |S_{N-1}|}{\mu_{n,N}} \sum_{j=1}^{\mu_{n,N}} a_j Y_{j,n}(\xi)$$

$$= \ \frac{C_n^\nu(1) |S_{N-1}|}{\mu_{n,N}} Y_n(\xi)$$

for $\xi \in S_{N-1}$. It also follows from the generating function for Gegenbauer polynomials given above that $C_n^\nu(1) = \frac{(N+n-3)!}{n!(N-3)!}$. (See [EMOT, p. 236] or [Ra, p.278]). So we conclude from (3.7), (3.10), and this last computation that

$$(3.12) \qquad Y_n(\xi) = \frac{\Gamma(\nu)(n + \nu)}{2\pi^{\nu+1}} \int_{S_{N-1}} C_n^\nu(\xi \cdot \eta) Y_n(\eta) dS(\eta)$$

for $\xi \in S_{N-1}$ where $\nu = \frac{N-2}{2}$.

The formula that we have just established in (3.12) is very useful in computing the principal-valued Fourier transform $\widehat{K}(y)$ of the Calderon-Zygmund kernel $K(x)$ of spherical harmonic type defined by (1.3)-(1.5) of Chapter 2 where
$$(3.13)$$
$$\widehat{K}(y) = \lim_{R \to \infty} \lim_{\varepsilon \to 0} (2\pi)^{-N} \int_{B(0,R)-B(0,\varepsilon)} e^{-iy \cdot x} K(x) dx \quad \text{ for } y \in \mathbf{R}^N.$$

In order to accomplish this computation, we need the orthogonality between surface spherical harmonics of different degrees, namely if $Y_j(\xi)$ and $Y_k(\xi)$ are surface spherical harmonics of degrees j and k, respectively, then

$$(3.14) \qquad \int_{S_{N-1}} Y_j(\xi) Y_k(\xi) dS(\xi) = 0 \quad \text{ for } j \neq k.$$

To establish (3.14), first we recall Green's second identity [Ru1, p. 297], which states that for $u, v \ \epsilon C^2[\overline{B}(0,1)]$,

$$(3.15) \qquad \int_{\partial B(0,1)} [u D_\mathbf{n} v - v D_\mathbf{n} u] dS = \int_{B(0,1)} [u \Delta v - v \Delta u] dx,$$

where $D_{\mathbf{n}}u = \nabla u \cdot \mathbf{n}$ and $\mathbf{n}$ is the outward pointing unit normal for $\partial B(0,1) = S_{N-1}$. Next, we use the expressions in (3.1) to obtain that

$$(3.16) \qquad\qquad r^j Y_j(\xi) = Q_j(x)$$

where $Q_j(x)$ is a spherical harmonic function. On S_{N-1}, we see from (3.16) that for $x = r\xi$,

$$D_{\mathbf{n}}Q_j(x) = jr^{j-1}Y_j(\xi) = jY_j(\xi) \quad \text{because } r = 1.$$

Consequently, on setting $u = Q_j$ and $v = Q_k$, we obtain from this last established fact, (3.15), and (3.16) that

$$(k-j) \int_{S_{N-1}} Y_j(\xi) Y_k(\xi) dS(\xi) = 0.$$

Hence, (3.14) is indeed valid because for $j \neq k$, $k - j \neq 0$.

We now return to the computation of $\widehat{K}(y)$ given in (3.13). It is clear from (1.4) of Chapter 2 that $\widehat{K}(0) = 0$. For $y \neq 0$, we shall prove the following theorem about $\widehat{K}(y)$:

Theorem 3.1 *Let $K(x) = Q_n(x)/|x|^{N+n}$ for $x \neq 0$ where Q_n is a spherical harmonic of degree n, $n \geq 1$. Then there exists a constant $\kappa_{n,N}$ depending only on n and the dimension N such that*

$$(3.17) \qquad\qquad \widehat{K}(y) = \kappa_{n,N} Q_n(y) |y|^{-n} \quad \text{for } y \neq 0,$$

where $\widehat{K}(y)$ is defined by the limit in (3.13).

Proof of Theorem 3.1.   To prove the theorem, we invoke the following identity from [Wa, p. 368]:

$$(3.18) \qquad e^{it\cos\theta} = 2^\nu \Gamma(\nu) \sum_{n=0}^{\infty} (\nu + n) i^n \frac{J_{\nu+n}(t)}{t^\nu} C_n^\nu(\cos\theta) \quad \text{for } t > 0,$$

where $\nu = \frac{N-2}{2}$ and $t \in \mathbf{R}$.

Next, with $y \neq 0$ and with $y = |y|\xi$ and $x = r\eta$ where $\xi, \eta \in S_{N-1}$ and $r = |x|$, we observe from (3.1) that $Q_n(x) = r^n Y_n(\eta)$ and hence $K(x) = Y_n(\eta)/r^N$. Consequently, the integral on the right-hand side of the equality in (3.13) can be written as

$$(3.19) \qquad\qquad \int_\varepsilon^R r^{-1} dr \int_{S_{N-1}} e^{-i|y|\xi\cdot r\eta} Y_n(\eta) dS(\eta).$$

Also, we see from (3.18) with $t = |y|r$ and $\xi \cdot \eta = \cos\theta$ that

$$e^{-i|y|\xi\cdot r\eta} = 2^\nu \Gamma(\nu) \sum_{k=0}^{\infty} (\nu + k)(-i)^k \frac{J_{\nu+k}(|y|r)}{(|y|r)^\nu} C_k^\nu(\xi \cdot \eta),$$

where we have used the fact that C_k^ν is an even or odd polynomial accordingly as k is even or odd.

From this last equality, we obtain that the inner integral in (3.19), namely, $\int_{S_{N-1}} e^{-i|y|\xi \cdot r\eta} Y_n(\eta) dS(\eta)$, is equal to

$$(3.20) \qquad 2^\nu \Gamma(\nu) \sum_{k=0}^{\infty} (\nu + k)(-i)^k \frac{J_{\nu+k}(|y|\, r)}{(|y|\, r)^\nu} \int_{S_{N-1}} C_k^\nu(\xi \cdot \eta) Y_n(\eta) dS(\eta).$$

Now as we have shown earlier, for each $\xi \in S_{N-1}$, $C_k^\nu(\xi \cdot \eta)$ is a surface spherical harmonic of degree k. Hence, it follows from the orthogonality relationship given in (3.14) that the summation in (3.20) reduces to

$$(3.21) \qquad 2^\nu \Gamma(\nu)(\nu + n)(-i)^n \frac{J_{\nu+n}(|y|\, r)}{(|y|\, r)^\nu} \int_{S_{N-1}} C_n^\nu(\xi \cdot \eta) Y_n(\eta) dS(\eta).$$

Next, using the special formula given by the equality in (3.12), we in turn have that the value of the expression in (3.21) is

$$(3.21') \qquad (-i)^n 2^{\nu+1} \pi^{\nu+1} \frac{J_{\nu+n}(|y|\, r)}{(|y|\, r)^\nu} Y_n(\xi).$$

Consequently,

$$(3.21'') \qquad \int_{S_{N-1}} e^{-i|y|\xi \cdot r\eta} Y_n(\eta) dS(\eta) = (-i)^n 2^{\nu+1} \pi^{\nu+1} \frac{J_{\nu+n}(|y|\, r)}{(|y|\, r)^\nu} Y_n(\xi).$$

So, from (3.19) and this last equality, we obtain that the integral on the right-hand side of the equality in (3.13) can be written as

$$(-i)^n 2^{\nu+1} \pi^{\nu+1} Y_n(\xi) \int_\varepsilon^R \frac{J_{\nu+n}(|y|\, r)}{(|y|\, r)^\nu} r^{-1} dr,$$

and therefore that

$$(3.22) \qquad \widehat{K}(y) = \lim_{R \to \infty} \lim_{\varepsilon \to 0} (2\pi)^{-N} (-i)^n 2^{\nu+1} \pi^{\nu+1} Y_n(\xi) \int_{\varepsilon|y|}^{R|y|} \frac{J_{\nu+n}(r)}{r^{\nu+1}} dr.$$

The Bessel function estimates given in (2.1) and (2.2), since $n \geq 1$, imply that the limits on the right-hand side of the equality in (3.22) exist. So $\kappa_{n,N}$ has the value

$$(3.23) \qquad \kappa_{n,N} = (2\pi)^{-N} (-i)^n 2^{\nu+1} \pi^{\nu+1} \int_0^\infty \frac{J_{\nu+n}(r)}{r^{\nu+1}} dr,$$

and the proof of the theorem is complete. ∎

Using the theory of hypergeometric functions, the integral in (3.23) is evaluated in [Wa, p. 391] and the following value is obtained:

$$(3.24) \qquad \int_0^\infty \frac{J_{\nu+n}(r)}{r^{\nu+1}} dr = \Gamma(n/2)/\Gamma(\nu + 1 + \frac{n}{2}) 2^{\nu+1}.$$

From Theorem 3.1, we can get a corollary that will give us the principal-valued Fourier coefficients of the periodic Calderon-Zygmund kernel $K^*(x)$. This periodic kernel $K^*(x)$ is defined in (1.8) of Chapter 2 to be

$$(3.25) \qquad K^*(x) = K(x) + \lim_{R \to \infty} \sum_{1 \le |m| \le R} [K(x + 2\pi m) - K(2\pi m)]$$

where $m \in \Lambda_N$. Also, the principal-valued Fourier coefficient of $K^*(x)$ is defined to be

$$(3.26) \qquad \widehat{K^*}(m) = \lim_{\varepsilon \to 0} (2\pi)^{-N} \int_{T_N - B(0,\varepsilon)} e^{-im \cdot x} K^*(x) dx,$$

and in (1.21) of Chapter 2, it is shown to have the following limit:

$$(3.27) \quad \widehat{K^*}(m) = \lim_{R \to \infty} \lim_{\varepsilon \to 0} (2\pi)^{-N} \int_{B(0,R) - B(0,\varepsilon)} e^{-im \cdot x} K(x) dx \quad \text{for } m \ne 0.$$

Corollary 3.2. *Let $K(x) = Q_n(x)/|x|^{N+n}$ for $x \ne 0$ where Q_n is a spherical harmonic of degree n, $n \ge 1$, and let $K^*(x)$ be its periodic analogue defined in (3.25). Also, let $\widehat{K^*}(m)$ be its principal-valued Fourier coefficint as defined in (3.26). Then*

$$\widehat{K^*}(m) = \widehat{K}(m) = \kappa_{n,N} Q_n(m) |m|^{-n} \text{ for } m \ne 0$$

where

$$(3.28) \qquad \kappa_{n,N} = (-i)^n 2^{-N} \pi^{-\frac{N}{2}} \Gamma(\frac{n}{2}) [\Gamma(\frac{n+N}{2})]^{-1}.$$

Proof of Corollary 3.2. To prove the corollary, we observe from (3.27), (3.13), and (3.17) as stated in Theorem 3.1 that

$$\widehat{K^*}(m) = \widehat{K}(m) = \kappa_{n,N} Q_n(m) |m|^{-n} \text{ for } m \ne 0.$$

Also, from (3.23) and (3.24) with $\nu = \frac{N-2}{2}$, we see that

$$\begin{aligned} \kappa_{n,N} &= (2\pi)^{-N} (-i)^n 2^{\nu+1} \pi^{\nu+1} \Gamma(n/2)/\Gamma(\nu + 1 + \frac{n}{2}) 2^{\nu+1} \\ &= (-i)^n 2^{-N} \pi^{-\frac{N}{2}} \Gamma(\frac{n}{2}) [\Gamma(\frac{n+N}{2})]^{-1}, \end{aligned}$$

and the proof of the corollary is complete.   ∎

In Appendix C, when we deal with the spherical harmonic expansion of a harmonic function in the interior of an N-Ball of radius R, the following proposition regarding Gegenbauer polynomials will prove very useful.

Proposition 3.3. *For $N \geq 3$ and $\nu = \frac{N-2}{2}$, with $0 \leq r < 1$ and $-1 \leq t \leq 1$, for Gegenbauer polynomials, the following formula prevails:*

$$(3.29) \qquad \frac{1-r^2}{(1-2rt+r^2)^{N/2}} = \sum_{n=0}^{\infty} [\frac{N-2+2n}{N-2}]C_n^\nu(t)r^n$$

where the series converges unifomly for $0 \leq r \leq r_0 < 1$ and $-1 \leq t \leq 1$.

It is clear that the left-hand side of (3.29) for $N = 2$ and $t = \cos\theta$ gives rise to the familiar Poisson kernel for the unit disk. The formula given here in (3.29) is going to play an analogous role for the unit N-ball. Also, since

$$(3.30) \qquad |C_n^\nu(t)| \leq C_n^\nu(1) = \frac{\Gamma(2\nu+n)}{\Gamma(2\nu)n!} \quad \text{for } -1 \leq t \leq 1,$$

(see [Ra, p.278 and p. 281]), it follows that the series in (3.29) converges uniformly for $0 \leq r \leq r_0 < 1$ and $-1 \leq t \leq 1$.

Proof of Proposition 3.3. We set $\rho = (1-2rt+r^2)^{1/2}$ and observe from the generating function for the Gegenbauer polynomials (see the paragraph below (3.3)) that

$$(3.31) \qquad \rho^{-(N-2)} = \sum_{n=0}^{\infty} C_n^\nu(t)r^n, \qquad \text{for } 0 < r < 1 \text{ and } -1 \leq t \leq 1.$$

Since $\partial\rho/\partial r = (-t+r)/\rho$, it is easy to see that

$$\frac{1-r^2}{(1-2rt+r^2)^{N/2}} = \frac{1}{\rho^{N-2}} + \frac{2r}{N-2}\partial\rho^{-(N-2)}/\partial r.$$

The equality in (3.29) follows immediately from (3.31) and this last formula. ∎

In closing this section, we want to establish a special case of the important formula (3.9) above, namely, when $N = 3$ and $\nu = 1/2$. In this case, we will have to deal with the Legendre polynomials, $\{P_n(t)\}_{n=0}^{\infty}$, and the associated Legendre functions, $\{P_n^k(t)\}_{n=0,k=1}^{\infty, n}$. The associated Legendre functions are defined as follows:

$$(3.32) \qquad P_n^k(t) = (1-t^2)^{k/2}d^k P_n(t)/dt^k \text{ for } k = 1, ..., n.$$

It is well-known (see [CH, p. 327] or [Sp, pp. 145-6] or [Pi, p. 229]) that these functions satisfy the following differential equation:

$$(3.33) \qquad [(1-t^2)y'(t)]' - \frac{k^2y(t)}{1-t^2} + n(n+1)y(t) = 0 \quad \forall t \in (-1, 1),$$

where $'$ denotes differentiation with respect to t.

Using the spherical coordinate notation introduced in Chapter 1, namely,

$$
\begin{aligned}
x_1 &= r\cos\theta \\
x_2 &= r\sin\theta\cos\phi \\
x_3 &= r\sin\theta\sin\phi,
\end{aligned}
$$

we see from (3.32) that

$$(3.34) \qquad r^n P_n^k(\cos\theta)\cos k\phi \text{ and } r^n P_n^k(\cos\theta)\sin k\phi \quad k = 1,...,n,$$

are homogeneous polynomials of degree n in the variables x_1, x_2, x_3. To do this, we need the observation that both $\cos k\phi$ and $\sin k\phi$ are homogeneous polynomials of degree k in the variables $\sin\phi$ and $\cos\phi$. This fact is easily proved by induction.

Also, using the Laplacian expressed in the spherical coordinates r, θ, ϕ, it is an easy matter using (3.33), to see that both

$$\Delta r^n P_n^k(\cos\theta)\cos k\phi = 0 \text{ and } \Delta r^n P_n^k(\cos\theta)\sin k\phi = 0$$

when $0 < r$, $0 < \theta < \pi$, and $0 \le \phi \le 2\pi$ (see [Sp, p. 145]). Since the set of functions in (3.34) are homogeneous polynomials of degree n in the variables x_1, x_2, x_3, we conclude that each of these functions are spherical harmonics. We will use this fact in the proof of the following theorem:

Theorem 3.4. *Let $S_2 \subset \mathbf{R}^3$ be the unit sphere and let n be any positive integer. Suppose $\{Y_{nj}(\xi)\}_{j=1}^{2n+1}$ is a set of surface spherical harmonics of degree n which are orthonormal in $L^2(S_2)$, i.e.,*

$$\int_{S_2} Y_{nj}(\xi)Y_{nk}(\xi)dS(\xi) = \delta_{jk} \qquad j, k = 1,...,2n+1.$$

where δ_{jk} is the Kronecker-δ. Then

$$P_n(\xi \cdot \eta) = \frac{4\pi}{2n+1}\sum_{j=1}^{2n+1} Y_{nj}(\xi)Y_{nj}(\eta) \quad \forall \xi, \eta \in S_2,$$

where $P_n(t)$ is the n-th Legendre polynomial.

Proof of Theorem 3.4. To prove the theorem, we let $\{Y_{nj}^{\Diamond}(\xi)\}_{j=1}^{2n+1}$ be another set of surface spherical harmonics of degree n which are orthonormal in $L^2(S_2)$. From (3.7), we have that the set of surface spherical harmonics of degree n is a vector space of dimension $2n + 1$. Consequently, the given set $\{Y_{nj}(\xi)\}_{j=1}^{2n+1}$ is a basis for the surface spherical harmonics of degree n, and we have that

$$(3.35) \qquad Y_{nj}^{\Diamond}(\xi) = \sum_{k=1}^{2n+1} a_{jk}Y_{nk}(\xi) \quad \forall \xi \in S_2,$$

where the a_{jk} are real constants and $j = 1, ..., 2n + 1$. Using the orthonormality of both sets in $L^2(S_2)$, it is easy to see that

$$\sum_{k=1}^{2n+1} a_{j_1 k} a_{j_2 k} = \delta_{j_1 j_2}.$$

Hence, the $(2n + 1) \times (2n + 1)$ matrix $O = (a_{jk})$ is an orthogonal matrix. So an easy computation using (3.35) shows that

$$(3.36) \qquad \sum_{j=1}^{2n+1} Y_{nj}(\xi) Y_{nj}(\eta) = \sum_{j=1}^{2n+1} Y_{nj}^{\Diamond}(\xi) Y_{nj}^{\Diamond}(\eta).$$

Next, we note that if $u(x)$ is harmonic in $\mathbf{R}^N$, so is $v(x) = u(Ox)$ where $O \in SO(3)$, the set of 3×3 real orthogonal matrices of determinant one (see [ABR, p.3]).

Consequently, it follows that $\{Y_{nj}(O\xi)\}_{j=1}^{2n+1}$ is a set of surface spherical harmonics of degree n that is also an orthonormal set in $L^2(S_2)$ for every orthogonal matrix O. We conclude from (3.36) that

(3.37)

$$\sum_{j=1}^{2n+1} Y_{nj}(\xi) Y_{nj}(\eta) = \sum_{j=1}^{2n+1} Y_{nj}(O\xi) Y_{nj}(O\eta) \quad \forall \xi, \eta \in S_2 \text{ and } \forall O \in SO(3).$$

Next, we observe that if $\xi, \eta, \widetilde{\xi}, \widetilde{\eta} \in S_2$ are four points with the property that $\xi \cdot \eta = \widetilde{\xi} \cdot \widetilde{\eta}$, then there exists an orthogonal matrix O such that

$$O\xi \cdot O\eta = \widetilde{\xi} \cdot \widetilde{\eta}.$$

Using this fact in conjunction with (3.37), we see that there is a function $g(t)$ defined for $-1 \leq t \leq 1$, which is also in $C([-1, 1])$, such that

$$(3.38) \qquad g(\xi \cdot \eta) = \sum_{j=1}^{2n+1} Y_{nj}(\xi) Y_{nj}(\eta) \quad \forall \xi, \eta \in S_2.$$

To obtain this function $g(t)$, reintroduce the spherical coordinate system above (3.34), set $\eta^* = (1, 0, 0)$, and observe that if $\xi = (\xi_1, \xi_2, \xi_3)$, then

$$\begin{aligned}
\xi_1 &= \cos \theta \\
\xi_2 &= \sin \theta \cos \phi \\
\xi_3 &= \sin \theta \sin \phi.
\end{aligned}$$

Next, observe that if $-1 \leq t \leq 1$, there exists θ_t with $0 \leq \theta_t \leq \pi$ such that $t = \cos \theta_t$. Set $\xi_t = (\cos \theta_t, \sin \theta_t, 0)$. Then $\xi_t \cdot \eta^* = t$ and define

$$g(t) = g(\xi_t \cdot \eta^*) = \sum_{j=1}^{2n+1} Y_{nj}(\xi_t) Y_{nj}(\eta^*).$$

For a given ξ and η, find ξ_t such that $\xi_t \cdot \eta^* = \xi \cdot \eta$ and $O \in SO(3)$ such that $\xi = O\xi_t$ and $\eta = O\eta^*$. Then $g(\xi \cdot \eta) = g(\xi_t \cdot \eta^*)$ and hence by this last set of equalities and (3.37),

$$g(\xi \cdot \eta) = \sum_{j=1}^{2n+1} Y_{nj}(\xi_t)Y_{nj}(\eta^*) = \sum_{j=1}^{2n+1} Y_{nj}(O\xi_t)Y_{nj}(O\eta^*) = \sum_{j=1}^{2n+1} Y_{nj}(\xi)Y_{nj}(\eta),$$

which gives the equality in (3.38).

We next use the spherical coordinate notation to define an orthonormal set in $L^2(S_2)$ of surface spherical harmonics as follows:

$$
\begin{aligned}
Y_{nj}^{\sharp}(\xi) &= b_{nj}P_n^j(\eta^* \cdot \xi)\cos j\phi && \text{for } j = 1, ..., n, \\
&= b_{nj}P_n^{j-n}(\eta^* \cdot \xi)\sin(j-n)\phi && \text{for } j = n+1, ..., 2n, \\
&= \left(\frac{2n+1}{4\pi}\right)^{1/2}P_n(\eta^* \cdot \xi) && \text{for } j = 2n+1
\end{aligned}
$$

where the b_{nj} are normalizing constants, the $P_n^j(t)$ are the associated Legendre functions introduced in (3.32), and $P_n(t)$ is the usual Legendre polynomial of order n. To be quite precise, the b_{nj} are constants defined so that

$$\int_{S_2} \left|Y_{nj}^{\sharp}(\xi)\right|^2 dS(\xi) = 1 \text{ for } j = 1, ..., 2n.$$

Observing from the spherical coordinate notation introduced above, that $\xi \cdot \eta^* = \cos\theta$, it is easy to check from well-known facts about Legendre polynomials (see [Ra, p. 175]) that

$$\int_{S_2} \left|Y_{n\,2n+1}^{\sharp}(\xi)\right|^2 dS(\xi) = 1.$$

Also, it is easy to see that $\{Y_{nj}^{\sharp}(\xi)\}_{j=1}^{2n+1}$ constitutes an orthogonal system in $L^2(S_2)$. Consequently, it follows from (3.36) and (3.38) that

$$(3.39) \qquad g(\xi \cdot \eta) = \sum_{j=1}^{2n+1} Y_{nj}^{\sharp}(\xi)Y_{nj}^{\sharp}(\eta) \quad \forall \xi, \eta \in S_2.$$

Next, we observe from (3.32) that $P_n^k(1) = 0$ for $k = 1, ..., n$. Hence, it follows from the definitions given above, that

$$Y_{nj}^{\sharp}(\eta^*) = 0 \text{ for } j = 1, ..., 2n.$$

But then we obtain from (3.39) and the definition of $Y_{n\,2n+1}^{\sharp}(\xi)$ given above that

$$g(\xi \cdot \eta^*) = \frac{2n+1}{4\pi}P_n(\eta^* \cdot \xi)P_n(1) \quad \forall \xi \in S_2.$$

Now $P_n(1) = 1$, and as we observed above, for every $t \in [-1,1]$, $\exists \xi_t \in S_2$ such that $\xi_t \cdot \eta^* = t$. So we conclude from this last equality that

$$g(t) = \frac{2n+1}{4\pi}P_n(t) \text{ for } t \in [-1,1],$$

and consequently that

$$g(\xi \cdot \eta) = \frac{2n+1}{4\pi} P_n(\xi \cdot \eta) \quad \forall \xi, \eta \in S_2.$$

This fact coupled with (3.38) gives the conclusion of the theorem. ∎

In Chapter 4, we shall need some further facts about the sequence of Gegenbauer polynomials $\{C_n^\nu(t)\}_{n=0}^\infty$. To handle these matters, given a real-valued measurable function f defined on the interval (-1,1), we say

$$f \in L_\mu^2[(-1,1)]$$

where $\mu(t) = (1-t^2)^{\nu-\frac{1}{2}}$ and $\nu = (N-2)/2$ with $N \geq 3$ provided the following prevails:

$$(3.40) \qquad \int_{-1}^1 |f(t)|^2 \mu(t) dt < \infty.$$

It is clear from the self-adjoint differential equation satisfied by the Gegenbauer polynomials stated just above (3.6) that

$$(3.41) \qquad \int_{-1}^1 C_j^\nu(t) C_k^\nu(t)(1-t^2)^{\nu-\frac{1}{2}} dt = 0 \quad \text{for } j \neq k.$$

Also, it is shown in [Ra, p. 278 and p. 281] that

$$(3.42) \qquad \int_{-1}^1 |C_n^\nu(t)|^2 (1-t^2)^{\nu-\frac{1}{2}} dt = \tau_n \quad \forall n,$$

where

$$(3.43) \qquad \tau_n = C_n^\nu(1) \frac{\Gamma(\frac{1}{2})\Gamma(\nu+\frac{1}{2})}{(\nu+n)\Gamma(\nu)}.$$

As a consequence of all this, we set

$$(3.44) \qquad a_n = \tau_n^{-1} \int_{-1}^1 C_n^\nu(t) f(t)(1-t^2)^{\nu-\frac{1}{2}} dt$$

and observe that the a_n are the Gegenbauer-Fourier coefficients for f.

In the next theorem, we will show that $\{C_n^\nu(t)\}_{n=0}^\infty$ is a complete orthogonal system. By this, we mean, given $f \in L_\mu^2[(-1,1)]$, then

$$(3.45) \qquad a_n = 0 \;\; \forall n \;\; \Longrightarrow f(t) = 0 \;\; \text{a.e. on } (-1,1),$$

where a_n is defined in (3.44).

Theorem 3.5. $\{C_n^\nu(t)\}_{n=0}^\infty$ *is a complete orthogonal system for* $L_\mu^2([-1,1])$ *where* $\mu = (1-t^2)^{\nu-\frac{1}{2}}$, $\nu = (N-2)/2$, *and* $N \geq 3$.

From Theorem 3.5, we can obtain the following corollary that will be useful in Chapter 4.

Corollary 3.6. *Let* $f \in C([-1,1])$ *with* a_n *defined by (3.44). Suppose that*

$$\lim_{n \to \infty} \sum_{j=0}^{n} a_j C_j^{\nu}(t) = g(t)$$

uniformly for $t \in [-1,1]$. *Then*

$$f(t) = g(t) \quad \forall t \in [-1,1].$$

Proof of Theorem 3.5. We are given $f \in L_{\mu}^2[(-1,1)]$ and that

$$\int_{-1}^{1} C_n^{\nu}(t) f(t)(1-t^2)^{\nu-\frac{1}{2}} dt = 0 \quad \forall n.$$

From the Rodrigues formula for $C_n^{\nu}(t)$ [AAR, p. 303], we see that $C_n^{\nu}(t)$ is a polynomial of degree n. Therefore, it follows from this last equality that

$$\int_{-1}^{1} t^n f(t)(1-t^2)^{\nu-\frac{1}{2}} dt = 0 \quad \forall n.$$

But then

$$(3.46) \qquad \int_{-1}^{1} P(t) f(t)(1-t^2)^{\nu-\frac{1}{2}} dt = 0,$$

for every polynomial $P(t)$.

It follows from the Weierstrass approximation that the analogue of (3.46) holds for every function $g \in C([-1,1])$. Using the Lebesgue dominated convergence theorem, we see that the same can then be said for every χ_I where $\chi_I(t)$ is the characteristic (indicator) function of a subinterval $I \subset [-1,1]$. Consequently, the same fact holds also for every simple function, and we conclude, using the Lebesgue dominated convergence theorem once again, that

$$(3.47) \qquad \int_{-1}^{1} h(t) f(t)(1-t^2)^{\nu-\frac{1}{2}} dt = 0,$$

for every function $h \in L^{\infty}([-1,1])$.

Next, we set

$$\begin{aligned} f_n(t) &= f(t) \quad \text{if} \quad |f(t)| \le n \\ &= 0 \quad \text{if} \quad |f(t)| > n. \end{aligned}$$

Then, $f_n \in L^{\infty}([-1,1])$, and it follows from (3.47) that

$$(3.48) \qquad \int_{-1}^{1} f_n(t) f(t)(1-t^2)^{\nu-\frac{1}{2}} dt = 0 \qquad \forall n.$$

Since $|f_n(t)f(t)| \le |f(t)|^2$ and

$$\lim_{n\to\infty} f_n(t) = f(t) \quad \text{for} \quad t \in (-1,1),$$

we conclude from (3.48) that

$$\int_{-1}^{1} |f(t)|^2 (1-t^2)^{\nu-\frac{1}{2}} dt = 0.$$

Hence, $f(t) = 0$ a.e. on $(-1,1)$. ∎

Proof of Corollary 3.6. Since the series that defines $g(t)$ converges uniformly for $t \in [-1,1]$, $g \in C([-1,1])$. Also, it follows from the orthogonality in (3.41) and the uniform convergence of this series that

$$a_n = \tau_n^{-1} \int_{-1}^{1} g(t)C_n^\nu(t)(1-t^2)^{\nu-\frac{1}{2}} dt \quad \forall n,$$

where we also have made use of (3.42) and (3.43).

But then it follows from (3.44) that

$$\int_{-1}^{1} [g(t) - f(t)]C_n^\nu(t)(1-t^2)^{\nu-\frac{1}{2}} dt = 0 \qquad \forall n,$$

and therefore from Theorem 3.5 that $f(t) = g(t)$ for every $t \in [-1,1]$. ∎

Exercises.

1. Given that the Jacobi polynomial $P^{(\nu-\frac{1}{2},\nu-\frac{1}{2})}(t) = \gamma_{n,\nu}C_n^\nu(t)$ where $\gamma_{n,\nu}$ is a positive constant and that the Jacobi polynomial $P^{(\alpha,\beta)}(t)$ satisfies the differential equation [AAR, p. 297]

$$\left(1-t^2\right) P_n^{(\alpha,\beta)\prime\prime}(t) + (\beta - \alpha + (\alpha + \beta + 2)t) P_n^{(\alpha,\beta)\prime}(t)$$

$$+ n(n + \alpha + \beta + 1) P_n^{(\alpha,\beta)}(t) = 0,$$

prove that $C_n^\nu(t)$ satisfies the differential equation:

$$(1-t^2)C_n^{\nu\prime\prime}(t) - 2t(\nu + \frac{1}{2})C_n^{\nu\prime}(t) = -n(n+2\nu)C_n^\nu(t) \quad \forall t \in (-1,1),$$

where $C_n^{\nu\prime}(t) = dC_n^\nu(t)/dt, \nu = \frac{N-2}{2}$, and $n \ge 1$.

2. Prove that, with the substitution of $t = \cos\theta$, the differential equation for $C_n^\nu(t)$ in Exercise 1 becomes

$$\frac{d}{d\theta}(\sin\theta)^{N-2}\frac{dC_n^\nu(\cos\theta)}{d\theta} = -n(n+N-2)(\sin\theta)^{N-2}C_n^\nu(\cos\theta)$$

for $\theta \in (-\pi, \pi)$.

3. With $\eta^* = (1, 0, ..., 0)$ and $\xi \in S_{N-1}$, using the spherical coordinate notation introduced in §3 of Chapter 1, prove that

$$r^3 C_3^\nu(\eta^* \cdot \xi)$$

is a homogeneous polynomial of degree 3 and is also a harmonic function in $\mathbf{R}^N$, $N \geq 3$, where $\nu = \frac{N-2}{2}$.

4. Set $\rho = \left(1 - 2rt + r^2\right)$ where $0 < r < 1$ nd $t \in [-1, 1]$. Prove that

$$\frac{1 - r^2}{(1 - 2rt + r^2)^{N/2}} = \frac{1}{\rho^{N-2}} + \frac{2r}{N-2} \partial \rho^{-(N-2)} / \partial r.$$

5. Using the differential equation for $C_n^\nu(t)$ given in Exercise 1, prove that

$$\int_{-1}^{1} C_j^\nu(t) C_k^\nu(t) (1 - t^2)^{\nu - \frac{1}{2}} dt = 0 \quad \text{for } j \neq k.$$

Real Analysis

1. Convergence and Summability

In this section, we shall prove various consistency theorems with respect to convergence and summability. In particular, the first theorem will show that the convergence of an integral to a given limit implies the Bochner-Riesz summability of that integral to the same limit.

Theorem 1.1. *Suppose $h(s) \in L^1(0, n)$ for every positive integer n. Set $H(R) = \int_0^R h(s)ds$ for R>0, and suppose*

$$\lim_{R \to \infty} H(R) = \gamma,$$

where γ is a finite real number. Then

$$(1.1) \qquad \lim_{R \to \infty} \int_0^R (1 - \frac{s^2}{R^2})^\alpha h(s)ds = \gamma,$$

for every $\alpha > 0$.

Proof of Theorem 1.1. We first of all observe from the Lebesgue dominated convergence theorem that

$$(1.2) \qquad \lim_{R \to \infty} \int_0^R (1 - \frac{s^2}{R^2})^\alpha e^{-s^2} ds = \pi^{1/2}/2.$$

Next, we set

$$h_1(s) = h(s) - \frac{2\gamma}{\pi^{1/2}} e^{-s^2},$$

and $H_1(R) = \int_0^R h_1(s)ds$. It is clear from the hypothesis of the theorem that

$$(1.3) \qquad \lim_{R \to \infty} H_1(R) = 0.$$

If we can show that this last limit implies that

$$(1.4) \qquad \lim_{R \to \infty} \int_0^R (1 - \frac{s^2}{R^2})^\alpha h_1(s)ds = 0,$$

then it follows from (1.2) and the definition of $h_1(s)$ that the limit in (1.1) holds. Therefore, to complete the proof of the theorem, it suffices to show that (1.3) implies (1.4). We now do this.

First of all, we observe after integrating by parts that

$$\int_0^R (1 - \frac{s^2}{R^2})^\alpha h_1(s)ds = 2\alpha \int_0^R H_1(s)(1 - \frac{s^2}{R^2})^{\alpha-1}\frac{s}{R^2}ds.$$

Hence, to establish (1.4), it is enough to show

$$(1.5) \qquad \lim_{R\to\infty} 2\alpha \int_0^R H_1(s)(1 - \frac{s^2}{R^2})^{\alpha-1}\frac{s}{R^2}ds = 0.$$

To do this, given $\varepsilon > 0$, use (1.3) to choose $R_0 > 0$, so that

$$(1.6) \qquad\qquad |H_1(R)| < \varepsilon \text{ for } R > R_0.$$

It is clear that

$$(1.7) \qquad \lim_{R\to\infty} \int_0^{R_0} (1 - \frac{s^2}{R^2})^{\alpha-1}\frac{s}{R^2}ds = 0.$$

Also, (1.6) implies that

$$2\alpha \left| \int_{R_0}^R H_1(s)(1 - \frac{s^2}{R^2})^{\alpha-1}\frac{s}{R^2}ds \right| \le \varepsilon \text{ for } R > R_0.$$

We conclude from (1.7) and this last inequality that

$$\limsup_{R\to\infty} 2\alpha \left| \int_0^R H_1(s)(1 - \frac{s^2}{R^2})^{\alpha-1}\frac{s}{R^2}ds \right| \le \varepsilon.$$

Since ε is an arbitrary positive number, we see that the limit in (1.5) is indeed valid, and the proof of the theorem is complete. ∎

Next, we show that convergence of a multiple series to a given finite limit implies Abel summability of the series to the same limit.

Theorem 1.2. *Let Λ_N designate the set of integral lattice points in $\mathbf{R}^N$, $N \ge 1$. Suppose (i) $\bar{a}_m = a_{-m}$ for $m \in \Lambda_N \backslash \{0\}$, (ii) $a_0 = 0$, and (iii) $\exists \beta > 0$ such that $|a_m| = O(|m|^\beta)$ as $|m| \to \infty$. Suppose, also,*

$$(1.8) \qquad \lim_{R\to\infty} \sum_{1 \le |m| \le R} a_m = \alpha \qquad \text{where } \alpha \text{ is a finite real number.}$$

Then,

$$(1.9) \qquad \lim_{t\to 0} \sum_{1 \le |m| < \infty} a_m e^{-|m|t} = \alpha.$$

Proof of Theorem 1.2. Set $S(r) = \sum_{0 < |m| \le r} a_m$. It is clear, with no loss in generality, we can suppose $\alpha = 0$. Then, for $r > 1$,

$$S(r) = \sum_{k=1}^{[r^2]} \sum_{|m|^2 = k} a_m,$$

where $[r^2]$ is the first integer less than or equal to r^2. Then,

$$\sum_{0<\,|m|\leq R} a_m e^{-|m|t} = \int_0^R e^{-rt} dS(r)$$

$$= S(R)e^{-Rt} + t\int_0^R e^{-rt} S(r)dr.$$

Consequently,

$$\sum_{1\leq\,|m|<\infty} a_m e^{-|m|t} = t\int_0^\infty e^{-rt} S(r)dr.$$

Given $\varepsilon > 0$, choose r_0 so large that $|S(r)| < \varepsilon$ for $r \geq r_0$. Then,

$$\left| \sum_{1\leq\,|m|<\infty} a_m e^{-|m|t} \right| \leq \left| t\int_0^{r_0} e^{-rt} S(r)dr \right| + \varepsilon t\int_{r_0}^\infty e^{-rt} dr,$$

and we conclude from this last inequality that

$$\limsup_{t\to 0} \left| \sum_{1\leq\,|m|<\infty} a_m e^{-|m|t} \right| \leq \varepsilon.$$

Since ε is an arbitrary positive number, we obtain from this last inequality that

$$\lim_{t\to 0} \sum_{1\leq\,|m|<\infty} a_m e^{-|m|t} = 0,$$

and the proof of the theorem is complete. ∎

2. Tauberian Limit Theorems

In this section, we deal with various limit theorems that are needed in tauberian theory. The first one is the following, which is evidently due to Landau, [Har, p. 176].

Theorem 2.1. *Suppose $f\in C^2(0,\infty)$. Suppose, also,*

$$(i) \lim_{t\to 0} f(t) = \alpha \ \text{where} \ \alpha \ \text{is} \ finite-valued,$$

$$(ii) f''(t) \geq -Ct^{-2} \text{for} \ t > 0 \ \text{where} \ C \ \text{is a positive constant.}$$

Then

(2.1) $$\lim_{t\to 0} t f'(t) = 0.$$

Proof of Theorem 2.1. The proof involves a subtle use of Taylor's theorem [Ru1, p. 111]. In particular, if t and $t + \eta$ are positive values with $\eta \neq 0$, it follows from Taylor's theorem that

$$f(t + \eta) - f(t) = \eta f'(t) + \frac{\eta^2}{2} f''[t + \theta(t, \eta)\eta]$$

where $0 < \theta < 1$. Therefore,

(2.2) $$f'(t) = [f(t + \eta) - f(t)]/\eta - \frac{\eta}{2} f''(t + \theta\eta).$$

Let $\varepsilon > 0$ be given, and choose δ so that $0 < \delta < 1$ and also so that

(2.3) $$\frac{C\delta}{2(1 - \delta)^2} < \varepsilon.$$

Take $\eta = \delta t$. Then $t + \eta = (1 + \delta)t$, and we see from (ii) and (2.2) that

$$
\begin{aligned}
tf'(t) &\leq \frac{f[(1 + \delta)t] - f(t)}{\eta} t + \frac{\eta}{2} \frac{C}{t} \\
&\leq \frac{f[(1 + \delta)t] - f(t)}{\delta} + \frac{C\delta}{2}.
\end{aligned}
$$

We consequently obtain from (i) and (2.3) that

$$\limsup_{t \to 0} tf'(t) \leq \frac{C\delta}{2} \leq \varepsilon.$$

Since ε is an arbitrary positive number, we have that

(2.4) $$\limsup_{t \to 0} tf'(t) \leq 0.$$

Next, we take $\eta = -\delta t$. Then $t - \eta = (1 - \delta)t$, and we see from (ii) and (2.2) that

$$
\begin{aligned}
tf'(t) &\geq \frac{f[(1 - \delta)t] - f(t)}{\eta} t + \frac{\eta}{2} \frac{C}{(t + \theta\eta)^2} \\
&\geq -\frac{f[(1 - \delta)t] - f(t)}{\delta} - \frac{C\delta}{2(1 - \delta)^2}.
\end{aligned}
$$

We consequently infer from (i) and (2.3) that

$$\liminf_{t \to 0} tf'(t) \geq -\varepsilon.$$

Hence,

$$\liminf_{t \to 0} tf'(t) \geq 0.$$

From this last inequality joined with the inequality in (2.4), we see that

$$\lim_{t \to 0} tf'(t) = 0,$$

which is the limit in (2.1).   ■

We also need the following Tauberian theorem due to Karamata [Ka].

Theorem 2.2. *Suppose $h(r)$ is a function defined on $(0,\infty)$ with the following properties: (i) $h(r)$ is continuous on $(0,\infty)\backslash E$ where $E=\bigcup_{j=1}^{\infty}\{r_j\}$ and*

$$0 < r_1 < r_2 < \cdots < r_n \to \infty;$$

(ii) $h(r)$ is continuous from the right; (iii) $h(r)=0$ for $0<r<r_1$; (iv) $h(r)$ is a nondecreasing function; (v) $\exists k>0$ such that $h(r)=O(r^k)$ as $r\to\infty$. Set

$$(2.5) \qquad f(t) = \int_0^{\infty} e^{-rt}dh(r) \quad for \ \ t>0,$$

and *suppose that $\lim_{t\to 0} tf(t)=\gamma$ where γ is finite-valued. Then*

$$\lim_{R\to\infty} \frac{h(R)}{R} = \gamma.$$

Proof of Theorem 2.2. We observe from the well-known Weierstrass theorem that if $G(r) \in C([0,1])$, then given $\varepsilon > 0$, there are two polynomials $P(r)$ and $p(r)$ such that $p(r) \leq G(r) \leq P(r)$ for $r \in [0,1]$ and

$$\int_0^1 [P(r) - G(r)]dr < \varepsilon \quad \text{and} \quad \int_0^1 [G(r) - p(r)]dr < \varepsilon.$$

A similar situation prevails for the function $g(r)$ defined as follows:

$$(2.6) \qquad \begin{aligned} g(r) &= 0 \quad \text{for } 0 \leq r \leq e^{-1} \\ &= r^{-1} \quad \text{for } e^{-1} < r \leq 1. \end{aligned}$$

Then, given ε with $0 < \varepsilon < \frac{1}{10}$, there are polynomials $P(r)$ and $p(r)$ such that

$$(2.7) \qquad p(r) \leq g(r) \leq P(r) \quad \text{for } r \in [0,1]$$

and

$$(2.8) \qquad \int_0^1 [P(r) - g(r)]dr < \varepsilon \quad \text{and} \quad \int_0^1 [g(r) - p(r)]dr < \varepsilon.$$

We show first that (2.7) and (2.8) hold with respect to $P(r)$. To do this, we define the continuous function $\phi(r)$ in the following manner:

$$\begin{aligned} \phi(r) &= 0 \quad \text{for } 0 \leq r \leq e^{-1} - \delta \\ &= e\frac{r - (e^{-1} - \delta)}{\delta} \quad \text{for } e^{-1} - \delta \leq r \leq e^{-1} \\ &= g(r) \quad \text{for } e^{-1} < r \leq 1, \end{aligned}$$

where $\delta = \varepsilon e^{-1}$. Since ϕ is a positive linear function in the interval

$$(e^{-1} - \delta, e^{-1}]$$

and equal to g elsewhere in the interval $[0,1]$, it is clear that $g \leq \phi$. It is also clear that

$$\int_0^1 [\phi(r) - g(r)]dr = \varepsilon/2.$$

We now apply the Weierstrass theorem to ϕ for the value $\varepsilon/2$ and obtain the $P(r)$ part of (2.7) and (2.8). A similar argument works for $p(r)$.

Next, with $P(r)$ and $p(r)$ as in (2.7) and (2.8), we see the following is true:

(2.9)

$$(i)\ \lim_{t\to 0} t \int_0^\infty e^{-rt}P(e^{-rt})dh(r) = \gamma \int_0^1 P(r)dr;$$

$$(ii)\ \lim_{t\to 0} t \int_0^\infty e^{-rt}p(e^{-rt})dh(r) = \gamma \int_0^1 p(r)dr.$$

Since both $P(r)$ and $p(r)$ are polynomials, it is enough to establish the limit in (2.9) for the special case of the polynomial r^n where n is a nonnegative integer, i.e,

(2.10)
$$\lim_{t\to 0} t \int_0^\infty e^{-rt}(e^{-rt})^n dh(r) = \gamma \int_0^1 r^n dr.$$

We observe from (2.5) that the left-hand side of the equality in (2.10) is

$$\lim_{t\to 0} tf[(n+1)t] = \gamma/(n+1).$$

An easy computation shows that the right-hand side of the equality in (2.10) is also $\gamma/(n+1)$. So (2.10) is established, and hence (2.9) is also valid.

To complete the proof of the theorem, we observe from (2.7), (2.8), and the hypothesis of the theorem that

$$\int_0^\infty e^{-rt}p(e^{-rt})dh(r) \le \int_0^\infty e^{-rt}g(e^{-rt})dh(r) \le \int_0^\infty e^{-rt}P(e^{-rt})dh(r)$$

for $t \in (0,\infty)\backslash E^\diamond$ where $E^\diamond = \bigcup_{j=1}^\infty \{r_j^{-1}\}$. Hence, it follows from (2.9) and this last set of inequalities that
(2.11)
$$\gamma \int_0^1 p(r)dr \le \limsup_{t\to 0, t\notin E^\diamond} t \int_0^\infty e^{-rt}g(e^{-rt})dh(r) \le \gamma \int_0^1 P(r)dr.$$

On the other hand, we see from (2.7) that

$$\gamma \int_0^1 p(r)dr \le \gamma \int_0^1 g(r)dr \le \gamma \int_0^1 P(r)dr.$$

Therefore, from (2.8) and (2.11) joined with this last set of inequalities, we obtain that

$$\left| \limsup_{t\to 0, t\notin E^\diamond} t \int_0^\infty e^{-rt}g(e^{-rt})dh(r) - \gamma \int_0^1 g(r)dr \right| \le 2\gamma\varepsilon.$$

But ε is an arbitrary positive number. Consequently,

$$\limsup_{t\to 0, t\notin E^\diamond} t \int_0^\infty e^{-rt}g(e^{-rt})dh(r) = \gamma \int_0^1 g(r)dr.$$

In a similar manner, we obtain

$$\liminf_{t\to 0, t\notin E^\diamond} t \int_0^\infty e^{-rt}g(e^{-rt})dh(r) = \gamma \int_0^1 g(r)dr.$$

From (2.6), we see that $\int_0^1 g(r)dr = 1$. So we conclude from these last two equalities that

(2.12)
$$\lim_{t\to 0, t\notin E^\diamond} t \int_0^\infty e^{-rt}g(e^{-rt})dh(r) = \gamma.$$

Referring back to (2.6), we see that for $t \in (0, \infty) \backslash E^\Diamond$,

$$\int_0^\infty e^{-rt} g(e^{-rt}) dh(r) = \int_0^{1/t} e^{-rt} g(e^{-rt}) dh(r)$$
$$= \int_0^{1/t} e^{-rt} e^{rt} dh(r)$$
$$= h(1/t).$$

Therefore, (2.12) gives us that

$$\lim_{t \to 0, t \notin E^\Diamond} th(1/t) = \gamma.$$

Setting $R = 1/t$ and observing that h is continuous from the right enables us to conclude that

$$\lim_{R \to \infty} \frac{h(R)}{R} = \gamma,$$

and the proof to the theorem is complete. ∎

3. Distributions on the N-Torus

$\mathcal{D}(T_N)$, called the class of test functions, is the class of real functions defined as follows:

$\mathcal{D}(T_N) = \{\phi : \phi \in C^\infty(\mathbf{R}^N), \ \phi \text{ is periodic of period } 2\pi \text{ in each variable}\}.$

The notion that

$$\phi_n \to 0 \ \text{ in } \mathcal{D}(T_N) \ \text{ for } \{\phi_n\}_{n=1}^\infty \subset \mathcal{D}(T_N)$$

is defined below (3.1) in Chapter 3.

S is called a distribution on T_N if it is a real linear functional on $\mathcal{D}(T_N)$ that meets the following condition:

(3.0) $\phi_n \to 0 \ \text{ in } \mathcal{D}(T_N) \implies S(\phi_n) \to 0.$

The class of distributions on the N-torus will be designated by $\mathcal{D}'(T_N)$. $S \in \mathcal{D}'(T_N)$ is defined initially as a real linear functional on $\mathcal{D}(T_N)$. We extend the definition of S by setting $S(\lambda + i\phi) = S(\lambda) + iS(\phi)$ for $\lambda, \phi \in \mathcal{D}(T_N)$. We then define $\widehat{S}(m)$ as follows:

$$(2\pi)^N \widehat{S}(m) = S(e^{-im \cdot x}) = S(\cos m \cdot x) - iS(\sin m \cdot x)$$

for $m \in \Lambda_N$.

The following theorem concerning $\widehat{S}(m)$ will be proved here.

Theorem 3.1. *Given $S \in \mathcal{D}'(T_N)$. There exists a positive integer J such that*

(3.1) $$\sum_{1 \leq |m|} \left|\widehat{S}(m)\right|^2 / |m|^{4J} < \infty.$$

Proof of Theorem 3.1. With no loss in generality, we can suppose from the start that $\widehat{S}(0) = 0$. So, suppose that the conclusion to the theorem is false, i.e.,

$$(3.2) \qquad \sum_{1 \leq |m|} \left| \widehat{S}(m) \right|^2 / |m|^{4n} = \infty \quad \text{for} \quad n = 1, 2, ...$$

We say that $\phi \in \mathcal{D}(T_N)$ is a trigonometric polynomial if there exists an $r_0 > 0$ such that

$$\widehat{\phi}(m) = 0 \quad \text{for} \quad |m| > r_0.$$

Also, we observe from the fact that S is a linear functional on $\mathcal{D}(T_N)$ that if ϕ is a trigonometric polynomial, then

$$S(\phi) = (2\pi)^N \sum_{1 \leq |m|} \widehat{S}(m) \widehat{\phi}(-m).$$

It consequently follows from (3.2) and elementary Hilbert space theory that there exists a sequence of trigonometric polynomials $\{\phi_n\}_{n=1}^{\infty}$ with $\widehat{\phi}_n(0) = 0 \ \forall n$ such that

$$(3.3) \qquad \left| \sum_{1 \leq |m|} \widehat{S}(m) \, \widehat{\phi}_n(-m) / |m|^{2n} \right| \geq n \left\| \phi_n \right\|_{L^2} \quad \text{for} \quad n = 1, 2, ...$$

We set

$$(3.4) \qquad \psi_n(x) = \sum_{1 \leq |m|} \widehat{\phi}_n(m) e^{im \cdot x} / |m|^{2n} \, n \left\| \phi_n \right\|_{L^2},$$

and observe that $\psi_n \in \mathcal{D}(T_N)$ is a trigonometric polynomial. Also, from (3.3), we have that

$$(3.5) \qquad |S(\psi_n)| \geq (2\pi)^N \quad \text{for} \quad n = 1, 2, ...$$

On the other hand, we see from (3.4) that

$$\left\| \Delta^k \psi_n \right\|_{L^2}^2 = (2\pi)^N \sum_{1 \leq |m|} \left| \widehat{\phi}_n(m) \right|^2 / |m|^{4(n-k)} \, n^2 \left\| \phi_n \right\|_{L^2}^2$$

for $k, n = 1, 2, ...$ Consequently, for each fixed k,

$$\lim_{n \to \infty} \left\| \Delta^k \psi_n \right\|_{L^2}^2 = 0.$$

We conclude that $\psi_n \to 0$ in $\mathcal{D}(T_N)$. But then

$$\lim_{n \to \infty} S(\psi_n) = 0.$$

This last limit is a direct contradiction to the inequalities in (3.5). Hence, the assumption in (3.2) must be false, and there must exist a positive integer J such that the inequality in (3.1) is true. $\blacksquare$

The next theorem we prove is the following:

Theorem 3.2. *Suppose $S \in \mathcal{D}'(T_N)$ and $\phi \in \mathcal{D}(T_N)$. Then*

$$(3.6) \qquad S(\phi) = (2\pi)^N \sum_{m \in \Lambda_N} \widehat{S}(m)\widehat{\phi}(-m)$$

where the series in (3.6) is absolutely convergent.

Proof of Theorem 3.2. With no loss in generality, we can suppose from the start, once again, that $\widehat{S}(0) = 0$. Next, let J be the positive integer that appears in (3.1) of Theorem 3.1 above. Then given $\phi \in \mathcal{D}(T_N)$, it is clear from the definition of $\mathcal{D}(T_N)$ given above that

$$\sum_{m \in \Lambda_N} \left|\widehat{\phi}(m)\right|^2 |m|^{4J} < \infty.$$

Therefore, by Schwarz's inequality,

$$\left(\sum_{1 \le |m|} \left|\widehat{S}(m)\widehat{\phi}(-m)\right|\right)^2 \le \sum_{1 \le |m|} \left|\widehat{S}(m)\right|^2 / |m|^{4J} \sum_{1 \le |m|} \left|\widehat{\phi}(m)\right|^2 |m|^{4J},$$

and it follows that the series given in (3.6) is absolutely convergent.

Next, set

$$\phi_n(x) = \sum_{|m|^2 \le n} \widehat{\phi}(m) e^{im \cdot x}.$$

Then because $\phi_n(x) \in \mathcal{D}(T_N)$ is a trigonometric polynomial,

$$S(\phi_n) = (2\pi)^N \sum_{1 \le |m|^2 \le n} \widehat{S}(m)\widehat{\phi}(-m).$$

Also, it is clear that $\phi_n \to \phi$ in $\mathcal{D}(T_N)$ as $n \to \infty$. Consequently,

$$\lim_{n \to \infty} S(\phi_n) = S(\phi),$$

and it follows that

$$\lim_{n \to \infty} (2\pi)^N \sum_{1 \le |m|^2 \le n} \widehat{S}(m)\widehat{\phi}(-m) = S(\phi),$$

which proves the theorem. $\blacksquare$

For $\lambda, \phi \in \mathcal{D}(T_N)$ and $S \in \mathcal{D}'(T_N)$, we define

$$\lambda S(\phi) = S(\lambda\phi),$$

and observe that $\lambda S \in \mathcal{D}'(T_N)$, i.e., λS is a real linear functional that meets (3.0) above.

Next, we establish the following proposition:

Proposition 3.3. *Given* $\lambda \in \mathcal{D}(T_N)$ *and* $S \in \mathcal{D}'(T_N)$,

$$(3.7) \qquad \widehat{\lambda S}(m) = \sum_{p \in \Lambda_N} \widehat{\lambda}(p)\widehat{S}(m-p) \quad \text{for } m \in \Lambda_N,$$

where the series in (3.7) is absolutely convergent.

Proof of Proposition 3.3. To establish (3.7), we see that $\widehat{\lambda S}(m) = (2\pi)^{-N} S(\lambda e^{-im \cdot x})$. Also, we define $\lambda_n(x)$ to be the trigonometric polynomial

$$(3.8) \qquad \lambda_n(x) = \sum_{|p|^2 \leq n} \widehat{\lambda}(p) e^{im \cdot x}.$$

So $\widehat{\lambda_n}(p) = \widehat{\lambda}(p)$ for $|p|^2 \leq n$ and $= 0$ otherwise. It is clear that $\lambda_n \to \lambda$ in $\mathcal{D}(T_N)$. Consequently,

$$(3.9) \qquad \lim_{n \to \infty} S(\lambda_n e^{-im \cdot x}) = S(\lambda e^{-im \cdot x}).$$

Now,

$$\lambda_n(x) e^{-im \cdot x} = \sum_{|p|^2 \leq n} \widehat{\lambda}(p) e^{-i(m-p) \cdot x}.$$

So, with m fixed in Λ_N, for $n > |m|^2$,

$$(2\pi)^{-N} S(\lambda_n e^{-im \cdot x}) = \sum_{|p|^2 \leq |m|^2} \widehat{\lambda}(p)\widehat{S}(m-p) + \sum_{|m|^2+1 \leq |p|^2 \leq n} \widehat{\lambda}(p)\widehat{S}(m-p).$$

Let J be the positive integer from Theorem 3.1. Then because $\lambda \in \mathcal{D}(T_N)$, it follows that

$$\sum_{p \in \Lambda_N} \left|\widehat{\lambda}(p)\right|^2 |m-p|^{4J} < \infty.$$

Then we obtain from Schwarz's inequality that the series in (3.7) is absolutely convergent.

Also, we see from this last equality that

$$(2\pi)^{-N} \lim_{n \to \infty} S(\lambda_n e^{-im \cdot x}) = \lim_{n \to \infty} \sum_{|p|^2 \leq n} \widehat{\lambda}(p)\widehat{S}(m-p).$$

But then from (3.9), it follows that

$$(2\pi)^{-N} S(\lambda e^{-im \cdot x}) = \lim_{n \to \infty} \sum_{|p|^2 \leq n} \widehat{\lambda}(p)\widehat{S}(m-p),$$

which establishes the proposition. ■

In §3 of Chapter 3, we introduce the class $\mathcal{A}(T_N) \subset \mathcal{D}'(T_N)$ as follows:

$$\mathcal{A}(T_N) = \{S \in \mathcal{D}'(T_N) : \left|\widehat{S}(m)\right| \text{ meets (3.4) and (3.5) in Chapter 3}\}.$$

In particular, we have $S \in \mathcal{A}(T_N)$ means that

(3.10)

$$(i) \exists c > 0 \text{ such that } \left|\widehat{S}(m)\right| \leq c \quad \forall m \in \Lambda_N,$$

$$(ii) \lim_{\min(|m_1|,\ldots,|m_N|)\to\infty} \left|\widehat{S}(m)\right| = 0.$$

Also, in §3 of Chapter 3, we define the class $\mathcal{B}(T_N) \subset \mathcal{A}(T_N)$ by replacing $(ii)(3.10)$ with

$$\lim_{|m|\to\infty} \left|\widehat{S}(m)\right| = 0.$$

In Chapter 3, we will need the following proposition:

Proposition 3.4. *Given $\lambda \in \mathcal{D}(T_N)$ and $S \in \mathcal{A}(T_N)$. Then $\lambda S \in \mathcal{A}(T_N)$. If, furthermore, $S \in \mathcal{B}(T_N)$, then $\lambda S \in \mathcal{B}(T_N)$.*

Proof of Proposition 3.4. It is sufficient just to prove the first part of the above proposition, since the second part follows in a very similar manner.

From the fact that $\lambda \in D(T_N)$, it follows that $\sum_{p\in\Lambda_N} \left|\widehat{\lambda}(p)\right| < \infty$. Consequently, we have from (3.7) and from (3.10)(i) that

$$\left|\widehat{\lambda S}(m)\right| \leq c \sum_{p\in\Lambda_N} \left|\widehat{\lambda}(p)\right| \quad \forall m \in \Lambda_N.$$

So to complete the proof of the proposition, it remains to show that

(3.11)

$$\lim_{\min(|m_1|,\ldots,|m_N|)\to\infty} \left|\widehat{\lambda S}(m)\right| = 0.$$

We will establish (3.11) by showing that given $\varepsilon > 0$,

(3.12)

$$\limsup_{\min(|m_1|,\ldots,|m_N|)\to\infty} \left|\widehat{\lambda S}(m)\right| \leq \varepsilon.$$

Since $\sum_{p\in\Lambda_N} \left|\widehat{\lambda}(p)\right| < \infty$, choose $r_0 > 0$ so that

$$\sum_{|p|>r_0} \left|\widehat{\lambda}(p)\right| \leq \varepsilon c^{-1}.$$

Then it follows from (3.7), (3.10)(i), and this last inequality that

$$\left| \widehat{\lambda S}(m) \right| \leq \varepsilon + \left| \sum_{|p| \leq r_0} \widehat{\lambda}(p) \widehat{S}(m - p) \right|.$$

Now the number of $p \in \Lambda_N$ such that $|p| \leq r_0$ is finite. Therefore, from (3.10)(ii) and this last inequality, we see that

$$\limsup_{\min(|m_1|,\ldots,|m_N|) \to \infty} \left| \widehat{\lambda S}(m) \right| \leq \varepsilon.$$

This establishes (3.12), and the proof of the proposition is complete. ∎

Exercises.

1. Suppose $S \in \mathcal{D}'(T_N)$ and

$$\sum_{1 \leq |m|} \left| \widehat{S}(m) \right|^2 / |m|^{12} = \infty.$$

Prove there exists a trigonometric polynomial ϕ such that

$$\left| \sum_{1 \leq |m|} \frac{\widehat{S}(m) \widehat{\phi}(-m).}{|m|^6} \right| \geq 6 \left\| \phi \right\|_{L^2}.$$

2. Let $\lambda \in \mathcal{D}(T_N)$ and $S \in \mathcal{D}'(T_N)$. Define

$$S_1(\phi) = S(\lambda \phi) \quad \text{for } \phi \in \mathcal{D}(T_N).$$

Prove $S_1 \in \mathcal{D}'(T_N)$.

4. $\mathbf{H}_j(x)$ and the C^α-Condition

In Chapter 6, we will need a theorem connecting the functions $H_j(x)$, $j = 1, \ldots, N$, previously introduced in Chapter 3 (see Lemma 1.4) with functions $u \in C^\alpha(T_N)$, $0 < \alpha < 1$. So here with $N \geq 2$, we establish the following result:

Theorem 4.1. *Suppose $f \in C(T_N)$. Set*

$$u_j(x) = (2\pi)^{-N} \int_{T_N} H_j(x - y) f(y) \, dy \quad \text{for } j = 1, \ldots, N.$$

Then $u_j(x) \in C^\alpha(T_N)$, $0 < \alpha < 1$.

Proof of Theorem 4.1. For ease of notation, we will assume that $j = 1$ and show that

$$u_1(x) \in C^\alpha(T_N), 0 < \alpha < 1.$$

A similar proof will prevail for the other values of j.

In order to accomplish this last fact, we will demonstrate the following:

there exists a positive constant c_1 such that given $x, x' \in \mathbf{R}^N$ with

$$|x - x'| \leq 10^{-1},$$

then

(4.1) $$\left| u_1(x) - u_1(x') \right| \leq c_1 \left| x - x' \right| \log \left| x - x' \right|^{-1}$$

where c_1 depends only on N and $\|f\|_{L^\infty(T_N)}$.

To establish (4.1), using the periodicity of f and H_1, we observe that

$$u_1(x) - u_1(x') = v_1(z) - v_1(0)$$

where $z = x - x'$,

$$v_1(z) = (2\pi)^{-N} \int_{T_N} H_1(z - y) f_1(y)\, dy,$$

and

$$f_1(y) = f(y + x') \quad \text{for } \forall y \in \mathbf{R}^N.$$

Since $\|f\|_{L^\infty(T_N)} = \|f_1\|_{L^\infty(T_N)}$, we see that to establish (4.1), it is sufficient to show that (4.1) holds for the special case when $x' = 0$ and $|x| \leq 10^{-1}$. So the proof of the theorem will be complete when we show the following:

(4.2) $$\exists c_1 > 0 \text{ such that } |u_1(x) - u_1(0)| \leq c_1 |x| \log |x|^{-1}$$

for $|x| \leq 10^{-1}$ where c_1 depends only on N and $\|f\|_{L^\infty(T_N)}$.

Using the fact that $H_1(x)$ is harmonic in $\mathbf{R}^N \setminus \cup_{m \in \Lambda_N} \{2\pi m\}$, we see that it is sufficient to establish the inequality in (4.2) when u_1 is replaced by w_1 where

(4.3) $$w_1(x) = (2\pi)^{-N} \int_{B(0,\frac{1}{2})} H_1(x - y) f(y)\, dy$$

and c_1 depends only on N and $\|f\|_{L^\infty(T_N)}$.

Also, we observe from Lemma 1.4 in Chapter 3 that there is a function $\eta_1(x)$, which is harmonic in $B(0, 1)$ such that

$$\eta_1(x) = H_1(x) + (2\pi)^N x_1 / |S_{N-1}| |x|^N \quad \text{for } x \in B(0, 1) \setminus \{0\}.$$

So using this $\eta_1(x)$ in conjunction with (4.3), we see that the inequality in (4.2) will be established if we can show

(4.4) $$\exists c_1 > 0 \text{ such that } |W_1(x) - W_1(0)| \leq c_1 |x| \log |x|^{-1}$$

for $|x| \leq 10^{-1}$ where c_1 depends only on N and $\|f\|_{L^\infty(B(0,\frac{1}{2}))}$ and

(4.5) $$W_1(x) = \int_{B(0,\frac{1}{2})} (x_1 - y_1) |x - y|^{-N} f(y)\, dy.$$

To show that (4.4) is valid, we write

$$B(0, 1/2) = A_{|x|} \cup B_{|x|}$$

where

$$A_{|x|} = B(0, 2\,|x|) \quad \text{and} \quad B_{|x|} = B\left(0, 1/2\right) \backslash B(0, 2\,|x|).$$

Then
(4.6)

$$W_1\left(x\right) - W_1\left(0\right) = [\int_{A_{|x|}} + \int_{B_{|x|}}][(x_1 - y_1)\,|x - y|^{-N} + y_1\,|y|^{-N}]f\left(y\right)dy.$$

Now,

$$\left|\int_{A_{|x|}}[(x_1 - y_1)\,|x - y|^{-N} + y_1\,|y|^{-N}]f\left(y\right)dy\right|$$

$$\leq 2\,\|f\|_{L^\infty\left(B\left(0,\frac{1}{2}\right)\right)}\int_{B(0,3|x|)}|y|^{-(N-1)}\,dy.$$

Since

$$\int_{B(0,3|x|)}|y|^{-(N-1)}\,dy = \gamma_N\,|x|,$$

where γ_N is a positive constant depending only on N, it follows from this last inequality and (4.6) that the inequality in (4.4) will be valid provided we show

$$(4.7) \quad \exists \gamma_N^* > 0 \text{ such that } \int_{B_{|x|}}|g\left(y\right) - g\left(y - x\right)|\ dy \leq \gamma_N^*\,|x|\log|x|^{-1}$$

for $|x| \leq 10^{-1}$ where γ_N^* depends only on N and

$$(4.8) \qquad\qquad\qquad g\left(y\right) = y_1/\left|y\right|^N.$$

From the mean-value theorem, for $y \in B(0, 1/2)$ with $|y| \geq 2\,|x|$ and $|x| \leq 10^{-1}$, we have $\exists \theta$ with $0 < \theta < 1$ such that

$$g\left(y\right) - g\left(y - x\right) = \sum_{j=1}^{N}\frac{\partial g}{\partial y_j}\left(y - \theta x\right)x_j.$$

But from (4.8), we see that

$$\left|\frac{\partial g}{\partial y_j}(y)\right| \leq \delta_j\,|y|^{-N} \text{ for } j = 1, ..., N$$

where δ_j is a constant depending only on N. Since $|y - \theta x| \geq |y|\,/2$, we conclude that

$$|g\left(y\right) - g\left(y - x\right)| \leq \delta_N^*\,|x|\,|y|^{-N}$$

for $y \in B(0, 1/2)$ with $|y| \geq 2\,|x|$ and $|x| \leq 10^{-1}$ where δ_N^* is a constant depending only on N.

Consequently,

$$\int_{B_{|x|}}|g\left(y\right) - g\left(y - x\right)|\ dy \leq \delta_N^*\,|x|\,\delta_N^{**}\int_{2|x|}^{1/2}r^{-1}dr$$

for $|x| \leq 10^{-1}$ where δ_N^{**} is another constant depending only on N.

The inequality in (4.7) follows immediately from this last inequality, and the proof of the theorem is complete.   ■

Exercises.

1. Let $H_1(x)$ be the function previously introduced in Lemma 1.4 of Chapter 3. Suppose that $f \in C(T_3)$. Define

$$u(x) = \int_{T_3 \backslash B(0,1)} H_1(x - y) f(y)\, dy.$$

Prove that $u \in C^\infty \left(B\left(0, \tfrac{1}{8}\right) \right)$.

APPENDIX C

Harmonic and Subharmonic Functions

1. Harmonic Functions

Given $\Omega \subset \mathbf{R}^N$, an open connected set, $N \geq 1$, we say $u(x)$ is harmonic in Ω if $u \in C^2(\Omega)$ and furthermore

$$\Delta u(x) = 0 \quad \forall x \in \Omega,$$

where

$$\Delta u(x) = \sum_{j=1}^{N} \frac{\partial u_j}{\partial x^j}(x).$$

The operator Δ is called the Laplace operator.

A well-known fact about a function u harmonic in Ω is that the mean-value theorem holds for u, namely, if for an $\varepsilon > 0$, $B(x, \rho + \varepsilon) \subset \Omega$, then for $\rho > 0$,

$$\frac{1}{|B(x,\rho)|} \int_{B(x,\rho)} u(y)\,dy = u(x)$$

where $|B(x,\rho)|$ stands for the N-volume of $B(x,\rho)$. (For the numerical value of $|B(x,\rho)|$, see (3.4) in Chapter 1. For the proof of the mean-value theorem for harmonic functions, see [ABR, Chapter 1]. The proof of this theorem can also be found in many other places.)

We shall need a number of theorems about harmonic functions that involve spherical harmonics. With $\mathcal{A}(R_1, R_2)$ designating the annular region

(1.1) $$\mathcal{A}(R_1, R_2) = \{x : R_1 < |x| < R_2\},$$

the first theorem of this nature that we prove is the following:

Theorem 1.1. *Suppose* $\{Q_n^\Diamond(x)\}_{n=1}^\infty$ *and* $\{Q_n^{\Diamond\Diamond}(x)\}_{n=1}^\infty$ *are two sequences of spherical harmonics of degree* n. *Suppose, furthermore,*

(1.2) $$\sum_{n=0}^\infty Q_n^\Diamond(x) = \sum_{n=0}^\infty Q_n^{\Diamond\Diamond}(x) \quad \forall x \in \mathcal{A}(R_1, R_2),$$

315

where *both series converge uniformly on compact subsets of* $\mathcal{A}(R_1, R_2)$, *the annular region given by* (1.1). Then

$$Q_n^\diamond(x) = Q_n^{\diamond\diamond}(x) \quad \forall x \in \mathbf{R}^N \text{ and } \forall n.$$

Proof of Theorem 1.1. Write $Q_n^\diamond(x) = r^n Y_n^\diamond(\xi)$ and $Q_n^{\diamond\diamond}(x) = r^n Y_n^{\diamond\diamond}(\xi)$ where $x = r\xi$ and $\xi \in \partial B(0,1)$. Then, because both series converge uniformly on $\partial B(0, \rho)$ when $\rho = (R_1 + R_2)/2$, it follows from (1.2) above and the orthogonality condition given in (3.14) of Appendix A and on the addition formula given in (3.9) of Appendix A that

$$\int_{\partial B(0,1)} [Q_n^\diamond(\rho\xi) - Q_n^{\diamond\diamond}(\rho\xi)] C_n^\nu(\eta \cdot \xi) dS(\xi) = 0 \quad \forall \eta \in \partial B(0,1) \text{ and } \forall n,$$

where $C_n^\nu(t)$ is the n-th Gegenbauer polynomial discussed in Appendix A, §3. Hence,

$$\int_{S_{N-1}} C_n^\nu(\xi \cdot \eta)[Y_n^\diamond(\xi) - Y_n^{\diamond\diamond}(\xi)] dS(\xi) = 0 \quad \forall \eta \in S_{N-1} \text{and } \forall n.$$

But then it follows from (3.12) in Appendix A that

$$Y_n^\diamond(\eta) = Y_n^{\diamond\diamond}(\eta) \ \forall \eta \in S_{N-1} \text{ and } \forall n,$$

which completes the proof of the theorem.   ∎

Theorem 1.2. *With $N \geq 2$, suppose $u(x)$ is harmonic in $B(0,R)$, $R > 0$. Then $\exists \{Q_n(x)\}_{n=0}^\infty$, where $Q_n(x)$ is a spherical harmonic of degree n, such that*

$$(1.3) \qquad u(x) = \sum_{n=0}^\infty Q_n(x) \ \forall x \in B(0, R).$$

Also, the series in (1.3) converges uniformly on compact subsets of $B(0,R)$.

Proof of Theorem 1.2. For $N = 2$, the proof of this theorem follows immediately from the well-known theory of holomorphic functions and the fact that $u = Re(f)$ in $B(0, R)$ where f is holomorphic in $B(0, R)$.

For the rest of the proof, we suppose that $N \geq 3$. Let $R_1 < R$. Then it follows from Theorem 1.1 that the theorem will be established, if we show the following:

(i) $\exists \{Q_n(x)\}_{n=0}^\infty$, where $Q_n(x)$ is a spherical harmonic of degree n, such that

$$(1.4) \qquad u(x) = \sum_{n=0}^\infty Q_n(x) \ \forall x \in B(0, R_1);$$

(ii) the series in (1.4) converges uniformly on compact subsets of $B(0, R_1)$.

To see that (1.4) is valid, we observe that $u \in C(\partial B(0, R_1))$. Hence, from the well-known technique for solving the Dirichlet problem for the closed ball $\overline{B}(0, R_1)$ (see [Ev, p. 41] or [Pi, p. 357]), we have that

$$(1.5) \qquad u(x) = \frac{R_1^2 - |x|^2}{|S_{N-1}| R_1} \int_{\partial B(0, R_1)} \frac{u(y)}{|x - y|^N} dS(y) \quad \forall x \in B(0, R_1).$$

With $y = R_1 \eta$ and $x = \rho \xi$ where $\xi, \eta \in S_{N-1}$, we observe from (3.29) in Appendix A that

$$\frac{R_1^2 - |x|^2}{|x - y|^N} = R_1^{-(N-2)} \frac{1 - (\frac{\rho}{R_1})^2}{[1 - 2\frac{\rho}{R_1}\xi \cdot \eta + (\frac{\rho}{R_1})^2]^{N/2}}$$

$$= R_1^{-(N-2)} \sum_{n=0}^{\infty} [\frac{N - 2 + 2n}{N - 2}] C_n^\nu(\xi \cdot \eta)(\frac{\rho}{R_1})^n,$$

where because of the bound in (3.30) of Appendix A, the series converges uniformly for $0 \leq \rho \leq R_2 < R_1$ and $\xi, \eta \in S_{N-1}$. As a consequence, we obtain from (1.5) that

$$(1.6) \qquad u(\rho\xi) = \sum_{n=0}^{\infty} \frac{N - 2 + 2n}{(N - 2)|S_{N-1}|}(\frac{\rho}{R_1})^n \int_{S_{N-1}} C_n^\nu(\xi \cdot \eta)u(R_1\eta)dS(\eta).$$

From the addition formula for surface spherical harmonics given in (3.9) of Appendix A, we see that the integral on the right-hand side of (1.6) represents a surface spherical harmonic of degree n, which we call $Y_n(\xi)$. Also, we have from (3.30) in Appendix A that there exists a constant c such that

$$(1.7) \qquad |Y_n(\xi)| \leq cn^{2\nu-1} \quad \forall \xi \in S_{N-1} \text{ and } \forall n.$$

Setting

$$Q_n(x) = \frac{N - 2 + 2n}{(N - 2)|S_{N-1}|}(\frac{\rho}{R_1})^n Y_n(\xi),$$

we have that $Q_n(x)$ is a spherical harmonic of degree n, and furthermore, from (1.7), that the series $\sum_{n=0}^{\infty} Q_n(x) = u(x)$ converges uniformly for $|x| \leq R_2 < R_1$. ∎

In Chapter 3, we shall need the following two theorems regarding point singularities for harmonic functions. We shall state them separately for dimension $N = 2$ and dimension $N \geq 3$. The two proofs will be different because in the former case, we'll use the theory of holomorphic functions.

Theorem 1.3. *With $N = 2$, suppose $u(x)$ is harmonic in $B(0,1)\backslash\{0\}$. Suppose, also, $u \in L^1(B(0,1))$. Then, $\exists$ constants b_0, b_1, and b_2 and a function $u_0(x)$ harmonic in $B(0,1)$ such that with $r = |x|$,*

$$(1.8) \quad u(x) = b_0 \log r + b_1 x_1 r^{-2} + b_2 x_2 r^{-2} + u_0(x) \quad \forall x \in B(0,1)\backslash\{0\}.$$

Theorem 1.4. *With $N \geq 3$, suppose $u(x)$ is harmonic in $B(0,1)\backslash\{0\}$. Suppose, also, $u \in L^1(B(0,1))$. Then, $\exists$ constants $b_0, b_1, ..., b_N$ and a function $u_0(x)$ harmonic in $B(0,1)$ such that with $r = |x|$,*

$$(1.9) \qquad u(x) = b_0 r^{-(N-2)} + \sum_{j=1}^{N} b_j x_j r^{-N} + u_0(x) \qquad \forall x \in B(0,1)\backslash\{0\}.$$

Proof of Theorem 1.3. Since we are going to use holomorphic function theory to establish this theorem, for convenience, we will set $x_1 = x$ and $x_2 = y$, and will designate the half-open line segment $\{(x,0) : 0 \leq x < 1\}$ by l_s. Then, for $(x,0)$ in the open line segment, we define

$$v(x,0) = -\int_{1/2}^{x} u_y(s,0)ds \quad \text{for } 0 < x < 1.$$

Next, we cut the unit disk, $B(0,1)$, with the half-open line segment l_s and define

$$v(x,y) = \int_{C_\gamma} u_x(s,t)dt - u_y(s,t)ds \quad \text{for } (x,y) \in B(0,1)\backslash l_s,$$

where C_γ is any piece-wise C^1 curve starting at $(1/2,0)$ and going to (x,y) in $B(0,1)\backslash l_s$ initially through the first quadrant. Also, letting $C(0,r)$ be the circle with center 0 and radius r, we see from the harmonicity of u in the punctured unit disk and from Green's theorem that there is a constant β such that

$$\int_{C(0,r)} u_x(s,t)dt - u_y(s,t)ds = \beta \quad \text{for } 0 < r < 1,$$

where the integration is in the counter-clockwise direction.

As a consequence of this last fact, we see that the function

$$v(x,y) - \beta\theta(x,y)/2\pi \text{ is both single-valued and harmonic in } B(0,1)\backslash\{0\},$$

where $\theta = \text{Im}(\log z)$, *i.e.*, $x = r\cos\theta$ and $y = r\sin\theta$. But then it follows that

$$u(x,y) - \beta \log r = \text{Re}(f(z)),$$

where $f(z)$ is holomorphic in $B(0,1)\backslash\{0\}$. So, from Laurent's theorem, we obtain that

$$u(x,y) - \beta \log r = \text{Re}(\sum_{n=-\infty}^{\infty} c_n r^n e^{in\theta})$$

where $\sum_{n=0}^{\infty} c_n r^n e^{in\theta}$ converges absolutely and uniformly on compact subsets of $B(0,1)$ and $\sum_{n=-\infty}^{-1} c_n r^n e^{in\theta}$ converges absolutely and uniformly on compact subsets of $B(0,1)\backslash\{0\}$. We conclude there exists a function $u_0(x,y)$ which is harmonic in B(0,1) such that

$$(1.10) \qquad u(x,y) - \beta \log r - u_0(x,y) = \sum_{n=1}^{\infty}(a_n \cos n\theta + b_n \sin n\theta)r^{-n},$$

where the series converges uniformly on compact subsets of $B(0,1)\backslash\{0\}$.

We set $w(x,y) = u(x,y) - \beta \log r - u_0(x,y)$ and let $k \geq 2$ be any fixed integer. Then

$$\int_0^{2\pi} w(r\cos\theta, r\sin\theta)\cos k\theta d\theta = \pi a_k r^{-k} \quad \text{for } 0 < r < 1/2.$$

Since $w \in L^1[B(0,1/2)]$, we obtain from this last equality after multiplying through by r^{-k} and integrating from 0 to ε that

$$\pi |a_k| \leq \varepsilon^{k-2} \int_{B(0,\varepsilon)} |w(x,y)|\, dxdy.$$

But $\int_{B(0,\varepsilon)} |w(x,y)|\, dxdy \to 0$, as $\varepsilon \to 0$. So we conclude from this last inequality that $a_k = 0$ for $k \geq 2$.

A similar technique shows that $b_k = 0$ for $k \geq 2$. Hence, we obtain from (1.10) that

$$u(x,y) - \beta \log r - u_0(x,y) = (a_1 \cos\theta + b_1 \sin\theta)r^{-1}$$

for $(x,y) \in B(0,1)\backslash\{0\}$, and the proof of the theorem is complete. ■

In order to prove Theorem 1.4, we shall need some lemmas.

Lemma 1.5. *With $N \geq 3$, suppose $u(x)$ is harmonic in $B(0,R_1)\backslash\{0\}$, where $R_1 > 0$. Suppose, also, $u \in L^\infty(B(0,R_1)\backslash\{0\})$. Then there exists a function $u_0(x)$ that is harmonic in $B(0,R_1/2)$ such that*

$$u(x) = u_0(x) \quad \forall x \in B(0,R_1/2)\backslash\{0\}.$$

What the conclusion of Lemma 1.5 states is that the singularity at the origin is removable.

Proof of Lemma 1.5. From a consideration of the function $v(x) = u(R_1 x)$, we see from the start, without loss in generality, 2nd we can assume that $R_1 = 1$.

To prove the lemma, we let $u_0(x)$ be the solution to the Dirichlet problem on $\overline{B}(0,1/2)$ for the boundary value $u(x)$ on $\partial B(0,1/2)$. We then set

$$w(x) = u(x) - u_0(x),$$

and see that the lemma will be established once we show

$$w(x) = 0 \quad \forall x \in B(0,1/2)\backslash\{0\}.$$

We will show

(1.11) $$w(x) \leq 0 \qquad \forall x \in B(0,1/2)\backslash\{0\}.$$

Similar reasoning will show the reverse inequality

$$w(x) \geq 0 \ \forall x \in B(0,1/2)\backslash\{0\}.$$

To establish the inequality in (1.12), we introduce the function

$$\Psi(x) = |x|^{-(N-2)} - 2^{N-2} \quad \forall x \in \mathbf{R}^N \backslash \{0\}.$$

Then $\Psi(x)$ is harmonic in $B(0, 1/2) \backslash \{0\}$ and also positive in this same region.

Let x_0 be a fixed but arbitrary point in $B(0, 1/2) \backslash \{0\}$, and let $\varepsilon > 0$, be given. If we can show

(1.12) $$w(x_0) \leq \varepsilon \Psi(x_0),$$

then by letting $\varepsilon \to 0$, we obtain $w(x_0) \leq 0$, and hence, the inequality in (1.11).

To establish the inequality in (1.12), we observe that there is a positive constant M such that

$$w(x) \leq M \quad \forall x \in B(0, 1/2) \backslash \{0\}.$$

Since $\varepsilon \Psi(x) \to \infty$, as $|x| \to 0$, we can find r_0 with $|x_0| > r_0 > 0$ such that

$$\varepsilon \Psi(x) \geq M \quad \forall x \in \partial B(0, r_0).$$

With $\mathcal{A}(r_0, 1/2)$ the annular region defined in (1.1) above, we see that $w(x) - \varepsilon \Psi(x)$ is harmonic in $\mathcal{A}(r_0, 1/2)$, continuous in the closed region $\overline{\mathcal{A}}(r_0, 1/2)$, 0 for $|x| = 1/2$, and ≤ 0 for $|x| = r_0$. Therefore, by the maximum principle for harmonic functions, $w(x) - \varepsilon \Psi(x) \leq 0$ for $x \in \mathcal{A}(r_0, 1/2)$. In particular,

$$w(x_0) - \varepsilon \Psi(x_0) \leq 0.$$

This gives the inequality in (1.12) and establishes the lemma. ∎

The next lemma that we need will involve the Kelvin transformation. In order to define this transformation, we first have to discuss inversions. Given $\Omega \subset \mathbf{R}^N \backslash \{0\}$ an open set, we define Ω^*, the inversion of Ω, to be

$$\Omega^* = \{\frac{x}{|x|^2}, \ x \in \Omega\}.$$

It is easy to check that if Ω is an open N-ball, then so is Ω^*. Also, it is easy to see that $y \in \Omega^* \Rightarrow \frac{y}{|y|^2} \in \Omega$ and consequently that $(\Omega^*)^* = \Omega$.

With Ω an open set as above and $u \in C(\Omega)$, we define $\mathcal{K}(u(x))$, the Kelvin transformation of u, as follows:

(1.13) $$\mathcal{K}(u(x)) = |x|^{2-N} u(\frac{x}{|x|^2}) \quad \forall x \in \Omega^*.$$

If $Q_n(x) = r^n Y_n(\xi)$ is a spherical harmonic of degree n, then

(1.14) $$\mathcal{K}(Q_n(x)) = r^{2-N-n} Y_n(\xi).$$

Using the spherical coordinate form of Δ given in (3.5) of Appendix A, it is easy to check that

$$r^{2-N-n}C_n^\nu(\cos\theta) \text{ is harmonic in } \mathbf{R}^N\backslash\{0\}.$$

Hence, it follows that as a function of r and ξ, $r^{2-N-n}C_n^\nu(\eta\cdot\xi)$ is harmonic in $\mathbf{R}^N\backslash\{0\}$ for every $\eta \in S_{N-1}$. Using this fact in conjunction with the formula involving the Riemann integral in (3.12) in Appendix A, we obtain that $r^{2-N-n}Y_n(\xi)$ is harmonic in $\mathbf{R}^N\backslash\{0\}$. Consequently, we have from (1.14) that

(1.15) $\mathcal{K}(Q_n(x))$ is harmonic in $\mathbf{R}^N\backslash\{0\}$,

for every spherical harmonic $Q_n(x)$ of degree n.

The next lemma we establish is the following:

Lemma 1.6. *With $N \geq 3$, suppose $u(x)$ is harmonic in Ω, where $\Omega \subset \mathbf{R}^N\backslash\{0\}$ is an open set. Then $\mathcal{K}(u(x))$, the Kelvin transformation of u, is harmonic in Ω^*.*

Proof of Lemma 1.6. It follows from (1.15) that if $f(x)$ is a harmonic polynomial, i.e., a finite linear combination of spherical harmonics, then $K(f(x))$ is harmonic in $\mathbf{R}^N\backslash\{0\}$.

To establish the lemma, it is sufficient to show the following:
(1.16)
$$B(x_0, 3r_0) \subset \Omega \text{ where } r_0 > 0 \Rightarrow \mathcal{K}(u(x)) \text{ is harmonic in } [B(x_0, r_0)]^*.$$

Applying Theorem 1.2 above to the N-ball $B(x_0, 3r_0)$, we see that there is a sequence of harmonic polynomials $\{f_n(x)\}_{n=1}^\infty$ such that

$$\lim_{n\to\infty} f_n(x) = u(x) \text{ uniformly } \forall x \in B(x_0, 2r_0).$$

Therefore,

(1.17) $\lim_{n\to\infty} \mathcal{K}[f_n(x)] = \mathcal{K}[u(x)]$ uniformly $\forall x \in [B(x_0, r_0)]^*$.

But as we observed above, $\mathcal{K}[f_n(x)]$ is harmonic in $[B(x_0, r_0)]^*$. Furthermore, the uniform limit of harmonic functions is harmonic. So the statement in (1.17) shows that the statement in (1.16) is indeed true, and the proof of the lemma is complete. ∎

Next, we establish the folowing lemma:

Lemma 1.7. *With $N \geq 3$ and $r_0 > 0$, let $\Omega = \{x : |x| > r_0\}$. Suppose $u(x)$ is harmonic in Ω, and*

(1.18) $$\limsup_{|x|\to\infty} |x|^{N-2}|u(x)| < \infty.$$

Then there exists a sequence $\{Y_n(\xi)\}_{n=0}^{\infty}$ of surface spherical harmonics of degree n such that

$$u(x) = \sum_{n=0}^{\infty} r^{2-N-n} Y_n(\xi) \quad \forall x \in \Omega,$$

where the series converges unifomly on compact subsets of Ω.

Proof of Lemma 1.7. From Lemma 1.6, we have that $\mathcal{K}[u(x)]$ is harmonic in the punctured N-ball, $B(0, r_0^{-1})\backslash\{0\}$, where $\mathcal{K}[u(x)]$ is given by (1.13). It follows from this formula for $\mathcal{K}[u(x)]$ and the *lim* sup in (1.18) that

$$\limsup_{|x|\to 0} |\mathcal{K}[u(x)]| < \infty.$$

Consequently, $\mathcal{K}[u(x)]$ is uniformly bounded in $B(0, r_0^{-1}/2)\backslash\{0\}$. Hence, by Lemma 1.5, the singularity at the origin is removable. Therefore, by Theorem 1.2, there exists a sequence $\{Y_n(\xi)\}_{n=0}^{\infty}$ of surface spherical harmonics of degree n such that

$$\mathcal{K}[u(x)] = \sum_{n=0}^{\infty} r^n Y_n(\xi) \quad \forall x \in B(0, r_0^{-1})\backslash\{0\},$$

where the series converges uniformly on compact subsets of $B(0, r_0^{-1})\backslash\{0\}$. But then applying the Kelvin transformation $\mathcal{K}$ to both sides of this last equality, we see that

$$(1.19) \qquad u(x) = \sum_{n=0}^{\infty} \mathcal{K}[r^n Y_n(\xi)] \quad \forall x \in \Omega,$$

where the series converges uniformly on compact subsets of Ω.

As we have observed previously,

$$\mathcal{K}[r^n Y_n(\xi)] = r^{2-N-n} Y_n(\xi).$$

Putting this value in the series on the right-hand side of the equality in (1.19) for each n completes the proof of the lemma. ■

The next lemma is the analogue of Theorem 1.1 for $\mathcal{K}[Q_n(x)]$, the Kelvin transform of a spherical harmonic.

Lemma 1.8. *Suppose $\{Y_n^{\diamond}(\xi)\}_{n=1}^{\infty}$ and $\{Y_n^{\diamond\diamond}(\xi)\}_{n=1}^{\infty}$ are two sequences of surface spherical harmonics of degree n. Suppose, in addition,*

$$(1.20) \qquad \sum_{n=0}^{\infty} r^{2-N-n} Y_n^{\diamond}(\xi) = \sum_{n=0}^{\infty} r^{2-N-n} Y_n^{\diamond\diamond}(\xi) \quad \forall r\xi \in \mathcal{A}(R_1, R_2),$$

where *both series converge uniformly on compact subsets of $\mathcal{A}(R_1, R_2)$, the annular region given by* (1.1). Then

$$Y_n^{\Diamond}(\xi) = Y_n^{\Diamond\Diamond}(\xi) \quad \forall \xi \in S_{N-1} \text{ and } \forall n \geq 0.$$

Proof of Lemma 1.8.   Because both series converge uniformly on $\partial B(0, \rho)$ when $\rho = (R_1 + R_2)/2$, it follows from (1.20) and the orthogonality condition given in (3.14) of Appendix A that

$$\rho^{2-N-n} \int_{\partial B(0,1)} [Y_n^{\Diamond}(\xi) - Y_n^{\Diamond\Diamond}(\xi)] C_n^{\nu}(\eta \cdot \xi) dS(\xi) = 0 \quad \forall \eta \in \partial B(0,1)$$

and $\forall n \geq 0$.

But then it follows from (3.12) in Appendix A that

$$Y_n^{\Diamond}(\eta) = Y_n^{\Diamond\Diamond}(\eta) \quad \forall \eta \in \partial B(0,1) \text{ and } \forall n \geq 0,$$

which completes the proof of the lemma.   ∎

The final lemma we need for the proof of Theorem 1.4 is the following:

Lemma 1.9.   *With $N \geq 3$, suppose $u(x)$ is harmonic in $B(0,1)\backslash\{0\}$. Then there exists $\{Y_n(\xi)\}_{n=0}^{\infty}$, a sequence of surface spherical harmonics of degree n, and a function $u_0(x)$, which is harmonic in $B(0,1/2)$ such that with $x = r\xi$,*

$$u(x) = u_0(x) - \sum_{n=0}^{\infty} r^{2-N-n} Y_n(\xi) \quad \forall x \in B(0,1/2)\backslash\{0\},$$

where the series converges uniformly on compact subsets of $B(0,1/2)\backslash\{0\}$.

Proof of Lemma 1.9.   We select a positive $R_1 < 1/2$, and apply Green's second identity in a standard manner (see [Ke, p. 215 and p. 261] or [Jo, p. 97]) in the annular region $\mathcal{A}(R_1, 1/2)$ to the functions $u(x)$ and $|x - y|^{2-N}$ and obtain

$$
\begin{aligned}
u(x) \;=\; & \frac{(N-2)^{-1}}{|S_{N-1}|} \Big[\int_{\partial B(0,\frac{1}{2})} (|x-y|^{2-N} \nabla u \cdot \mathbf{n} - u \nabla |x-y|^{2-N} \cdot \mathbf{n}) dS(y) \\
& - \int_{\partial B(0,R_1)} (|x-y|^{2-N} \nabla u \cdot \mathbf{n} - u \nabla |x-y|^{2-N} \cdot \mathbf{n}) dS(y) \Big]
\end{aligned}
$$

$\forall x \in \mathcal{A}(R_1, 1/2)$ where $|S_{N-1}|$ is the volume of the unit $(N-1)$-sphere and $\mathbf{n}$ is the outward pointing unit normal. We call the first expression on the right-hand side of the above equality $u_0(x)$ and the second $v(x)$. So

(1.21) $$u(x) = u_0(x) - v(x)$$

where $u_0(x)$ is harmonic in $B(0, 1/2)$ and $v(x)$ is harmonic for $|x| > R_1$. An easy calculation shows

$$\lim_{|x| \to \infty} |x|^{N-2} v(x) = |S_{N-1}|^{-1} \int_{\partial B(0, R_1)} \nabla u \cdot \mathbf{n} \, dS(y).$$

Therefore, it follows from Lemma 1.7 that there exists a sequence $\{Y_n(\xi)\}_{n=0}^{\infty}$ of surface spherical harmonics of degree n such that

$$v(x) = \sum_{n=0}^{\infty} r^{2-N-n} Y_n(\xi) \quad \forall x \in \mathcal{A}(R_1, 1/2)$$

where the series converges uniformly on compact subsets of the annulus $\mathcal{A}(R_1, 1/2)$. Consequently, we see from (1.21) that

$$(1.22) \qquad u(x) = u_0(x) - \sum_{n=0}^{\infty} r^{2-N-n} Y_n(\xi) \quad \forall x \in \mathcal{A}(R_1, 1/2).$$

To see that this last equality is valid in all of $B(0, 1/2) \backslash \{0\}$, let x_0 be an arbitrary but fixed point in $B(0, 1/2) \backslash \{0\}$, with $|x_0| < R_1$. We select an $R_1' > 0$ with $R_1' < |x_0| < R_1$. Then we proceed exactly as above and obtain a sequence of surface spherical harmonics $\{Y_n^{\Diamond}(\xi)\}_{n=0}^{\infty}$ such that

$$u(x) = u_0(x) - \sum_{n=0}^{\infty} r_n^{2-N-n} Y_n^{\Diamond}(\xi) \quad \forall x \in \mathcal{A}(R_1', 1/2).$$

But then it follows from (1.22) that

$$\sum_{n=0}^{\infty} r^{2-N-n} Y_n(\xi) = \sum_{n=0}^{\infty} r_n^{2-N-n} Y_n^{\Diamond}(\xi) \quad \forall x \in \mathcal{A}(R_1, 1/2),$$

and consequently from Lemma 1.6 that

$$Y_n^{\Diamond}(\xi) = Y_n(\xi) \quad \forall \xi \in S_{N-1}.$$

We conclude that the equality in (1.22) holds also at x_0 and therefore $\forall x \in B(0, 1/2) \backslash \{0\}$.

In a similar manner, we conclude that the series in (1.22) converges uniformly on compact subsets of $B(0, 1/2) \backslash \{0\}$. ∎

Proof of Theorem 1.4. Since our given $u(x)$ is harmonic in $B(0, 1) \backslash \{0\}$, we invoke Lemma 1.9 and obtain a sequence of surface spherical harmonics $\{Y_n(\xi)\}_{n=0}^{\infty}$ and a function $u_0(x)$ harmonic in $B(0, 1/2)$ such that

$$(1.23) \qquad u(x) - u_0(x) = -\sum_{n=0}^{\infty} r^{2-N-n} Y_n(\xi) \quad \forall x \in B(0, 1/2) \backslash \{0\},$$

where the series in this last equality converges uniformly on compact subsets of $B(0, 1/2) \backslash \{0\}$.

With $0 < r < 1/4$, we multiply both sides of (1.23) by $C_k^\nu(\xi \cdot \eta)$, integrate over S_{N-1}, and use formulas (3.12) and (3.14) in Appendix A to obtain

$$(1.24) \quad r^{N+k-2} \int_{S_{N-1}} [u(r\xi) - u_0(r\xi)]C_k^\nu(\xi \cdot \eta)dS(\xi) = -\frac{2\pi^{\nu+1}}{\Gamma(\nu)(k+\nu)}Y_k(\eta)$$

$\forall \eta \in S_{N-1}$.

Taking $k \geq 2$, we integrate the absolute value of both sides of (1.24) again, this time from 0 to ε to obtain

$$\varepsilon|Y_k(\eta)| \leq c\varepsilon^{k-1} \int_{B(0,\varepsilon)} |u(x) - u_0(x)|\,dx$$

$\forall \eta \in S_{N-1}$ where c is a constant independent of ε. Since $u \in L^1(B(0,1))$ and u_0 is harmonic in $B(0,1/2)$, it is clear from this last inequality that

$$|Y_k(\eta)| \leq \varepsilon^{k-2}o(1) \text{ as } \varepsilon \to 0 \quad \forall \eta \in S_{N-1}.$$

We conclude that $Y_k(\eta) = 0 \; \forall \eta \in S_{N-1}$ for $k \geq 2$.

As a consequence, we obtain from (1.24) that

$$u(x) = u_0(x) - \frac{Y_0(\xi)}{r^{N-2}} - \frac{Y_1(\xi)}{r^{N-1}} \quad \forall x \in B(0,1/2)\backslash\{0\}.$$

The conclusion to the theorem follows easily from this last equality. ∎

We will need the following theorem in Chapter 3.

Theorem 1.10. *Let $u(x)$ be harmonic in $\mathbf{R}^N$, $N \geq 2$. Suppose there is a positive integer n and a constant $c > 0$, such that*

$$(1.25) \qquad\qquad |u(x)| \leq c|x|^n \quad \text{for } 1 \leq |x| < \infty.$$

Then $u(x)$ is a polynomial of degree at most n.

Proof of Theorem 1.10. It follows from Theorem 1.2 above with $x = r\xi$, $\xi \in S_{N-1}$, that

$$(1.26) \qquad\qquad u(x) = \sum_{k=0}^{\infty} r^k Y_k(\xi) \quad \forall x \in \mathbf{R}^N,$$

where $Y_k(\xi)$ is a surface spherical harmonic of degree k and the series in (1.26) converges uniformly on compact subsets of $\mathbf{R}^N$.

We let $j > n$ be a positive integer, multiply both sides of the equality in (1.26) by $C_j^\nu(\xi \cdot \eta)$, and integrate over S_{N-1} to obtain

$$r^j Y_j(\xi) = b_j \int_{S_{N-1}} u(r\eta)C_j^\nu(\xi \cdot \eta)dS(\eta) \quad \forall \xi \in S_{N-1},$$

where b_j is a constant, and we have invoked formula (3.12) in Appendix A.

It follows from this last equality and (1.25) that for $r > 1$,

$$|Y_j(\xi)| \leq c r^{n-j} b_j \int_{S_{N-1}} |C_j^\nu(\xi \cdot \eta)| \, dS(\eta) \quad \forall \xi \in S_{N-1}.$$

Letting $r \to \infty$, we see that

$$Y_j(\xi) = 0 \quad \forall \xi \in S_{N-1} \text{ and } j > n,$$

and hence from (1.26) that

$$u(x) = \sum_{k=0}^{n} r^k Y_k(\xi) \quad \forall x \in \mathbf{R}^N.$$

But $r^k Y_k(\xi)$ is a spherical harmonic of degree k. Hence, $r^k Y_k(\xi)$ is a homogeneous polynomial of degree k in the variables $x_1, ..., x_N$. So the conclusion to the theorem follows from this last equality. ∎

Exercises.

1. Let $x_0 \in \mathbf{R}^3$ with $x_0 \neq 0$. Also, let $S(x_0, r_0)$ be the 2-sphere, which is the boundary of the 3-ball $B(x_0, r_0)$ where $0 < r_0 < |x_0|$. Set

$$[S(x_0, r_0)]^* = \left\{ y : y = \frac{x}{|x|^2} \text{ for } x \in S(x_0, r_0) \right\}.$$

Prove that $[S(x_0, r_0)]^*$ is also a 2-sphere.

2. With $\eta^* = (1, 0, ..., 0)$ and $\xi \in S_{N-1}$, where S_{N-1} is the unit $N-1$-sphere in $\mathbf{R}^N$, using the spherical coordinate notation introduced in §3 of Chapter 1, prove that for $n \geq 1$,

$$r^{2-N-n} C_n^\nu(\eta^* \cdot \xi)$$

is a harmonic function in $\mathbf{R}^N \setminus \{0\}$, where $N \geq 3$ and $\nu = \frac{N-2}{2}$.

2. Subharmonic Functions

Subharmonic functions in N-space are defined in a similar manner to convex functions in 1-space. Recall that $f(t) \in C(\alpha, \beta)$ is convex in the open interval (α, β) if the following is valid for every finite closed sub-interval $[\alpha_1, \beta_1]$ of (α, β): Let $l(t)$ be the line going through the two points $(\alpha_1, f(\alpha_1))$ and $(\beta_1, f(\beta_1))$. Then

$$f(t) \leq l(t) \quad \forall t \in [\alpha_1, \beta_1].$$

Subharmonic functions are defined similarly. Let $\Omega \subset \mathbf{R}^N$ be an open set. We say $u(x) \in C(\Omega)$ if $u(x)$ is a continuous real-valued function in Ω. Then $u(x) \in C(\Omega)$ is subharmonic in Ω if the following holds for every closed ball $\overline{B}(x_0, r) \subset \Omega$, $r > 0$.

Let $v(x) \in C(\overline{B}(x_0,r))$ and harmonic in $B(x_0,r)$. Suppose also that $v(x) = u(x)$ for $x \in \partial\overline{B}(x_0,r)$. Then

$$u(x) \leq v(x) \ \forall x \in \overline{B}(x_0,r).$$

To handle the properties of subharmonic functions that are useful in Fourier analysis, we next introduce the concept of a generalized Laplacian.

Let $G(x)$ be a function in L^1 in a neighborhood of the point x_0. Then we use the following notation for the volume mean of G at x_0:

$$(2.1) \qquad G_{[r]}(x_0) = |B(x_0,r)|^{-1} \int_{B(x_0,r)} G(x)dx,$$

where $|B(x_0,r)|$ stands for the N-dimensional volume of $B(x_0,r)$.

We define the upper generalized Laplacian of G at x_0, designated by $\Delta^*G(x_0)$, in the following manner:

$$(2.2) \qquad \Delta^*G(x_0) = 2(N+2)\limsup_{r\to 0} \frac{G_{[r]}(x_0) - G(x_0)}{r^2}.$$

$\Delta_*G(x_0)$, the lower generalized Laplacian, is defined similarly using the lim inf. It is easy to prove that if G is in class C^2 in a neighborhood of x_0, then $\Delta^*G(x_0) = \Delta_*G(x_0) = \Delta G(x_0)$ where $\Delta G(x_0)$ is the usual Laplacian of G at x_0.

We want, ultimately, to show that if $u(x) \in C(\Omega)$ and if $\Delta^*u(x) \geq 0$ for all $x \in \Omega$, then $u(x)$ is subharmonic in Ω. In order to do this, we first need the following theorem:

Theorem 2.1. *Let $\Omega \subset \mathbf{R}^N$ be an open set where $N \geq 1$. Suppose that $u \in C(\Omega)$ and that u has property P at every $x \in \Omega$, where property P is defined in (3.3) below. Then u is subharmonic in Ω.*

u has property P at x means the following: $\exists \{r_n\}_{n=1}^\infty$ *with $r_n > 0 \ \forall n$ and $\lim_{n\to\infty} r_n = 0$ such that*

$$(2.3) \qquad u_{[r_n]}(x) - u(x) > 0 \quad \forall n.$$

Proof of Theorem 2.1. Suppose to the contrary, that u is not subharmonic in Ω. Then there exists a closed ball $\overline{B}(x_0,r) \subset \Omega$, $r > 0$, and $v \in C(\overline{B}(x_0,r))$ and harmonic in $B(x_0,r)$ with $v(x) = u(x)$ for $x \in \partial\overline{B}(x_0,r)$ and there is an $x^* \in B(x_0,r)$ such that

$$(2.4) \qquad u(x^*) > v(x^*).$$

Set $w(x) = u(x) - v(x)$ and let $\gamma = \max_{x\in\overline{B}(x_0,r)} w(x)$. Then $\gamma > 0$ because of (2.4). Also, set $M = \{x \in \overline{B}(x_0,r) : w(x) = \gamma\}$. Then M is a closed set contained in the open ball $B(x_0,r)$.

Let η designate the distance from M to $\partial\overline{B}(x_0,r)$. Then $\eta > 0$. We know there exists $x^{**} \in M$ such that the distance from x^{**} to $\partial\overline{B}(x_0,r)$ is exactly

η. Hence, every ball of the form $B(x^{**}, \rho)$ with $0 < \rho < \eta$ contains points not in M. Therefore,

$$(2.5) \qquad w_{[\rho]}(x^{**}) < w(x^{**}) \quad \text{for } 0 < \rho < \eta.$$

On the other hand, u has property P at x^{**}, and $v_{[\rho]}(x^{**}) = v(x^{**})$ for $0 < \rho < \eta$. So,

$$w_{[\rho]}(x^{**}) = u_{[\rho]}(x^{**}) - v(x^{**}) \quad \text{for } 0 < \rho < \eta.$$

For $n > n_0$, the r_n that are involved in the definition of property P at x^{**} are smaller than η. Consequently,

$$\begin{aligned} w_{[r_n]}(x^{**}) &= u_{[r_n]}(x^{**}) - v(x^{**}) \\ &> u(x^{**}) - v(x^{**}). \end{aligned}$$

Therefore,

$$(2.6) \qquad w_{[r_n]}(x^{**}) > w(x^{**}) \quad \text{for } n > n_0.$$

The inequalities in (2.5) and (2.6) are mutually contradictary. Hence, u is indeed subharmonic in Ω. ∎

Theorem 2.2. *Let $\Omega \subset \mathbf{R}^N$ be an open set where $N \geq 1$. Suppose that $u \in C(\Omega)$ and that*

$$(2.7) \qquad \Delta^* u(x) \geq 0 \qquad \forall x \in \Omega.$$

Then $u(x)$ is subharmonic in Ω.

Proof of Theorem 2.2. For j, a positive integer, set

$$(2.8) \qquad u_j(x) = u(x) + |x|^2 / j.$$

Then it follows from the inequality in (2.7) that $\Delta^* u_j(x) \geq 2N/j$ for every $x \in \Omega$. But then we see from the definition given in (2.2) that each $u_j(x)$ has property P for every $x \in \Omega$. Hence, we obtain from Theorem 2.1 that

$$(2.9) \qquad u_j(x) \text{ is subharmonic in } \Omega \text{ for every } j.$$

Let $\overline{B}(x_0, r) \subset \Omega$, where $r > 0$, and suppose $v(x)$ is the function harmonic in $B(x_0, r)$ that agrees with $u(x)$ for $x \in \partial \overline{B}(x_0, r)$ and $v_j(x)$ is the function harmonic in $B(x_0, r)$ that agrees with $u_j(x)$ for $x \in \partial \overline{B}(x_0, r)$. Now the following facts are clear from (2.8) and the maximum principle for harmonic functions:

$$(i) \lim_{j \to \infty} u_j(x) = u(x) \text{ uniformly for } x \in \overline{B}(x_0, r);$$

$$(ii) \lim_{j \to \infty} v_j(x) = v(x) \text{ uniformly for } x \in \overline{B}(x_0, r).$$

From (2.9), we have that

$$u_j(x) \leq v_j(x) \quad \text{for } x \in B(x_0, r) \quad \text{and } \forall j.$$

So from (i) and (ii) above, it follows that

$$u(x) \le v(x) \quad \text{for } x \in B(x_0, r).$$

Hence, indeed, u is subharmonic in Ω. ∎

Exercises.

1. Suppose $G(x) \in C^2(B(0,1))$ where $B(0,1)$ is the unit 3-ball in $\mathbf{R}^3$. With $\Delta^* G(x_0)$ defined in (2.2) above, prove

$$\Delta^* G(x_0) = \frac{\partial^2 G}{\partial x_1^2}(x_0) + \frac{\partial^2 G}{\partial x_2^2}(x_0) + \frac{\partial^2 G}{\partial x_3^2}(x_0) \quad \text{for } x_0 \in B(0,1).$$

2. Suppose $f(t)$ is an increasing convex function for $0 < t < \infty$. Suppose also that $f(t) \in C^2((0,\infty))$.Z Set

$$G(x) = f(x_1^2 + x_2^2 + 1) \quad \text{for } x \in \mathbf{R}^2.$$

Prove that $G(x)$ is a subharmonic function in $\mathbf{R}^2$.

Bibliography

[Ash]   J. M. Ash, *Multiple trigonometric series,* in *Studies in harmonic analysis,* edited by J. M. Ash, Math. Assoc. of America, Studies in Mathematics, **13** (1976), 76-96.

[AW1]   J. M. Ash and G. Wang, *A survey of uniqueness questions in multiple trigonometric series,* in Contemporary mathematics: A Conference in Harmonic Analysis and Nonlinear Differential Equations in Honor of Victor L. Shapiro, **208** (1997), 35-71.

[AW2]   J. M. Ash and G. Wang, *Sets of uniqueness for spherically convergent multiple trigonometric series,* Trans. Amer. Math. Soc. **354** (2002), 4769-4788.

[AW3]   J. M. Ash and G. Wang, *Uniqueness questions for multiple trigonometric series,* in Contemporary mathematics: Advances in ergodic theory and harmonic analysis (to appear).

[AAR]   G. E. Andrews, R. Askey, and R. Roy, Special functions, Cambridge University Press, Cambridge, 1999.

[ABR]   S. Axler, P. Bourdon, and W. Ramey, Harmonic function Theory, Springer-Verlag, New York, 2001.

[Bec]   B. Beckman, Codebreakers, Amer. Math. Soc., Providence, 2002.

[Beu]   A. Beurling, *Sur les ensembles exceptionels, Acta math.* **72** (1940), 1-13.

[Boc1]   S. Bochner, *Summation of multiple Fourier series by spherical means,* Trans. Amer. Math. Soc. **40** (1936), 175-207.

[Boc2]   S. Bochner, *Boundary values of analytic functions in several variables and almost periodic functions,* Ann. of Math. **45** (1944), 708-722.

[Bou]   J. Bourgain, *Spherical summation and uniqueness of multiple trigonometric series,* Internat. Math. Res. Notices (1996), 93-107.

[BN]   H. Brezis and L. Nirenberg, *Characterizations of ranges of some nonlinear operators and applications to boundary value problems,* Ann. Scuo. Norm Sup. Pisa **5** (1978), 225-336.

[CZ1]   A. P. Calderon and A. Zygmund, *On the existence of certain singular integrals,* Acta Math. **88** (1952), 85-139.

[CZ2]   A. P. Calderon and A. Zygmund, *Singular integrals and periodic functions,* Studia Math. **14** (1954), 249-271.

[CZ3]   A. P. Calderon and A. Zygmund, *Local properties of solutions of elliptic partial differential equations,* Studia Math. **20** (1961), 171-225.

[Coh]   P. J. Cohen, *Topics in the theory of uniqueness of trigonometric series,* doctoral thesis, University of Chicago, Chicago, IL, 1958.

[Co]   R. Cooke, *A Cantor -Lebesgue theorem in two dimensions,* Proc. Amer. Math. Soc. **30** (1971), 547-550.

[CH]   R. Courant and D. Hilbert, Methods of mathematical physics, Vol 1, John Wiley & Sons, New York, 1966.

[Da]   J. W. Dauben, George Cantor, Harvard University Press, Cambridge, 1979.

[DE]   R. H. Dyer and D. E. Edmunds, *Removable singularities of solutions of the Navier-Stokes equations,* J. London Math. Soc. **2** (1970), 535-538.

[DG]   D. G. De Figueiredo, and J. P. Gossez, *Nonlinear perturbations of a linear elliptic problem near its first eigenvalue*, J. Differential Equations **30** (1978), 1-19.

[Du]   P. Duren, A century of mathematics in America, Part III, American Mathematical Society, Providence, 1989.

[EK]   L. Edelstein-Keshet, Mathematical models in biology, Random House, New York, 1988.

[EMOT]   A. Erdelyi, W. Magnus, F. Oberhettinger, and F. G. Tricomi, Higher transcendental functions, Vol 2, McGraw-Hill Book Co., New York, 1953.

[Ev]   L. C. Evans, Partial differential equations, American Mathematical Society, Providence, 1998.

[Fal]   K. Falconer, Fractal geometry, John Wiley & Sons, New York, 1990.

[Ga]   G. P. Galdi, An Introduction to the mathematical theory of the Navier-Stokes equations, Vol II, Springer-Verlag, New York, 1994.

[Ge]   C. Gerhardt, *Stationary solutions to the Navier-Stokes equations in dimension four, Math. Zeit.* **165** (1979), 193-197.

[Har]   G. H. Hardy, Divergent series, Oxford University Press, Oxford, 1949.

[HL]   G. H. Hardy and J. E. Littlewood, *Two theorems concerning Fourier series,* J. London Math. Soc. **1** (1926), 19-25.

[HeLo]   H. Helson and D. Lowdenschlager, *Prediction theory and Fourier series in several variables,* Acta Math. **99** (1958), 165-200.

[Ho]   K. Hoffman, Banach spaces of analytic functions, Prentice-Hall, Englewood Cliffs, N .J., 1962.

[Jo]   F. John, Partial differential equations, fourth edition, Springer Verlag, New York, 1982.

[Ka]   J. Karamata, *Neuer Beweis und Verallgemeinerung der Tauberschen Satse, welche die Laplaschen und Stieltjessche Transformation betreffen,* J. Reine und Angewandte Math. **164** (1931), 27-39

[Ke]   O. D. Kellogg, Foundations of potential theory, Ungar, NewYork, 1929.

[KW]   J. L. Kazdan and F. W. Warner, *Remarks on some quasilinear elliptic equations,* Comm. Pure Appl. Math. **28** (1975), 567-597.

[La]   O. A. Ladyzhenskya, The mathematical theory of viscous incompressible flow, Gordon and Breach, New York, 1969.

[LL]   E. M. Landesman and A. C. Lazer, *Nonlinear perturbations of linear elliptic boundary value problems at resonance,* J. Math. Mech. **19** (1970), 609-623.

[LefS]   L. E. Lefton and V. L. Shapiro, *Resonance and quasilinear parabolic partial differential equations,* J. Diff. Equations, **101** (1993), 148-177.

[LS]   G. E. Lippman and V. L. Shapiro, *Capacity and absolute Abel summability of multiple Fourier series,* J. Appox. Theory, **10** (1974), 313-323.

[Man]   B. B. Mandelbrot, The fractal geometry of nature, W. H. Freeman and Company, New York, 1982.

[Mur]   J. D. Murray, Mathematical Biology, Springer-Verlag, New York, 1990.

[Mar]   J. Marcinkiewicz, *Sur le multiplicateurs des series de Fourier multiples,* Studia Math. **8** (1938), 78-91.

[Mu]   O. R. Musin, *The kissing problem in three dimensions,* Discrete & Computational Geometry **35** (2006), 375-384.

[Pi]   M. A. Pinsky, Partial differential equations and boundary-value problems with applications, 2nd ed., McGraw-Hill, Inc., New York, 1991.

[PZ]   F. Pfender and Gunter M. Ziegler, *Kissing numbers, sphere packings, and some unexpected proofs,* Notices Amer. Math. Soc. **51** (2004), 873-883.

[Ra]   E. D. Rainville, Special functions, The Macmillan Company, New York, 1960.

[Ru1]   W. Rudin, Principles of mathematical analysis, 3rd ed., McGraw-Hill Book Co., New York, 1976.

[RR]   F. Riesz and M. Riesz, *Uber die Randverte einer analytischen function*, Compte Rendu Quatrieme Congres des Mathematiciens Scandinavia tenu a Stockholm, 1916, 27-44.

[Ru2]   W. Rudin, *Uniqueness theory for Laplace series*, Trans. Amer. Math. Soc. **68** (1950), 287-303.

[Ru3]   W. Rudin, Fourier analysis on groups, John Wiley & Sons, New York, 1962.

[Sa]   S. Saks, Theory of the integral, 2nd ed., Hafner Publishing Company, New York, 1937.

[Sch]   I. J. Schoenberg, *Positive definite functions on spheres*, Duke Math. J. **9** (1942), 96-107.

[Se]   R. T. Seeley, *Spherical harmonics*, Amer. Math. Monthly **73** (1966), 115-121.

[Sh1]   V. L. Shapiro, *Fourier series in several variables*, Bull. Amer. Math. Soc. **70** (1964), 48-93.

[Sh2]   V. L. Shapiro, *Uniqueness of multiple trigonometric series*, Annals of Math. (2) **66** (1957), 467-480.

[Sh3]   V. L. Shapiro, *Fractals and distributions on the N-torus*, Proc. Amer. Math. Soc. **131** (2004), 3431-3440.

[Sh4]   V. L. Shapiro, *Poisson integrals and nontangential limits*, Proc. Amer. Math. Soc. **134** (2006), 3181-3189.

[Sh5]   V. L. Shapiro, *Algebraic integers and distributions on the N-torus*, J. Functional Analysis **13** (1973), 138-153.

[Sh6]   V. L. Shapiro, *Sets of uniqueness on the 2-torus*, Trans. Amer. Math. Soc. **165** (1972), 127-147.

[Sh7]   V. L. Shapiro, *Singular integrals and spherical convergence*, Studia Math. **44** (1972), 253-262.

[Sh8]   V. L. Shapiro, *On Green's theorem*, J. London Math. Soc. **32** (1957), 261-269.

[Sh9]   V. L. Shapiro, Topics in Fourier and geometric analysis, Memoir 39, American Math. Soc., Providence, 1961.

[Sh10]   V. L. Shapiro, *Generalized and classical solutions of the nonlinear stationary Navier-Stokes equations*, Trans. Amer. Math. Soc. **216** (1976), 61-79.

[Sh11]   V. L. Shapiro, *Singular quasilinear heat equations*, Indiana Univ. Math. J. **58** (2009), 443-477.

[Sh12]   V. L. Shapiro, *Resonance and quasilinear ellipticity*, Trans. Amer. Math. Soc. **294** (1986), 567-584.

[Sh13]   V. L. Shapiro, *Quasilinear ellipticity on the N-torus*, nonlinear and convex analysis, Lecture Notes Pure and Appl. Math. **107**, edited by B. L. Lin and S. Simons, 203-209.

[Sh14]   V. L. Shapiro, Singular quasilinearity and higher eigenvalues, Memoir 726, American Math. Soc., Providence, 2001.

[Sh15]   V. L. Shapiro, *One-sided conditions for the Navier-Stokes equations*, J. Diff. Equations, **94** (1982), 413-439.

[Sh16]   V. L. Shapiro, *Removable sets for pointwise solutions of the generalized Cauchy-Riemann equations*, Ann. Math **92** (1970), 82-101.

[Sh17]   V. L. Shapiro, *Harmonic functions in the unit disk*, Amer. Math. Monthly, **109** (2002), 37-45.

[Sh18]   V. L. Shapiro, *Bounded generalized analytic functions on the torus*, Pacific Jour. of Math. **14** (1964), 1413-1422.

[Sh19]   V. L. Shapiro, *The uniqueness of solutions of the heat equation in an infinite strip*, Trans. Amer. Math. Soc. **125** (1966), 326-361.

[Sh20]   V. L. Shapiro, *Nonlinear ellipticity on unbounded domains*, Rocky Mountain J. Math., 25 pages, to appear in 2011.

[Sh21]   V. L. Shapiro, *Isolated singularities in steady-state fluid flow*, Siam J. Math. Anal. **7** (1976), 577-601.

[Sh22]   V. L. Shapiro, *A Counter-example in the Theory of planar viscous incompresible flow,* J. Diff. Equations, **22** (1976), 164-179.

[Sh23]   V. L. Shapiro, *Isolated singularities for solutions of the nonlinear stationary Navier-Stokes equations,* Trans. Amer. Math, Soc **187** (1974), 335-363.

[Sh24]   V. L. Shapiro, *Pressure, capacity, and the Navier-Stokes equations,* J. Math. and Applications, **65** (1978), 236-246.

[Sh25]   V. L. Shapiro, *Positive Newtonian Capacity and stationary viscous incompressible flow,* Indiana Univ. Math. J. **24** (1974), 1-16.

[ShWe]   V. L. Shapiro and G. V. Welland, *Sobolev spaces, the Navier-Stokes equations and capacity,* Proc. Symp. Pure Math., **35** (1979), 369-375.

[Sm]   J. Smoller, Shock waves and reaction-diffusion equations, New York, 1983.

[Sp]   M. R. Spiegel, Fourier analysis, McGraw-Hill Publishing Co., New York, 1974.

[St]   M. Struwe, Variational methods, Fourth ed., Springer-Verlag, Berlin, 2008.

[SW]   E. M. Stein and G. Weiss, Introduction to Fourier analysis in Euclidean spaces, Princeton Univ. Press, Princeton, 1971.

[Te]   R. Temam, Navier-Stokes equations and nonlinear functional analysis, SIAM, Philadelphia, 1983.

[Ti1]   E. C. Titchmarsh, The theory of functions, Oxford Univ. Press, Oxford, 1950.

[Ti2]   E. C. Titchmarsh, Introduction to the theory of Fourier integrals, 2nd ed., Oxford Univ. Press, Oxford, 1948.

[Wa]   G. N. Watson, A treatise on the theory of Bessel functions, 2nd ed., Cambridge Univ. Press, Cambridge, 1944.

[Zy1]   A. Zygmund, Trigonometric Series,Vol I, Cambridge Univ. Press, Cambridge, 1959.

[Zy2]   A. Zygmund, Trigonometric Series, Vol II, Cambridge Univ. Press, Cambridge, 1959.

Index